Advancing Physics A2

British Library Cataloguing in Publication Data
A catalogue record for this book is available from the British Library

First published June 2000
Revised edition published 2008
by Institute of Physics Publishing, Dirac House, Temple Back, Bristol BS1 6BE, UK

ISBN 978-0-7503-0781-9

Printed and bound in Spain by Dédalo Offset.

Revised and first edition authors and contributors (books and CD-ROMs):

Steve Adams,
Susan Aldridge
Michael Barnett
Christopher Bishop
Richard Boohan
David Brabban
Jim Breithaupt (A–Z)
Michael Brimicombe
Philip Britton
Ian Brown
James Butler
Peter Campbell
Geoff Camplin
Simon Carson
Simon Collins
Mark Cramoysan
John Cullerne
Kathleen Davies
Laurence Dickie
Ken Dobson

Ingrid Ebeyer
Tony Egan
Diana Emes
Jonathan Foyle
Richard Field
Mike Gidlow
David Grace
Sarah Grant
Stephen Hall
Stephen Hearn
Lawrence Herklots
David Homer
Julian Hoult
Neil Hutton
Mike Kearney
Ian Lawrence
Allan Mann
Rick Marshall
John Mascall
John Miller

John Mitchell
David Morland
Andrew Morrison
Jon Ogborn (project director)
Simon Petts
Steve Pickersgill
Andrew Raw
Helen Reynolds
Laurence Rogers
David Sang
Clare Sansom
Allan Seago
Chris Shilliday
Robert Strawson
Janet Taylor
Kevin Walsh
Elizabeth Weiser
Catherine Wilson
Ken Zetie

Revised edition production:

Susannah Bruce, Laura Churchill, Kate Gardner, Andrew Giaquinto, Jane Henley, Kerry Hopkins, Anastasia Ireland, Huw Johnson, Evie Palmer, Cee Pike, Teresa Ryan, Andrew Stevens, Fred Swist, Jamie Symonds, Alison Tovey

First edition production:

Penelope Barber, Alan Evans, Huw Johnson, Hayley Liddle, Kevin Lowry, Bridget Pairaudeau, Angela Gage, Andrew Stevens, Jamie Symonds, Alison Tovey, Brenda Trigg, Lucy Williams

ADVANCING PHYSICS A2

Revised edition edited by
Jon Ogborn and Rick Marshall

First edition edited by
Jon Ogborn and Mary Whitehouse

CD-ROM edited by
Ian Lawrence, Rick Marshall and Jon Ogborn

CD-ROM first edition edited by
Ian Lawrence and Mary Whitehouse

IOP Publishing

Contents

How to use this book and CD-ROM

The book and CD-ROM go together. Both are essential to the Advancing Physics course.

The book provides, in each chapter:
- **Narrative text** telling the story of part of physics, putting it in context and explaining to you why it is worth understanding
- **Key summaries** that contain all the essential ideas, showing their structure in a visual form
- Short **Quick check** questions at the end of each section, to start you off being able to make calculations and arguments for yourself
- **Links to the CD-ROM** at the end of each section, pointing out further questions for practice, activities to try, key terms to look up in the A–Z, further readings for interest and key items to use for revision
- At the end of the chapter a **Summary check-up** lists all the essential ideas you should have learned, with some further **Questions** to try

The CD-ROM provides, for each chapter:
- **Activities**, including some software on the CD-ROM, and many experiments to do
- **Questions** at different levels, from warm-up and practice exercises to questions at A2 examination level
- **Readings** that extend the material in the textbook or provide interesting alternative views of the physics in the course and its background, with an emphasis on the people involved
- **Images** to look at that add to those in the textbook, and **Key summary** diagrams to study
- **Computer files** for use with a variety of computer software, notably spreadsheets and modelling packages
- **Revision checklists** of things you should have learned to be able to do
- **An A–Z of physics** covering all the ideas in the course in a form useful for revision
- **Help with skills** of experimenting and data handling

Studying a section of a chapter ✓

Each chapter is divided into numbered sections. When studying a section you should:
- First read the section through, get a sense of what it is all about
- Work through lab activities on the CD-ROM, guided by your teacher
- Go back and study the Key summaries, which display ideas pictorially. You will have to work at these to get their full message. Look carefully at how the logic of arguments and relationships between ideas are laid out graphically. Try reconstructing the diagrams for yourself – this can teach you a lot
- Work through all of the Quick check questions
- Follow the Links to the CD-ROM for more questions to try, activities to look at, key terms to check out and other items for further interest or revision
- Use the A–Z to look up ideas you don't understand well enough, or to get a view different from the one in the book
- Personalise your CD-ROM by creating your own "shadow file" of notes, comments, summaries, etc.

Revising for tests or examinations ✓

- Review your "shadow file" notes
- Go to the CD-ROM revision checklist for each chapter and follow the links to relevant revision terms
- Browse the A–Z terms listed for each section of a chapter
- Review the relevant Key summary diagrams in the book or on the CD-ROM
- Try more questions on the CD-ROM
- Stretch yourself with some further readings
- Try further activities including ones using software

The Revision CD that is also available provides
- Checklists of things you need to know and be able to do
- Revision notes and diagrams
- Previous examination questions
- Worked examples

Links between the student's book and CD-ROM

At the end of each section of each chapter you will find a list of links to the CD-ROM that provide one-click access to:

- questions to practise with
- activities to try out
- key terms to look up in the A–Z
- readings to go further for interest
- the revision checklist and other revision resources

Book page

Quick check

CD links

Book summary check-up page

CD-ROM page

10 Creating models

Models are simple, artificial worlds created to give insight into how real systems work and to predict what they might do. We start with simple models of objects and their behaviour, and go on to model variations between one quantity and another. We will give examples of models of:

- radioactive decay
- decay of charge on a capacitor
- oscillating systems and how they can resonate

The potential spread of a forest fire can be modelled.

10.1 What if...?

Walking or driving through the countryside, you see signs warning against starting fires by accident – one tree set alight and soon a whole forest can be ablaze. What if the trees were planted less densely? At what density might fires be unable to spread? This is an example of a "What if...?" question, the sort that physicists are often good at tackling. One important way of answering such a question is to build a model on a computer – to make a **computational model**.

One way to make a forest fire model is to divide the forest area into squares and plant computer-generated trees in a number of those squares. Then put a computer-made burning tree in one place and see how the fire spreads. To do that computer trees must know what to do. They need a rule for their behaviour, such as "if a tree is next to a burning tree it becomes a burning tree".

What if the tree density is varied? When will a fire spread right across the forest, from side to side? The result is striking. Below a certain density, fires never spread right through the forest. Above that density they always do. And the range of densities over which they may, or may not, spread is quite narrow. There is a critical density for the continuous spread of fires.

The same problem is well known in oil prospecting. Oil is found underground trapped

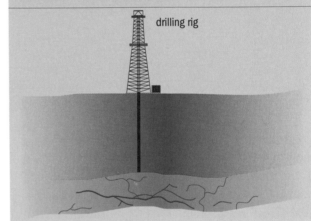
Can the oil be extracted?

drilling rig

A layer of cracked rock may contain oil that can be extracted.

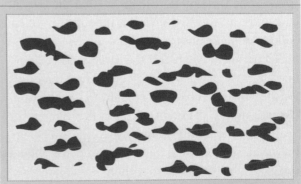
A "conducting insulator"?

Will it conduct? This illustration shows graphite grains in an insulating ceramic. At this density there is no conducting path across the material so it can't conduct.

Computer model of forest fires

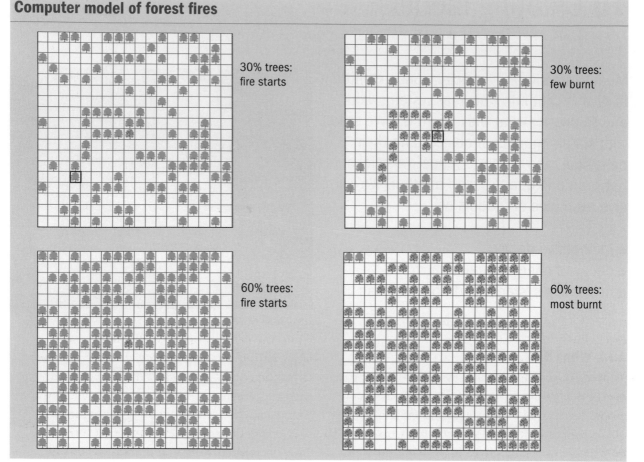

30% trees: fire starts

30% trees: few burnt

60% trees: fire starts

60% trees: most burnt

This model shows that a fire in 30% woodland does not spread everywhere. A fire in 60% woodland spreads across the entire forest.

in cracked rocks. Only if the cracks make a continuous open path through the rock can oil work its way through the rock layer to be extracted. How are the two problems the same? For "tree" read "open crack", for "burning tree" read "oil-filled crack", and now the rule becomes that if an open crack is next to a filled crack, the open crack becomes filled. Because it was for oil percolating through cracked rock that the critical density result was first noticed, the model is often called the percolation model.

Electrical engineers use the same model. They need to make hard-wearing conducting materials for the brushes that carry current to the rotors of electric motors. A good way to do this is to embed grains of conducting graphite in a hard-wearing but insulating ceramic matrix. How dense should the grains be to ensure that there is a conducting path right through the material? You guessed it – above exactly the same critical density as that for oil-filled cracks and the spread of forest fires.

Models made with objects and rules

The percolation model illustrates several things about such computational models:

- the model is an imagined, artificial world with objects and rules that you create – you "play God" in making the model;
- there are just a few objects and rules – they are all there is to the model;
- the model is a very simplified, "stripped down" version of reality – the trick is to find the right essentials to include;
- the model may have an unexpected behaviour (e.g. the existence of a critical density);
- one form of model may be adapted to a number of very different problems.

Models like these have been used to study economic change and share prices, the flocking of birds and the behaviour of ants, the cracking of materials, the flow of powders, and the occurrence of earthquakes. Physicists who know how to make models to answer "What if...?" questions

Computer model of rabbits breeding and spreading on an island

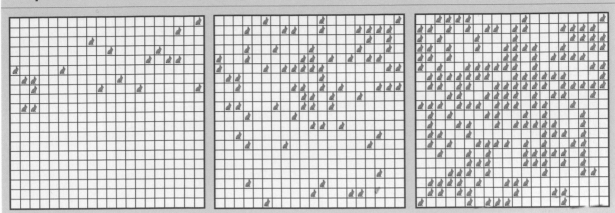

In this model the birth rate is 12% and the death rate is 10%, so that the rabbit population increases.

can turn their hands to a vast array of problems. The financial world calls it rocket science, and rewards it well. When you have a model, you can study its behaviour. You can try it under various conditions. In this way, you discover its properties. Some of these properties will be necessary consequences of the very nature of the model – any model of the same kind must behave in a certain way. In this way you are starting to do mathematics; you are studying the mathematical nature of a type of model.

Rabbits and radioactivity

From time to time, populations explode. One summer there may be a plague of insects; in other summers hardly any. Rabbits introduced onto an empty island can multiply rapidly – breeding like rabbits – until they eat all the grass, the population cannot be sustained and numbers collapse.

Why can populations be unstable like this? It's a kind of feedback. More rabbits means more babies, meaning more rabbits still, and so on. It's easy to make a simple computer model with just one object, rabbits, and two rules:
- a rabbit breeds a new rabbit, with a certain probability;
- a rabbit dies, with a certain probability.

The rules are applied to each rabbit. Then they are applied all over again to the rabbits that are left and to all the new ones. And so on, repeatedly.

Notice the simple, stripped down nature of the model. There are no males and females, there is no period of pregnancy, just a chance that a rabbit

produces a new one, and a chance that a rabbit dies. These are the bare essentials.

If you increase the birth probability a little above the death probability, the screen rapidly fills with rabbits. The more rabbits there are the faster the population grows. Such changes are called exponential. A change is exponential when the rate of change of something is proportional to the amount of that something that there is. Now decrease the birth probability below the death probability and the screen empties as the population collapses. If the probabilities are equal, any level of population is possible. Random fluctuations drive it up or down.

Rabbits don't really breed or die purely at random. Radioactive decay, however, is an example of a genuinely random event. Understanding radioactive decay is important for dealing with the radioactive waste from power generation, industry and hospitals. It matters how active the material is and how long it lasts: the faster it decays the more active it is but the less time it lasts. This is equally important in handling the life-saving radioactive materials that are used to destroy cancers, for example.

The decay of a nucleus is a typical quantum event (chapter 7, *Advancing Physics AS*). A particle, such as an alpha particle or an electron (beta particle) has a small probability of being emitted by the nucleus in a given time. The event is absolutely and essentially random, affected by nothing outside the nucleus. This was initially hard to believe. Early workers on radioactivity, like Rutherford, tried heating,

Computer model of radioactive decay

● Unstable nucleus ● Stable product

In this model each nucleus has a 2% chance of decay in each time unit. The numbers remaining (unstable nuclei) and numbers decayed (stable product) are shown at time intervals of 0, 20 and 40 units.

cooling, subjecting materials to pressure, and so on to try to affect the rate of decay – all to no avail. It is just a flow of randomly timed events.

Models of radioactivity are therefore very simple. A nucleus can be given just one rule: to decay with a fixed probability in a given time. That's all. The result: a population that declines, but ever more slowly and slowly, as the number left to decay gets less and the number decaying, proportional to the number left, similarly reduces. This is why the graph against time of the number of active nuclei always falls, but also why its rate of fall gets less and less as the number left gets smaller.

Tracing substances through the body

An overactive thyroid gland accumulates iodine. Doctors can diagnose such a condition by tracing substances as they pass through the body. One way to do this is to attach a small amount of a radioactive material to the substance and see where the active material goes.

A doctor needs a radioactive material that decays quite quickly, for two reasons. First, it needs to have enough activity to detect. Activity, measured in becquerel (Bq), is simply the number of decays or counts per second. The second reason is to make sure the substance lasts in the body only as long as needed by the investigation. For example, the tracer iodine-123 decays so that on average half of it is gone every 13 hours; another commonly used tracer is technetium-99, half of

which on average decays every six hours (p7).

The activity of a source is linked to the probability of decay. If, in a short interval of time, each nucleus has a probability p to decay, and there are N nuclei at that moment, then:

$$\text{number decaying} = pN$$

Random decays happen on average at a definite overall rate that you can hear with a Geiger counter. Waiting for a time Δt, the number of decays you observe is on average proportional to the interval Δt. The longer you wait, the more chance there is to get a decay. That is, the probability p of decay in time Δt is proportional to Δt:

$$p = \lambda \Delta t$$

where the constant λ is called the **decay constant**. It is just the probability that a nucleus will decay in one second. Thus the activity a of the source is:

$$\text{activity } a = \text{number decaying per second}$$

$$a = \frac{pN}{\Delta t} = \frac{\lambda \Delta t N}{\Delta t}$$

Thus:

$$a = \lambda N$$

If you can measure the activity then, knowing how

Key summary: smoothed-out radioactive decay

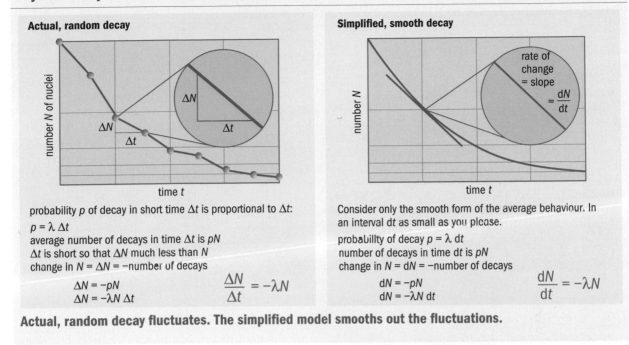

Actual, random decay

probability p of decay in short time Δt is proportional to Δt:

$p = \lambda \, \Delta t$

average number of decays in time Δt is pN

Δt is short so that ΔN much less than N

change in $N = \Delta N = -$number of decays

$$\Delta N = -pN$$
$$\Delta N = -\lambda N \, \Delta t$$

$$\frac{\Delta N}{\Delta t} = -\lambda N$$

Simplified, smooth decay

Consider only the smooth form of the average behaviour. In an interval dt as small as you please.

probability of decay $p = \lambda \, dt$

number of decays in time dt is pN

change in $N = dN = -$number of decays

$$dN = -pN$$
$$dN = -\lambda N \, dt$$

$$\frac{dN}{dt} = -\lambda N$$

Actual, random decay fluctuates. The simplified model smooths out the fluctuations.

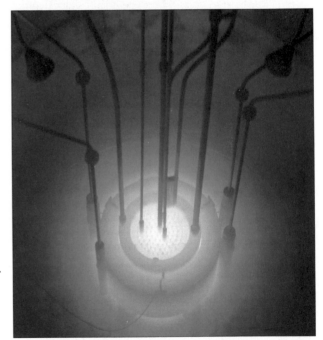

This is a photograph of radioactive material being stored under water; the blue light comes from Cerenkov radiation.

many nuclei there are, you can calculate the decay constant. Of course, the random decays fluctuate in number around a steady average. A description of the decay will often ignore the fluctuations, smoothing them out.

To measure the activity of a source is to aim at a moving target. As the nuclei decay, there are fewer left to decay so, although the probability of decay doesn't change, the activity falls. If the source lasts thousands of years, as many do, this is no problem. However, if the lifetime is short you have to count decays for long enough to even out the fluctuations and get a reliable average, but not for so long that the activity has changed while you are measuring it.

The gradual, smoothed-out fading away of radioactivity can be described in a very simple way – the number of nuclei present at any moment decreases at a rate equal to the activity:

$$\frac{dN}{dt} = -\lambda N \ (\text{with } \lambda N = a)$$

The term $\dfrac{dN}{dt}$, called the **derivative** or differential of N with respect to time, is the limit of the ratio $\dfrac{\Delta N}{\Delta t}$ as the time interval Δt is made shorter and shorter. Actually, if you reduce the time interval more and more you get down to single isolated decays, not to a smooth rate of decay, so this

equation is another simplified model. It is stripped down to leave out individual events. It just captures the smooth overall behaviour of the decay, which is common to many other problems, so the model is very useful. An equation like this, involving the rate of change of a quantity, is called a differential equation.

The spread of farming

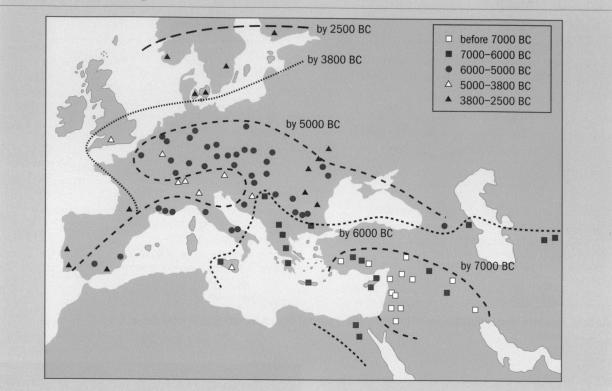

The map shows sites where remains of crops first domesticated in the fertile crescent have been radiocarbon dated. Sites more distant from the sources in the fertile crescent are generally of later date. Note the initial spread into Europe through the Balkans and the rapid spread through the plains north of the Alps. The average rate of spreading is about 1 km a year.

Key summary: clocking radioactive decay

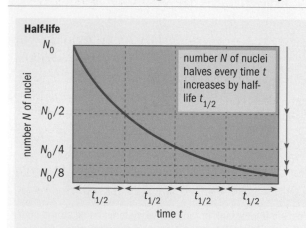

Half-life

number N of nuclei halves every time t increases by half-life $t_{1/2}$

In any time t the number N is reduced by a constant factor

In one half-life $t_{1/2}$ the number N is reduced by a factor 2

In L half-lives the number N is reduced by a factor 2^L

(e.g. in 3 half-lives N is reduced by the factor $2^3 = 8$)

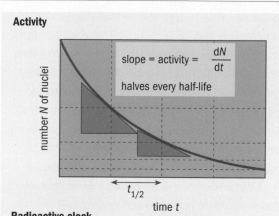

Activity

slope = activity = $\dfrac{dN}{dt}$

halves every half-life

Radioactive clock

Measure activity. Activity proportional to number N left

Find factor F by which activity has been reduced

Calculate L so that $2^L = F$

$L = \log_2 F$

age $= t_{1/2} L$

The half life of a radioactive isotope can be used as a clock

Radioactive clocks: tracking the ancient origins of farming

Farming (invented independently in China, the Pacific, the Americas and in Africa), seems first to have started about ten thousand years ago. It was in the "fertile crescent" of the Middle East, today occupied by Israel, Jordan, Syria and Iraq, that human beings first abandoned hunting and gathering and settled to grow domesticated food crops. These crops, especially wheat, barley and peas, are today part of your everyday diet. If you eat lamb, it's also thanks to these original farmers, who were the first to domesticate sheep and goats.

It took only three to four thousand years for agricultural sites to spring up all over Europe – as far afield as Spain, Britain and Scandinavia. Starting in 8000 BC in the fertile crescent, the outward wave of innovation can be seen in the ages of the earliest archaeological traces of farming. This spread has been tracked by using radioactive decay to date the seeds found at archaeological sites. The radioactive decay of carbon in the seeds is used to calculate their age through a process called radiocarbon dating.

Radiocarbon dating works like this. The carbon dioxide that plants take in and use to make tissue is slightly radioactive. Some of the stable isotope carbon-12 in the atmosphere is turned into the radioactive isotope carbon-14 through the action of cosmic rays. While alive, the plants have a constant ratio of the two isotopes in their tissue, as you do in yours. After they die the fraction of radioactive carbon-14 starts to fall, because it decays and is no longer replenished. So these radioactive nuclei clock the time since the plant died through their constant probability of decay in a fixed time.

For carbon-14 the timescale is a few thousand years: half of what there is decays on average in 5730 years. In the next 5730 years, half of what is left decays, leaving a quarter, and so on. This time, 5730 years, is called the **half-life** of carbon-14. The half-life $t_{1/2}$ of a radioactive material is just the average time for the number of nuclei to be reduced by a factor of two. The half-life is not half of the total life, but is the average time for half of what is there to go away.

To get a radiocarbon date you have to destroy some of the material to measure the activity of

How to get a carbon date

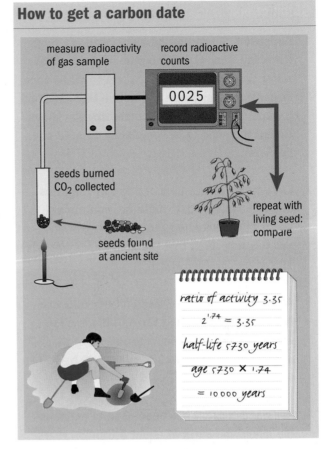

measure radioactivity of gas sample

record radioactive counts

0025

seeds burned CO_2 collected

seeds found at ancient site

repeat with living seed: compare

ratio of activity 3.35

$2^{1.74} = 3.35$

half-life 5730 years

age 5730 × 1.74

= 10 000 years

the carbon-14 in it. There is a trade-off between how much material you use and the accuracy of the estimate of the age because of the random fluctuations in the activity.

The 5730-year half-life of carbon-14 can be measured in a few minutes. So can the 4.5 billion-year half-life of uranium-238. Impossible? Not at all: you just measure the activity of a sample containing a known number of the radioactive nuclei. The greater the activity, the shorter the half-life. From the activity and the number of nuclei you can calculate the decay constant λ (p5) and from that the average time for half the nuclei to decay. It is given by:

$$t_{1/2} = \frac{0.693}{\lambda} = \frac{\ln 2}{\lambda}$$

The half-life is a measure of how long a radioactive substance will last. The decay constant tells you how rapidly it decays. Both quantities are useful, though they say the same thing in different ways. Notice that they are yet another example of a reciprocal relationship in physics.

Quick check

1. Suppose a quick-growing weed invades a pond. Each day, each tiny leaf on the surface of the water is replaced by two new leaves of the same size. After a few weeks you notice that the pond is half covered by weed. Explain why you should expect the pond to be completely covered one day later.

2. The activity of a sample of a few micrograms of a radioactive waste material, containing about 10^{15} nuclei, is 10^6 Bq (disintegrations per second). Show that the decay constant λ is $10^{-9}\,\mathrm{s}^{-1}$.

3. Radiocarbon dating is tried on 500–600 year-old material. Given that the half-life is 5730 years, show that roughly a twentieth of the original radioactive carbon nuclei will have decayed.

4. Radioactive decay has been used to estimate the age of the Earth at about 4.5 billion years. Explain why the isotopes used must have half-lives of about one billion years, and why the activity detected is necessarily very small.

5. The half-life of a radioactive waste material from a nuclear reactor is six months. Show that the activity will decrease by a factor of about 1000 in five years and by a factor of about one million in ten years.

6. Suggest conditions under which the following changes might be approximately exponential: breeding of mosquitoes in a marsh; growth of money in a bank account.

Links to the *Advancing Physics* CD-ROM

Practise with these questions:

5C   Comprehension *First steps in mathematical modelling*

10C Comprehension *Disposal of radioactive waste*

20S Short answer *Randomness and half-life*

30S Short answer *Decay in theory and practice*

40S Short answer *Model growth and sample decay*

Try out these activities:

10S Software-based *Models of forest fires and percolation*

20S Software-based *Models of rabbit populations*

50E Experiment *A model of radioactive decay using dice*

60S Software-based *WorldMaker models of radioactive decay*

A-Z

Look up these key terms in the A–Z:
Exponential decay processes; half-life; models; radioactive decay; random processes

Go further for interest by looking at:

10T Text to read *The Turin shroud – relic or fake?*

✓

Revise using the revision checklist and:

10O OHT *Smoothed out radioactive decay*

20O OHT *Radioactive decay used as a clock*

10.2 Stocks and flows

When you run out of food at home you buy more, but you get annoyed if the supermarket runs out. The manager has to plan to replace each item at the same average rate as people are buying it. Similarly, when water stored in reservoirs is used for drinking water there is a problem if rain doesn't replenish it as fast as it is taken out.

Capacitors and electric charge

Modern cameras usually have an electronic flash for taking pictures at night. In the camera, opposite electric charges are placed on the conducting plates of a capacitor by connecting them to a large potential difference. The capacitor is then discharged through the flash tube. The discharge is so rapid that motion in front of the camera is "frozen" in the resulting photo. But you may have to wait a while after taking a picture for the charge separation to be built up again and for the flash to work. It's like slowly filling up a large tank and then letting the water out in a rush.

A similar natural phenomenon is lightning: air currents carry charged ice crystals in storm clouds, slowly building up a large potential difference between the top and bottom of the cloud. Then suddenly the air conducts and a lightning flash discharges the cloud – one of nature's capacitors.

A capacitor is very simple – just a pair of electrical conductors close together. Because charges come in pairs, positive and negative, "charging" a capacitor means pulling those charges apart and getting the positive charge on one conductor and the negative charge on the other. Capacitors are often made of sheets of metal foil with an insulating layer between them.

Water clocks

Greek and Roman law courts used a water clock to keep advocates from talking too much. A bowl full of water dripped into another bowl underneath. When the top bowl (called a clepsydra, meaning "water thief") was empty, the speaker's time was up. For centuries, until the invention of clockwork, water or sand clocks and candles or incense sticks marked with time graduations were the only ways (in addition to sundials) to measure time.

As the water in a water clock runs out, the

Key summary: water and electric charge

Water

reservoir filled with water

pressure difference increases as amount of water behind dam increases

Electric charge

conducting plates with opposite charges concentrated on them

$-Q$ $+Q$

potential difference V increases as amount of charge separated increases

Define capacitance:
$C = Q/V$
charge separated per volt

to calculate Q or V:
$Q = CV$
$V = Q/C$

Units:
charge Q
potential difference V
capacitance C

coulomb C
volt V
farad $F = C\,V^{-1}$

Capacitors keep opposite charges separated. The charge is proportional to the capacitance at a given potential difference.

This one-sixth size working model of a Chinese wooden water clock dates from AD 1080. It works like a water wheel, turning at a steady rate. The original mechanism was used to drive an astronomical clock.

An early Egyptian clock, 1415–1380 BC. This is a plaster cast of the original found at Karnak Temple in 1904. The original is made of alabaster. In use it was filled with water which slowly leaked out through a small hole in the bottom.

Key summary: exponential water clock

What if...
volume of water per second flowing through outlet tube is proportional to pressure difference across tube, and the tank has uniform cross section?

volume of water V

pressure difference p

height h

fine tube to restrict flow

flow rate = dV/dt

Pressure difference proportional to height h. Constant cross section so height h proportional to volume of water V.

$$p \propto V$$

Rate of flow of water proportional to pressure difference

$$f = dV/dt \propto p$$

flow of water decreases water volume
rate of change of water volume proportional to water volume

$$dV/dt \propto -V$$

V

t

time to half empty is large if tube resists flow and tank has large cross section

Water level drops exponentially if the rate of flow is proportional to pressure difference and the cross section of the tank is constant

Key summary: exponential decay of charge

What if...
current flowing through resistance is proportional to potential difference and potential difference is proportional to charge on each plate of the capacitor?

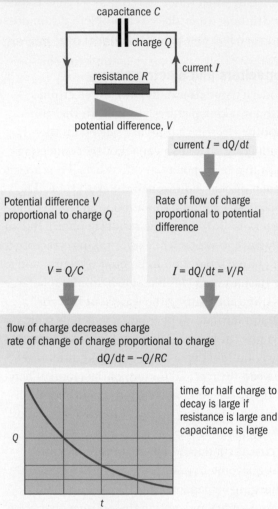

capacitance C

charge Q

current I

resistance R

potential difference, V

current $I = dQ/dt$

Potential difference V proportional to charge Q

Rate of flow of charge proportional to potential difference

$$V = Q/C$$

$$I = dQ/dt = V/R$$

flow of charge decreases charge
rate of change of charge proportional to charge

$$dQ/dt = -Q/RC$$

Q

t

time for half charge to decay is large if resistance is large and capacitance is large

Charge decays exponentially if the current is proportional to potential difference and the capacitance C is constant

drips from it come more slowly, because there is less pressure from the water that is left helping to push the drops out. The pressure depends on how deep the water is in the clock, so it could be that the fall rate of the water level in the clock is proportional to the level itself. If so, the water level would fall exponentially.

Before the advent of digital timing, ingenious physicists sometimes used capacitors as electric water clocks to time rapid events such as the

flight of a bullet, judging the time interval by how much charge had flowed between the plates of a capacitor. The larger the product RC of resistance R and capacitance C, the slower the charge decays. To measure short times, R and C both need to be kept small.

You can see that the way a capacitor discharges through an ohmic resistor (so that the rate of flow of charge is proportional to the potential difference driving the flow) is exactly analogous to the behaviour of an exponential water clock. Actually, it's harder to get a real water clock to behave exponentially because the rate of flow of water isn't necessarily proportional to the pressure difference, especially if the water rushes and swirls as it comes out.

Exponential change: a fundamental pattern

We have given you several examples of the same exponential pattern, both growth and decay. All are fundamentally similar. The things they have in common are simply:

- at any instant, the rate of change of a quantity Q is proportional to the amount of that quantity present;
- $dQ/dt = kQ$, where the constant k may be positive (growth) or negative (decay);
- the quantity Q grows or decays so that its amount changes by a constant factor in equal intervals of time. *Adding* to the time *multiplies* the quantity.

There is a number, another of those unending numbers like π, which is the natural base to use for exponential change. It is the number $e = 2.718...$ You can see it on your calculator if you enter 1 and press the function key for e^x. The number e arises naturally if you consider the simplest possible case of exponential decay. This case is just $dQ/dt = -Q$, with $k = -1$. The relationship between Q and t is then $Q/Q_0 = e^{-t}$. In unit time, Q falls by the factor e. To find the actual value of e you have to do some arithmetic. You cannot get a "final right answer" – just better and better approximations. The way shown in "Key summary: steps to the value of e" is to divide the unit time into ever smaller slices. In each slice you assume – but never quite correctly – that the rate of change is given by the value of Q at the start of the interval. Then you draw a straight line of that slope for that interval. Starting from the initial value of Q, you get a graph which, like the exponential decay curve, slopes down but ever less steeply. Unlike the exponential decay curve, it is made of straight line pieces, each sloping down a bit too steeply. So the value of Q after unit time is always a bit too small, and the value of e you obtain is a bit too big. However, the narrower the time slices you choose, the nearer you get to $e = 2.718...$, and to $\frac{1}{e} = 0.3678...$ for the ratio of Q after unit time to its value at the start.

Key summary: steps to the value of e

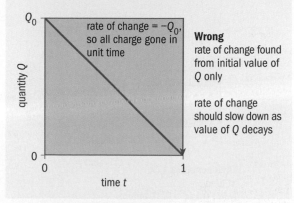

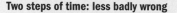

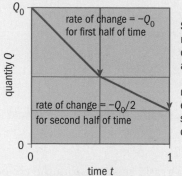

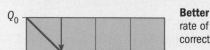

The decay curve can be approximated by a series of short steps

Key summary: radioactive decay times

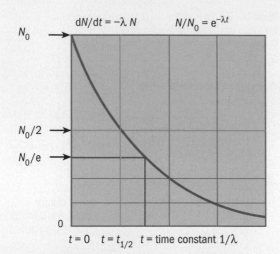

$$\frac{dN}{dt} = -\lambda N \qquad N/N_0 = e^{-\lambda t}$$

$t = 0 \quad t = t_{1/2} \quad t = \text{time constant } 1/\lambda$

Time constant $1/\lambda$
at time $t = 1/\lambda$
$N/N_0 = 1/e = 0.37$ approx.
$t = 1/\lambda$ is the time constant of the decay

Half-life $t_{1/2}$
at time $t_{1/2}$ number N becomes $N_0/2$
$N/N_0 = \frac{1}{2} = -\exp(-\lambda t_{1/2})$
$\ln \frac{1}{2} = -\lambda t_{1/2}$
$t_{1/2} = \frac{\ln 2}{\lambda} = \frac{0.693}{\lambda}$

The half-life $t_{1/2}$ is related to the decay constant λ

The step-by-step graph gets you from the recipe for how a quantity changes – the differential equation that says: "If it's like this now, then what happens next is…" – to how the quantity changes gradually over time. The great Swiss-born mathematician Leonhard Euler (1707–1783) invented this simple step-by-step method. The number e is sometimes known as Euler's number.

For capacitor discharge, the differential equation is:

$$\frac{dQ}{dt} = -\frac{Q}{RC}$$

with the constant k equal to $-\frac{1}{RC}$. The separated charge then changes with time as

$$\frac{Q}{Q_0} = e^{-t/RC}$$

often written instead as

$$\frac{Q}{Q_0} = \exp\left(-\frac{t}{RC}\right)$$

The product RC has units of time (you can check this). After time $t = RC$, the ratio $\frac{t}{RC}$ is 1 and Q has been reduced to $\frac{1}{e}$ of its original value. The product RC is known as the **time constant** of the discharge circuit.

Fast flashes of energy

It's the night of a big film première. Photographers' camera flashes go off in bursts as each celebrity arrives. Like lightning, a camera flash is a reminder that energy is stored when electric charge is separated. The camera contains a capacitor, which it charges to a potential difference of say 200 V, enough to trigger the ionisation of the xenon gas in the flash tube. The capacitor in your camera may have a capacitance of, say, 2 millifarad (mF).

Since $C = \frac{Q}{V}$, the charge separated is:

$$Q = CV = 2 \times 10^{-3} \, \text{F} \times 200 \, \text{V} = 0.4 \, C$$

If a steady supply of 200 V supplied a charge of $0.4 \, C$, the energy delivered would be $E = QV = 0.4 \, C \times 200 \, \text{V} = 80 \, \text{J}$. But that's not true for a capacitor because, as it gives up its charge, the potential difference drops as there is less charge separated between its plates. So the energy per unit of charge falls during the discharge. The result is that the energy stored and provided during discharge is:

$$E = \frac{1}{2}QV$$

The energy stored in your flash gun capacitor is therefore:

$$\frac{1}{2} \times 0.4 \, C \times 200 \, \text{V} = 40 \, \text{J}$$

Typically, the bright flash lasts about 1 ms, "freezing" movement. Although the energy of the flash is the same as that from a not very bright 40 W lamp for one second, the power of the flash is much greater because it is all delivered in such a short time:

$$\text{power of flash} = \frac{\text{energy delivered}}{\text{time}}$$

Key summary: energy stored by a capacitor

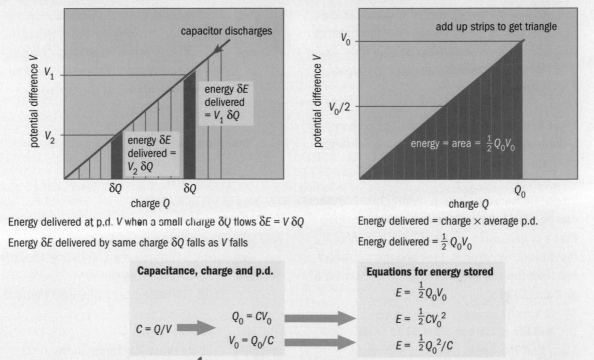

Energy delivered at p.d. V when a small charge δQ flows $\delta E = V\,\delta Q$

Energy δE delivered by same charge δQ falls as V falls

Energy delivered = charge × average p.d.

Energy delivered = $\tfrac{1}{2}Q_0 V_0$

Capacitance, charge and p.d.

$$C = Q/V \quad\Longrightarrow\quad \begin{array}{l} Q_0 = CV_0 \\[4pt] V_0 = Q_0/C \end{array}$$

Equations for energy stored

$$E = \tfrac{1}{2}Q_0 V_0$$
$$E = \tfrac{1}{2}CV_0^{\,2}$$
$$E = \tfrac{1}{2}Q_0^{\,2}/C$$

The energy stored by a capacitor is $\dfrac{1}{2}QV$

In a lightning flash a faint "leader" seeks out a path from the cloud towards the ground, taking about 50 ms and carrying a charge of about 5 coulombs at a potential as high as 100 million volts. This leader is met by an upward-moving streamer that produces the powerful "return stroke" as the lightning travels from ground to cloud in less than 0.1 ms, delivering a peak current of about 30 000 amperes.

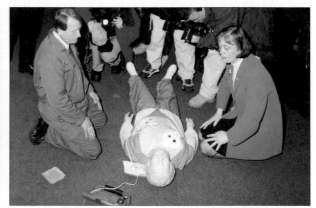

Energy stored on a capacitor can save lives. Heart attacks occur when the heart muscles start rapid random contractions. A controlled electric shock delivered from a charged capacitor via electrodes on the chest can restore the regular heartbeat, but this needs to be done quickly. It is important to have small portable heart defibrillators which can be taken quickly to the patient. This picture shows a demonstration on a dummy.

$$= \frac{40\,\text{J}}{10^{-3}\,\text{s}} = 40\,\text{kW}$$

To get such a short flash, the resistance of the ionised xenon in the flash tube must be very small. The time constant RC must be of the order the flash time 1 ms, so that

$RC = R \times 2 \times 10^{-3}\,\text{F} = 10^{-3}\,\text{s}$, giving $R = 0.5\,\Omega$. The charge $0.4\,C$ takes much longer, perhaps 20 s, to be pulled apart onto the plates of the capacitor by the camera's charging circuit. That's an average charging current of only 20 mA. This is why you have to wait some time between taking photographs using a flash.

Quick check

1. You are designing a circuit to measure a time of about 1 ms by observing the decay of charge on a capacitance of 1 μF during this time. Show that you should choose a resistor of roughly 1 kΩ resistance.

2. Show that the time taken to reduce a charge to a tenth of its original value is about 2.3 times the time constant *RC*.

3. Power for memory backup of integrated circuits can be provided by a charged capacitor. Show that a capacitance of about 1 mF is needed if the current required is 1 μA and the p.d. of 5 V must not drop by more than 10% in a quarter of an hour (1000 s).

4. The half-life of carbon-14 is 5730 years. Show that the decay constant λ is about $1.2 \times 10^{-4}\,\text{yr}^{-1}$ and that this is equivalent to about $4 \times 10^{-12}\,\text{s}^{-1}$ (1 year = $3.1 \times 10^{7}\,\text{s}$).

5. Use the data from question 4 to show that a mass of about 6 μg of carbon-14 is needed to give an activity of $10^{6}\,\text{Bq}$.

6. A nuclear fusion device (Tokamak) is required to deliver 1 MJ of energy to a gas discharge, using capacitors discharged through the gas. Show that a capacitance of about 0.02 F is required if the largest workable potential difference is 10 kV.

Links to the *Advancing Physics* CD-ROM

Practise with these questions:
60S   Short answer *Charging capacitors*
65S   Short answer *Separating charges*
70S   Short answer *Discharge and time constants*
80S   Short answer *Discharging a capacitor*
110S Short answer *Energy stored by capacitors*
120S Short answer *Energy to and from capacitors*

Try out these activities:
140S Software-based *Modelling the Euler algorithm graphically*
150S Software-based *Stepwise through decay*

Look up these key terms in the A–Z:
Capacitance; differential equation; exponential decay processes; rates of change; *RC* circuits

Go further for interest by looking at:
20T   Text to read *Modelling conquers the Atlantic – eventually*

Revise using the revision checklist and:
300   OHT *Analogies between charge and water*
400   OHT *Half-life and time constant*
500   OHT *Energy stored by a capacitor*

10.3 Clockwork models

Are you always on time, or often late? Timetables rule schools and colleges, businesses and travel. Workers clock in and out; paid by the hour, their time has become money. Yet in the story of humankind, accurate clocks and careful timekeeping are relatively recent. Mechanical clocks first appeared in Europe in the 14th century, but it was Huygens who, in 1656–7, made the first really good pendulum clock. Galileo had the idea as early as 1581, inspired – so the story goes – by timing with his pulse a chandelier swinging in the cathedral at Pisa. Galileo found that the time of swing did not seem to depend on how big the swing was. Huygens found how to make this constant timekeeping property control a working clock.

Huygens' first clock was accurate to about one minute a day. The swinging pendulum kept the time – its "tick-tock" sound coming from mechanical pulses turning the clock hands. To keep it going, the pendulum took energy from a descending weight. Today, you probably think nothing of a cheap watch able to time events to a hundredth of a second, and which gains or loses barely a fraction of a second in a day: one part in 10^5. The satellite Global Positioning System relies on clocks that are stable to one part in 10^{13}.

About 300 years ago Galileo took a huge step forward, now taken completely for granted. He introduced the time variable t into scientific thinking. The very idea of rates of change $\frac{dQ}{dt}$, invented by Newton and Leibniz in the 17th century, is Galileo's grandchild.

In those 300 years, "clockwork" has become a metaphor for smooth, reliable, predictable running. The regular rhythms of the heavens had for millennia ruled sowing and reaping by farmers. Now the idea grew up that these had their origin in the universe itself being a kind of clockwork (chapter 11). No idea has been more important to western, and now world, culture. The universe became imagined as a clockwork machine, its behaviour decided by mechanical principles. Fates, spirits and gods lost their powers, and science acquired its reputation for reducing everything to basic mechanical principles. Hated by some, gloried in by others, this shift in the imagination still drives thought today. All down to clocks? Maybe.

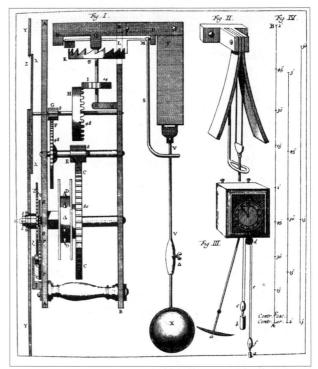

An illustration of the movement of one of Huygens' first pendulum clocks. The curved pieces of metal at the top of the pendulum guided it into a cycloidal arc that helped to achieve isochronicity.

The "pips" arrive in person. Early attempts to standardise timekeeping used a trustworthy clock set to London time, to which distant clocks were synchronised. Miss Ruth Belville, shown here, continued the practice. She called at the Royal Greenwich Observatory every Monday until the 1930s to check her chronometer against Greenwich time and would then visit the principal clockmakers in London to set their timekeepers. The Greenwich Time Signal now performs this function.

Different oscillators: different clocks

Huygens' clock was revolutionary because it used the regular swing of a pendulum to keep time. The crucial property was that the length of swing did not depend on the amplitude of the oscillations (to a good approximation, for small amplitudes). So the pendulum provided a steady time pulse, despite the clock running down. The swings of a pendulum are thus said to be (approximately) isochronous (from the Greek for "same time").

Key summary: language to describe oscillations

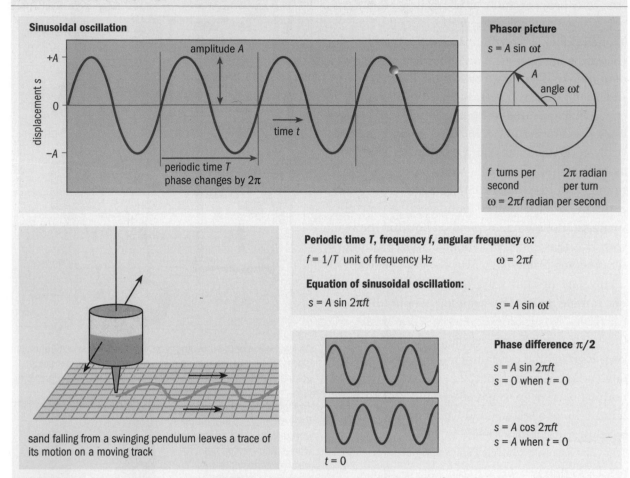

Sinusoidal oscillation

amplitude A

displacement s

+A

0

−A

time t

periodic time T
phase changes by 2π

Phasor picture

$s = A \sin \omega t$

A

angle ωt

f turns per second

2π radian per turn

$\omega = 2\pi f$ radian per second

sand falling from a swinging pendulum leaves a trace of its motion on a moving track

Periodic time T, frequency f, angular frequency ω:

$f = 1/T$ unit of frequency Hz

$\omega = 2\pi f$

Equation of sinusoidal oscillation:

$s = A \sin 2\pi f t$

$s = A \sin \omega t$

Phase difference $\pi/2$

$s = A \sin 2\pi f t$
$s = 0$ when $t = 0$

$s = A \cos 2\pi f t$
$s = A$ when $t = 0$

$t = 0$

A sinusoidal oscillation has an amplitude A, periodic time T, frequency $f = \dfrac{1}{T}$ and a definite phase

This is a property shared by a large number of oscillating systems called harmonic oscillators.

Every mechanical harmonic oscillator has the following features:

- it is accelerated back towards the equilibrium position by a spring-like force that always pulls it back towards the equilibrium position;
- at the equilibrium position there is no net force; here the velocity of the mass continues unchanged;
- it stores energy. The energy goes back and forth, all being stored by some kind of "spring" (potential energy) at the extremes of the oscillation, and all being carried by the motion (kinetic energy) as it passes through the equilibrium position;
- resistive forces gradually take energy from the oscillator so that its amplitude decreases unless

energy is fed back in to compensate;

- its time trace is a sinusoidal curve, oscillating at the natural frequency of the oscillator.

The next big advance in clockmaking was the result of developing sea trade. Ships were often lost through inaccurate navigation. A good clock makes it possible to find longitude. You note the difference in the time of midday between where you are and a standard place, say Greenwich. If you have a clock set to Greenwich time, and the Sun reaches its highest point two hours after noon by that clock, you know that you are $\frac{2}{24}$ of the way round the world from Greenwich, or 30 degrees. But Earth turns quickly, and an error of eight seconds in the Greenwich time of your travelling clock puts you (at the equator) two nautical miles off position. Enough to hit the rocks.

Which is what happened on a foggy autumn

John Harrison holding a copy of his best and smallest chronometer. Over his left shoulder can be seen a long case clock with a gridiron pendulum designed not to expand when the temperature rises – another of his inventions.

This is John Harrison's third attempt at a marine chronometer. Two balance wheels, controlled by spiral springs, oscillate back and forth, counter-rotating to compensate for the motion of a ship. The final design was much more compact than this huge machine and bore a closer resemblance to the timepieces we know today. Captain James Cook took a copy on his voyages.

night in 1707. A Royal Navy flotilla under Admiral Sir Clowdisley Shovell mistook its longitude and struck rocks off the Scilly Isles, with the loss of four ships and about 2000 men. The resulting public outcry gave urgency to finding a better method of establishing longitude at sea. The government set up a Board of Longitude, which offered a prize of £20000 (a big lottery win in today's money) for a solution. After much dispute, and against the wishes of the Astronomer Royal, the prize was finally given grudgingly to a Yorkshire carpenter turned clockmaker called John Harrison (1693–1776).

Harrison knew that pendulum clocks would be no good in a ship rolling on the ocean swell so he invented new kinds of oscillator and made their time of oscillation as constant as possible. In all he made four clocks, each better than the last. One, shown on this page, used a pair of balance wheels, twisted back and forth by spiral springs. A time trace of the motion of Harrison's wheels would need to plot angle against time, not displacement against time. But this trace would be harmonic, a sinusoidal curve mirroring that of a pendulum.

Modern quartz crystal timing devices use masses and springs provided free by nature – the masses are quartz atoms and the springs are the forces between them. A slice of pure quartz crystal oscillates with a very stable frequency. Quartz is used because it is piezo-electric (also used in ultrasound generators, chapter 1, *Advancing Physics AS*). This means that if a potential difference is applied across the crystal, the crystal is stretched or squashed. So an alternating potential difference can make it oscillate. As it oscillates the motion is picked up as an oscillating potential difference

across the crystal, which can be used to control digital circuits that count and display the time.

Today, using atomic clocks, time and frequency are the quantities physicists can measure with the greatest accuracy. Even distance is measured by time – the distance light travels in a defined time.

Modelling harmonic oscillators

To understand the essentials of a harmonic oscillator, we ask you, as with earlier models (p2), to imagine a very stripped down case. This is not because only simple cases can be modelled; it is because more complicated examples just add extra features on top of the essentials. So we look at the essentials first. One of Harrison's clocks used Hooke's principle that the force exerted by a stretched or squashed spring is proportional to the deformation. We shall show you that this, with Newton's laws of motion, is all you need to model harmonic oscillators.

Think of a mass, held between two rigid walls by a pair of springs. Forget about the need to support the mass: imagine perfect wheels or a super air cushion. Forget about friction. These can be put back in later.

Imagine holding the mass to one side. The spring with the largest extension exerts the largest force. Now let go. The mass is accelerated towards the centre, speed builds up and the mass travels past the centre. Now the situation is reversed so that the mass slows down until it stops with a net force in the opposite direction. The whole sequence of actions repeats in reverse, taking the

Key summary: motion of a harmonic oscillator

Everything about harmonic motion follows from the restoring force being proportional to minus the displacement

mass back to where it started, ready to start a new cycle of the motion.

Predicting the time trace

The velocity of the mass is changing all the time, which makes predicting where it will get to next just a bit tricky. Air-traffic controllers have the same problem: they must observe an aircraft's velocity v now and predict where it will be next – you don't want to end up with two planes in the same place at the same time. This can be done, approximately, by taking a short time slice δt and saying that the aircraft will travel through a displacement $v\delta t$. This works only if the velocity doesn't change much in time δt. Then they must either observe the new velocity and do it all over again for the next time interval, or they must have a flight plan saying what the velocity is going to be. This is how approximate step-by-step numerical predictions of the paths of moving objects can be made.

Key summary: dynamics of a harmonic oscillator

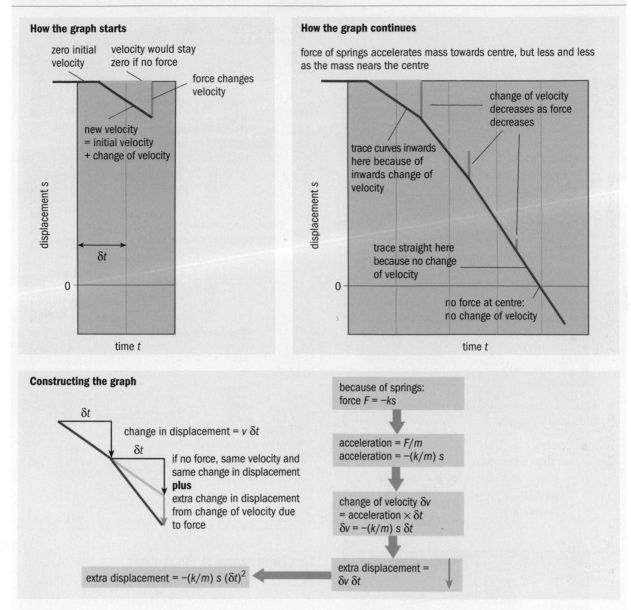

How the graph starts

zero initial velocity

velocity would stay zero if no force

force changes velocity

new velocity = initial velocity + change of velocity

δt

displacement s

0

time t

How the graph continues

force of springs accelerates mass towards centre, but less and less as the mass nears the centre

change of velocity decreases as force decreases

trace curves inwards here because of inwards change of velocity

trace straight here because no change of velocity

no force at centre: no change of velocity

displacement s

0

time t

Constructing the graph

δt

change in displacement = $v\,\delta t$

δt

if no force, same velocity and same change in displacement
plus
extra change in displacement from change of velocity due to force

extra displacement = $-(k/m)\,s\,(\delta t)^2$

because of springs:
force $F = -ks$

acceleration = F/m
acceleration = $-(k/m)\,s$

change of velocity δv
= acceleration $\times \delta t$
$\delta v = -(k/m)\,s\,\delta t$

extra displacement =
$\delta v\,\delta t$

Health warning! This simple (Euler) method has a flaw. It always changes the displacement by too much at each step. This means that the oscillator seems to gain energy.

Nature has already filed the "flight plan" for the harmonic oscillator (as for a projectile, chapter 9, *Advancing Physics AS*). With mass m and spring constant k the acceleration is given by:

$$a = \frac{F}{m} \text{ and } F = -ks \text{ so } a = -\frac{k}{m}s$$

Newton's laws let you calculate the acceleration, given the mass and the spring force at a given displacement s. From the acceleration, you find the change in velocity. From the velocity you find the new displacement. Starting from a known position and velocity, you can track the motion of an oscillating mass, moment by moment.

The time trace of the displacement always curves inward. This is because the spring always pulls the mass back to the centre. You get a sinusoidal curve when the acceleration is proportional to the displacement and opposite in direction to it.

Key summary: changing rates of change

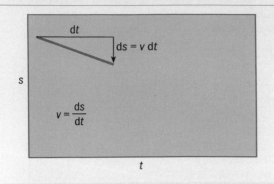

slope = rate of change of displacement
= velocity v

rate of change of slope = rate of change of velocity

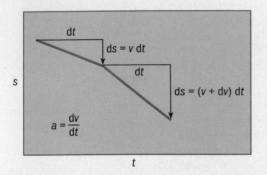

new slope = new rate of change of displacement
= new velocity $(v + dv)$

new $ds = (v + dv)\,dt$
$dv = a\,dt$

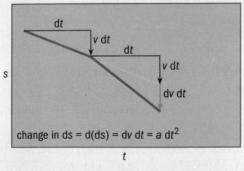

change in $ds = d(ds) = dv\,dt = a\,dt^2$

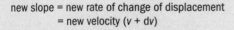

change in $ds = d(ds) = dv\,dt$
$= a\,dt^2$

$$\frac{d}{dt}\left(\frac{ds}{dt}\right) = \frac{d^2s}{dt^2} = a$$

The first derivative $\dfrac{ds}{dt}$ says how steeply the graph slopes. The second derivative $\dfrac{d^2s}{dt^2}$ says how rapidly the slope changes.

What is the period of a pendulum?

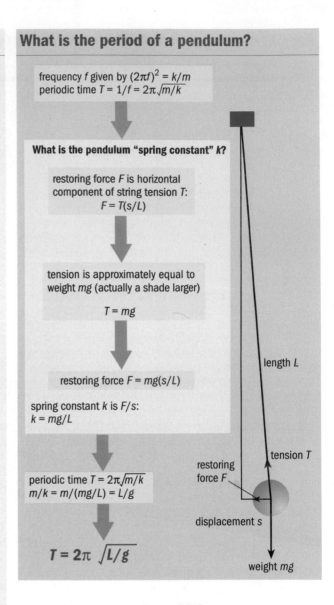

frequency f given by $(2\pi f)^2 = k/m$
periodic time $T = 1/f = 2\pi\sqrt{m/k}$

What is the pendulum "spring constant" k?

restoring force F is horizontal component of string tension T:
$F = T(s/L)$

tension is approximately equal to weight mg (actually a shade larger)
$T = mg$

restoring force $F = mg(s/L)$

spring constant k is F/s:
$k = mg/L$

periodic time $T = 2\pi\sqrt{m/k}$
$m/k = m/(mg/L) = L/g$

$$T = 2\pi\sqrt{L/g}$$

length L

tension T

restoring force F

displacement s

weight mg

The time trace is sinusoidal

Here is a way to show that the time trace of the displacement of a harmonic oscillator is in fact sinusoidal. Start by supposing that it is, and show that this works.

$$\text{Suppose } s = A\sin 2\pi ft$$

$$\text{Then } v = \frac{ds}{dt} = 2\pi fA\cos 2\pi ft$$

$$\text{and } a = \frac{dv}{dt} = -(2\pi f)^2 A\sin 2\pi ft = -(2\pi f)^2 s$$

So the acceleration a is proportional to $-s$, as required.

The argument shows that if you assume that the displacement is sinusoidal, the acceleration is a sinusoidal curve of the opposite sign to the

Key summary: force, acceleration, velocity and displacement

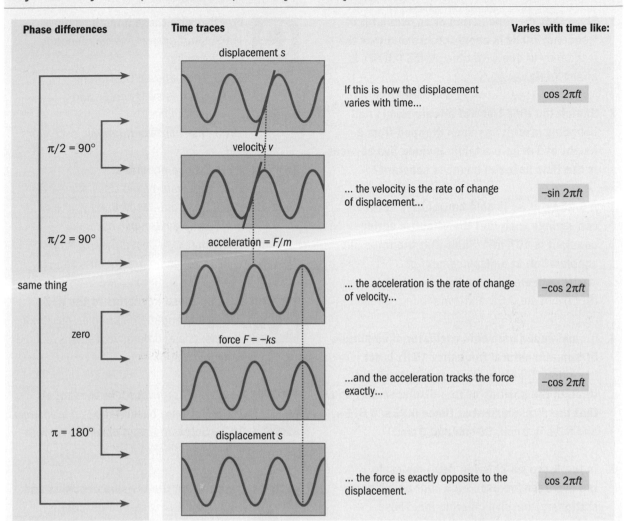

Graphs of displacement, velocity, acceleration and force against time have similar shapes but differ in phase

displacement, which is exactly what the equation

$$\frac{\mathrm{d}^2 s}{\mathrm{d}t^2} = a = -\frac{k}{m}s$$

says it must be. And you get extra information: the frequency of oscillation f is given by

$$(2\pi f)^2 = \frac{k}{m} \text{ or } 2\pi f = \sqrt{\frac{k}{m}}$$

Such simple models

The two highly idealised, very simplified models we have described to you are vital building bricks for making models in physics. They appear over and over again. Exponential changes are important in nuclear chain reactions, fire prevention, the transmission of signals along cables, motion with friction. Harmonic oscillations occur in the swaying of skyscrapers in the wind, sounds of musical instruments, vibrations of atoms in solids and tuning circuits in radios. Other examples of both abound.

Put the two models together and you have something more still: oscillations that die away gradually, like real-world oscillations generally do. So what started looking almost too simple can soon be made very practical. Combining exponential decay and harmonic motion, you can now model the important practical case of exponentially decaying harmonic oscillations.

Quick check

1. Show that the time period of an oscillator of frequency 40 Hz is about 0.025 s and that the frequency of one with time period 0.070 s is about 14 Hz.

2. Sketch the time trace of a table-tennis ball bouncing after it has been dropped from a height of 1 m onto a table. Include five bounces. Is the time between bounces constant?

3. A 0.25 kg mass is held horizontally between two springs such that the effective spring constant is $10 \, \text{N m}^{-1}$. Show that the force and acceleration at a displacement of
   (a) 100 mm are 1 N and $4 \, \text{m s}^{-2}$;
   (b) 50 mm are 0.5 N and $2 \, \text{m s}^{-2}$.

4. An undamped harmonic oscillator of amplitude 50 mm and natural frequency 10 Hz is set in motion. Timing is started when it passes through the position of zero displacement. Show that the displacement at times 0.05 s, 0.075 s and 0.1 s is 0 mm, 50 mm and 0 mm.

5. In the hydrogen chloride (HCl) molecule the hydrogen ion oscillates near an almost stationary, massive chlorine ion. These molecules oscillate at a frequency of approximately 9×10^{13} Hz. Show that this is in the infrared part of the spectrum and that the spring constant of the HCl bond is about $550 \, \text{N m}^{-2}$ (mass of hydrogen atom = 1.7×10^{-27} kg).

6. Part of a building structure is oscillating sinusoidally in the wind with a period of 10 s and amplitude 1 m. Show that the maximum speed is about $0.6 \, \text{m s}^{-1}$ and that the maximum acceleration is about $0.4 \, \text{m s}^{-2}$.

Links to the *Advancing Physics* CD-ROM

Practise with these questions:
150S Short answer *Revisiting motion graphs*
160S Short answer *Oscillators*
170S Short answer *Energy and pendulums*
190S Short answer *Harmonic oscillators*

Try out these activities:
250S Software-based *Oscillating freely*
270S Software-based *Build your own simple harmonic oscillator*
280S Software-based *Step by step through an oscillation*

Look up these key terms in the A–Z:
Differential equation; harmonic oscillator; models; phase difference; radians; simple harmonic motion

Go further for interest by looking at:
290S Software-based *Slopes and models*
300S Software-based *Making links with mathematics*

Revise using the revision checklist and:
600 OHT *A language to describe oscillations*
700 OHT *Snapshots of the motion of a simple harmonic oscillator*
800 OHT *Step by step through the dynamics*
900 OHT *Rates of change*
1000 OHT *Graphs of simple harmonic motion*

10.4 Resonating

A tragic and apparently unlikely accident happened in Angers, France, in 1850. A group of soldiers was marching in step across a bridge when the bridge began to vibrate dramatically. The well-drilled soldiers did not break step but continued the regular beat of their march. The vibration increased until the bridge collapsed, sending more than 200 men to their deaths. This bleak story is an example of resonance. The tragedy occurred because the frequency of the marching feet matched a natural frequency of oscillation of the bridge.

It's the same as when you push a child on a swing. Push regularly in time with the movement of the swing and gentle pushes soon build up a large amplitude. Energy is being fed into the oscillations of the swing, and the oscillations build up until energy is "leaking out" of the motion of

the swing as fast as you feed it in. The oscillating swing itself acts as a store of energy. That energy goes from kinetic energy of the fast-moving child and swing as they pass through the lowest point of the motion, to potential energy stored as the swing goes up in the gravitational field of the Earth and then back again.

Energy stored in a spring

Ancient armies relied heavily on storing energy in springy things. Roman soldiers had catapults able to hurl stones at the enemy; the bowmen at Agincourt had yew bows that needed great strength to pull back; crossbows stored even more energy by having a thicker bow that had to be wound back mechanically. A modern example is the spring suspension of a car, designed to absorb the energy of shocks to the wheels from bumps in the road.

When you stretch a catapult the first little bit is

"No person in charge of a troop or body of persons exceeding twenty in number shall permit such troop or body of persons while on the Bridge and between the piers of the Bridge to proceed in regular step."

(Byelaws for regulating the traffic on the Clifton Suspension Bridge, Bristol... in pursuance of Section 46 of the Clifton Suspension Bridge Act, 1952)

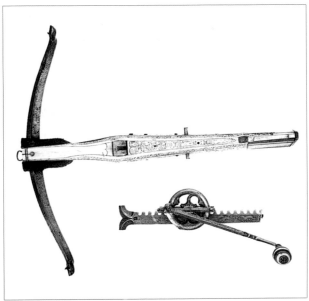

This photograph shows a steel crossbow from mid-16th century Germany. The cord is missing. The bow was tensioned by the ratchet mechanism shown.

Key summary: energy stored in a spring

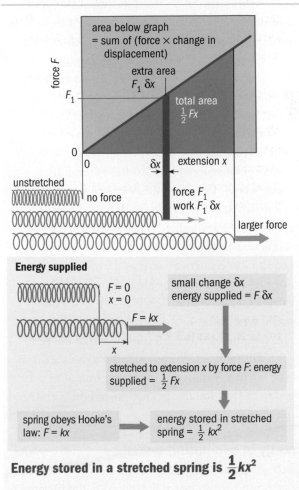

Energy stored in a stretched spring is $\frac{1}{2}kx^2$

This suspension bridge over the Tacoma Narrows near Seattle, Washington state, US, collapsed soon after it was built. In a strong wind, the bridge oscillated. It was known locally as Galloping Gertie. One day in 1940, when the wind induced resonant oscillations in the bridge, the oscillations built up enough to destroy the bridge.

easy: you don't have to pull hard, and you store little energy in the elastic. Let go, and the stone drops feebly from it. The small force from your hand did little work in extending it. But stretch the catapult some more and the job gets harder. The force to stretch it is larger, and for the same extra extension you store more energy than before. Pull even harder, store even more energy for the extra stretch, and a stone flies off impressively, even dangerously, when you let go.

The force F to extend a spring by x is $F = kx$, if the force is proportional to the extension (Hooke's law). But the energy stored is not simply force×displacement, $Fx = kx^2$. It isn't, because the force grows steadily larger as the spring is stretched. The average force is half as much, $\frac{F}{2}$, since the force starts at zero and finishes at F.

As a result, the energy stored in a stretched spring $E_{\text{potential}} = \frac{1}{2} Fx$. Since $F = kx$, this can also be written as $E_{\text{potential}} = \frac{1}{2} kx^2$. The squared term accounts for the increasing extra energy stored for each additional equal extra extension of the spring. The form of the argument is very similar to that for the energy stored on a charged capacitor (p13).

Energy in oscillating systems

The energy stored in the stretched cables of a suspension bridge just stays there for the lifetime of the bridge. But the energy stored in a mechanical oscillator is continually shifting back and forth, from energy carried by the motion of the moving parts (kinetic energy) to energy stored in the restraining springs (potential energy). The energy in an oscillator continually passes back and forth between these two parts of the system. The energy associated with the vibration gradually leaks out and, as the energy stored gets less and less, the oscillations die away.

In a harmonic oscillator, the stretch x of the spring is the displacement s of the oscillator. So the potential energy is $\frac{1}{2} ks^2$ and varies as the displacement changes. But the potential energy is always positive, while the displacement passes through zero and changes sign twice every cycle. So the potential energy reaches a maximum twice per cycle, not once. It doesn't matter from the energy storage point of view whether the springs are stretched or squashed in one direction or in the opposite direction.

It is the same with the kinetic energy, $\frac{1}{2} mv^2$. The velocity reaches a maximum twice per cycle, first in one direction and then again coming back. The kinetic energy reaches its maximum just as the potential energy reaches its minimum, and vice versa, so the kinetic and potential energies have exactly opposite phases. The sum of the two is constant at all times.

Key summary: energy flow in an oscillator

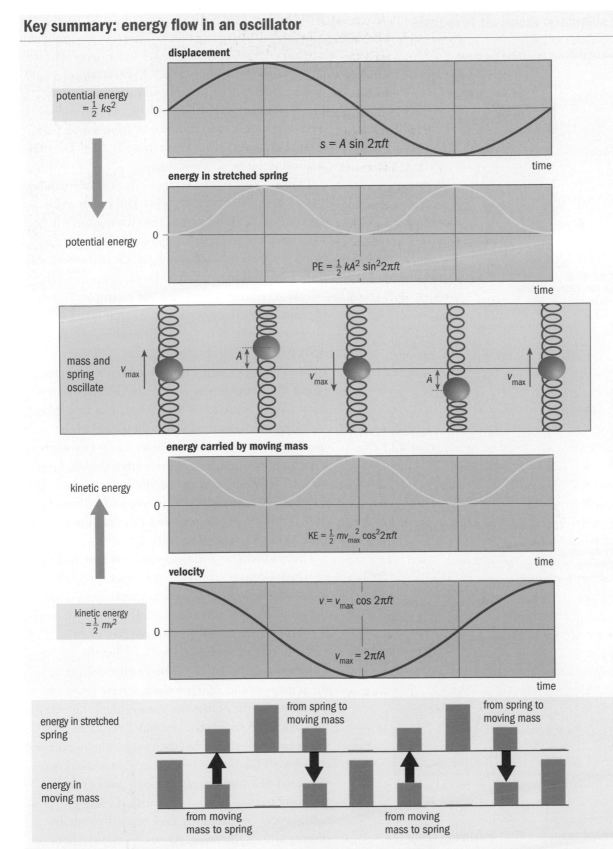

potential energy
$= \frac{1}{2} k s^2$

potential energy

displacement

$s = A \sin 2\pi f t$

time

energy in stretched spring

$PE = \frac{1}{2} k A^2 \sin^2 2\pi f t$

time

mass and spring oscillate

v_{max} A v_{max} $\dot{A}$ v_{max}

kinetic energy

energy carried by moving mass

$KE = \frac{1}{2} m v_{max}^2 \cos^2 2\pi f t$

time

kinetic energy
$= \frac{1}{2} m v^2$

velocity

$v = v_{max} \cos 2\pi f t$

$v_{max} = 2\pi f A$

time

energy in stretched spring

from spring to moving mass

from spring to moving mass

energy in moving mass

from moving mass to spring

from moving mass to spring

The energy stored in an oscillator goes back and forth between stretched spring and moving mass, between potential and kinetic energy

Key summary: resonant response

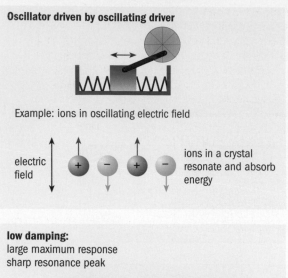

Oscillator driven by oscillating driver

Example: ions in oscillating electric field

electric field

ions in a crystal resonate and absorb energy

low damping:
large maximum response
sharp resonance peak

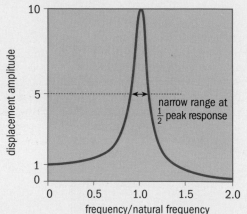

narrow range at $\frac{1}{2}$ peak response

(displacement amplitude vs frequency/natural frequency)

more damping:
smaller maximum response
broader resonance peak

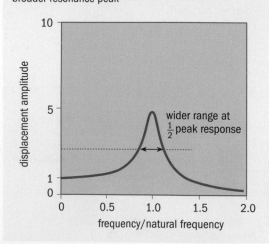

wider range at $\frac{1}{2}$ peak response

(displacement amplitude vs frequency/natural frequency)

Resonant response is at maximum when the frequency of a driver is equal to the natural frequency of an oscillator

Resonance everywhere

You are riding a moped or scooter. Annoyingly, at certain engine speeds the rear-view mirror vibrates strongly, blurring your view. Or, on a trip in a car or bus, there is a distracting sound from a body panel vibrating or rattling only at certain speeds. This is resonance: a natural vibration set off with a large amplitude by a force that varies at just that natural frequency.

This is how it works. Any oscillating force makes whatever it affects vibrate a bit. But much more happens if the frequency of the driving force matches a natural frequency of oscillation of the thing it affects. Then the amplitude of the response builds up, and may get very large.

Take a tall metal chimney, for example. Winds blowing on it create eddies of air that break away alternately from either side of the chimney. Each eddy (vortex) gives the chimney a small sideways push as it breaks away. But what if this happens at just the frequency at which the chimney sways naturally? Then each little push from a new eddy breaking away is timed just right to add to the oscillation produced by the last. Again and again the chimney is made to move with a slightly larger amplitude. In the end the oscillations can build up so much that the chimney breaks and collapses. (The spiral strips you see round such chimneys are there to discourage such eddies.)

The resonant response depends on how rapidly energy leaks away from the oscillator – on the amount of damping. Damping can be measured by the number of oscillations until the vibration has died away after a sudden impulse (say to $\frac{1}{e}$ of the initial amplitude). This is called the Q-factor (quality factor). The less the damping, the narrower and sharper the resonant response; the more the damping, the greater the range of frequencies to which the resonator responds. But the greater the damping, the smaller the maximum response. This is why, to remove that annoying vibration of a body panel in a car, you can use wadding packed behind it to absorb energy.

Resonance is extremely important throughout physics and engineering. Atoms resonate to light of the right frequency. Buildings, towers and bridges resonate to vibrations caused by wind. Tuning in the radio uses resonance. Resonance is everywhere.

Resonance, resonance, resonance

Empty bottles and ancient amphitheatres

If you blow across the neck of a bottle you get a low-pitched sound. The mass of air in the neck of the bottle moves in and out, with the air in the body acting like a spring. Ancient open-air amphitheatres sometimes had large narrow-necked bottles, like those used to store oil and wine, placed under the stone seats because they improved the acoustics of the theatre by resonating.

Music in the home

Small loudspeakers may sound "tinny" but hi-fi speakers are often mounted in large cabinets, difficult to fit in with the furniture. They need a large volume of air behind the loudspeaker designed to resonate at low frequencies around 30–40 Hz, boosting the response of the loudspeaker at those low frequencies.

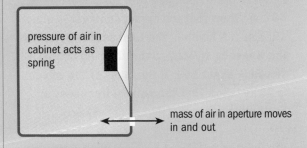

pressure of air in cabinet acts as spring

mass of air in aperture moves in and out

Put it in the microwave

Water molecules flip back and forth in liquid water at frequencies of the order 20 GHz. If your microwave were tuned to this resonant frequency, the absorption of energy would be so strong that the food would only be heated to a depth of 10 mm or so, leaving the interior cold. So the microwave frequency is set lower (typically 2.5 GHz) so that the food is heated throughout. Note that the resonant frequency for water molecules in ice is higher still, even further from resonance. This is why defrosting food in a microwave takes so long.

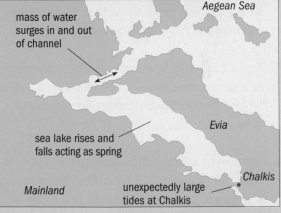

Mysterious tides

At Chalkis in Greece, large tides in the almost tideless Mediterranean Sea puzzled Aristotle. Resonance is the explanation. Chalkis is at the bottom of a "bottle" – a large sea lake with a narrow channel opening at the far end. The spring is the rising and falling of water in the lake under gravity; the mass is water surging in and out of the channel. The natural frequency is very close to the 12.4-hour frequency with which the Moon's tidal tug varies. The lake resonates and the water at Chalkis rises and falls much more than expected.

Aegean Sea

mass of water surges in and out of channel

Evia

sea lake rises and falls acting as spring

Mainland

unexpectedly large tides at Chalkis

Chalkis

Resonance in musical instruments

The xylophone, originating in South-East Asia, is an instrument with wooden bars that are struck to make a note. A version called the marimba taken to South America via Africa has hollow gourds or metal tubes hung below the wooden bars, each tuned to resonate to a bar. The resonance strengthens and enriches the sound.

Resonance wins the radar war

In 1940 two British scientists flew to America with an invention that helped win the Second World War. It was the cavity magnetron, generating powerful short-wavelength radar waves. Inside, cavities in a copper cylinder resonated to electromagnetic waves of just the desired frequencies, building up a large power.

Quick check

1. A spring balance reading up to 10 kg has a scale 100 mm long. Show that the spring stores about 5 J of energy when it carries the maximum load.

2. A 50 kg person sitting on the wing of a car deflects the suspension over that wheel by 50 mm. Show that the spring constant is about $10^4 \, \text{N m}^{-1}$. A bump in the road suddenly deflects the wheel by 100 mm. Show that the energy stored is about 50 J. If the mass of the wheel is 25 kg, show that the natural frequency of oscillation is about 3 Hz.

3. You want to construct a vibration-free table for an experiment by laying a massive slab on springs. The resonant frequency is to be lower than 5 Hz. If the mass of the slab is 10 kg, show that the spring constant of the springs must be less than $10^4 \, \text{N m}^{-1}$.

4. A paper loudspeaker cone has a mass of 5 g and is deflected by 1 mm by a load of 100 g laid on the speaker cone. Show that the resonant frequency of the cone is about 70 Hz.

5. Explain in terms of resonance what can happen to a suspension bridge if it sways with a natural frequency equal to the frequency of gusts of wind on the bridge.

6. One end of a long narrow springy steel strip is attached to a vibration generator. The generator is driven at constant amplitude at a range of frequencies. Sketch a graph of the amplitude of the displacement at the far end of the strip. Pay attention to how it oscillates at a very low frequency.

Links to the *Advancing Physics* CD-ROM

Practise with these questions:
210D Data handling *Energy in a simple oscillator*
220S Short answer *Bungee jumping*
240S Short answer *Oscillator energy and resonance*
250S Short answer *Resonance in car suspension systems*

Try out these activities:
370S Software-based *Energy in oscillators*
360S Software-based *Simulating resonance*

Look up these key terms in the A–Z:
Harmonic oscillator; resonance and damping

Go further for interest by looking at:
380S Software-based *Modelling chaos*
30T   Text to read *The Tacoma Narrows Bridge collapse*
40T   Text to read *Tacoma Narrows: Re-evaluating the evidence*

Revise using the revision checklist and:
1200 OHT *Elastic energy*
1300 OHT *Energy flow in an oscillator*
1400 OHT *Resonance*

Summary check-up

Exponential changes ✓

- In exponential change the rate of change of a quantity is proportional to that quantity
- Radioactive decay is exponential, with $\frac{dN}{dt} = -\lambda N$
- The differential equation for discharge of a charge Q on a capacitance C through a resistance R is $\frac{dQ}{dt} = -\frac{Q}{RC}$
- The solution of the differential equation for discharge of a capacitor is $\frac{Q}{Q_0} = e^{-\frac{t}{RC}}$. The corresponding solution for radioactive decay is $\frac{N}{N_0} = e^{-\lambda t}$.
- The half-life $t_{1/2}$ of radioactive decay is equal to $t_{1/2} = \frac{\ln 2}{\lambda}$, where λ is the decay constant
- The time constant RC is the time for the charge on a capacitor to reduce to $\frac{1}{e} \approx 0.37$ of its former value

Harmonic oscillators ✓

- The period of a harmonic oscillator T is independent of its amplitude. $T = 2\pi \sqrt{\frac{m}{k}}$
- The acceleration of a harmonic oscillator is proportional to its displacement and always acts towards the equilibrium (zero displacement) position
- The differential equation for a harmonic oscillator has the form $\frac{d^2 s}{dt^2} = -\frac{k}{m} s$
- The variation of displacement of a harmonic oscillator with time is sinusoidal, having the general form $s = A \sin(2\pi f t + \phi)$ where ϕ is a phase angle
- There are fixed phase relationships between the variations of displacement, force, acceleration and velocity

Stored energy ✓

- The energy stored by a capacitor is $\frac{1}{2}QV = \frac{1}{2}CV^2 = \frac{1}{2}Q^2C$
- The energy stored in a stretched spring at extension x is $\frac{1}{2}kx^2$ with stretching force $F = kx$
- The energy of a mechanical oscillator is the sum of its potential energy and its kinetic energy

Resonance ✓

- An oscillator driven by a sinusoidally varying force resonates with large amplitude if the force varies at or close to its natural frequency

Questions

1. The data in table 1 give the average rate of decay in disintegrations per minute from a radioactive sample, measured over a period of one minute at successive times.

Table 1

Time/hour	Rate
0	8190
1	6050
2	4465
3	3300
4	2430
5	1800
6	1330
7	980
8	720
9	535
10	395

Rate = rate of disintegration in counts per minute

(a) Approximately how many decays are there in the first hour?
(b) Plot a graph of the rate of disintegration against time. How would you test whether it is exponential in form?
(c) Use the graph to estimate the half-life.
(d) Find the radioactive decay constant. Give the units.
(e) How many nuclei decay per second for each million nuclei in the sample?
(f) Approximately how many undecayed nuclei were there at the start?

2. Technetium-99 is a medical tracer used to diagnose heart and muscle function. It is prepared in an excited state, which decays with a half-life of 6 hours emitting a gamma-ray of energy 140 keV. ($1 eV = 1.6 \times 10^{-19}$ J.) Suppose that a decay rate of $1000 s^{-1}$ is required for diagnosis.
(a) What is the probability of decay, per second, of an excited technetium-99 nucleus?
(b) How many technetium-99 atoms need to enter the patient's heart to give an initial decay rate of $1000 s^{-1}$?
(c) In the end, all these nuclei will decay. What is the total energy in joules released in the patient?
(d) The mass of a technetium atom is 99 atomic mass units (a.m.u.), where 1 a.m.u. $= 1.66 \times 10^{-27}$ kg. What mass of technetium-99 needs to be used in the tracer?
(e) What activity will remain after two days?

3. A $10 000 \mu F$ capacitor is charged to 6 V. It is discharged through a 6000Ω resistor. Then it is recharged, and discharged through a 6 V lamp rated at 1 W.

(a) What is the initial charge on the capacitor?
(b) Estimate the initial current when it starts discharging through the 6000Ω resistor.
(c) Use the answer to (b) to estimate the fraction of the charge that has gone in the first ten seconds.
(d) Calculate the product RC. What are its units? What is its significance?
(e) What energy is stored on the capacitor?
(f) What resistance has the 6 V, 1 W lamp?
(g) Approximately how long will the capacitor take to discharge through the lamp?

4. A scene in a movie shows a car dropped on a flat surface bouncing on its suspension springs. The time of oscillation is about 0.5 s. The mass of the car might be about 500 kg. Take $g = 10 N kg^{-1}$.
(a) Estimate the spring constant of the suspension.
(b) By how much does the car compress the suspension when it is at rest?
(c) If the drop initially compresses the springs by 100 mm beyond the equilibrium point, estimate the energy initially going into the oscillations.
(d) Estimate the maximum velocity of the car as it oscillates.

5. The diagram shows the absorption of infrared radiation by a thin slice of sodium-chloride crystal, plotted against the wavelength of the radiation. This is a resonance effect: the electric field of the radiation makes the charged sodium and chlorine ions oscillate at near to their natural frequency.

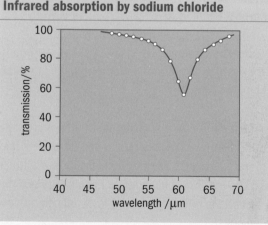

Infrared absorption by sodium chloride

(a) At what frequency is there maximum absorption?
(b) What is the natural frequency of oscillation of ions in sodium chloride?
(c) Taking the mass of an oscillating ion to be of the order 30 a.m.u., with 1 a.m.u. $= 1.66 \times 10^{-27}$ kg, estimate the spring constant for these oscillations.

11 Out into space

From the beginning, people have looked at the sky, imagining what might be out there. Reaching out into space, first in the imagination and now in reality, has changed human beings' understanding of their place in the universe. We will tell the stories of:

- how different models of the solar system arose: the "clockwork universe"
- the idea of universal gravitation
- the idea of momentum and its conservation
- how space probes and satellites are sent into space

11.1 Rhythms of the heavens

The regular cycles of day and night, the phases of the Moon, and seasonal changes in the path of the Sun across the sky set the rhythms of all life on Earth. They have naturally formed the basis of calendars: days, months and years.

There are also less frequent events in the sky. Eclipses of the Moon or Sun used to be seen as omens, almost always bad. Comets came and went

As this time-lapse photograph shows, the stars follow curved paths across the sky. Many stars rise and set. If you can see the Pole Star, other stars near it are in view all night.

These ancient astronomical instruments in Jaipur, India, were used to make accurate observations of the signs of the zodiac.

This radio telescope started operating at Arecibo in Puerto Rico in 1963. Made by reshaping a natural valley to support a metal mesh 300 m in diameter, it is still the largest dish in the world.

Time-lapse simulation of planetary motion

○ Mercury
○ Venus
○ Mars
○ Saturn
○ Jupiter

The word planet means wanderer. As this time-lapse simulation shows, the five planets visible to the naked eye seem to move relative to the fixed stars. Compared to the Sun or Moon, the planets have additional, special motions. They progress at varying speeds and some at times seem to loop backwards before continuing a forward path (retrograde motion).

Night skies through the year

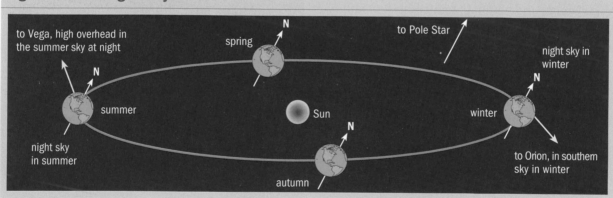

to Vega, high overhead in the summer sky at night

N

spring

N

to Pole Star

night sky in winter

N

summer

Sun

N

winter

night sky in summer

N

to Orion, in southern sky in winter

autumn

without warning. Such events were disturbing, until they too became predictable. Myths tried to explain what was going on "out there".

Early civilisations valued priests and astronomers who understood the rhythm of the heavens and could predict its events. Seed for food crops was irreplaceable and had to be sown at the right time each spring. It was vital to know when to harvest and to move herds. Using the most advanced technologies available at that time, great astronomical observatories were erected for measurement and for ritual. These are some of the largest and most costly engineering projects ever

undertaken. Some took decades to build; their construction depended on being able to house and feed temporary cities of workers and their families. Today's observatories are no less impressive.

Pictorial representations of celestial events come from all cultures and periods, with symbolic as well as practical value. You too have been told stories about how the spinning of the Earth explains night and day, and how the orbiting of a spinning, tilted Earth around the Sun explains the seasons, the changing path of the Sun across the sky and the different constellations of stars seen at different times of the year.

The battle between gods and demons is a central theme in Hindu mythology. At a banquet of the gods, the demon Rahu stole a sip of the elixir of eternal life. Seeing this, the Sun and Moon reported him. The god Vishnu decapitated Rahu and threw his head, which had gained eternal life, up into the sky. In revenge for decapitation, Rahu's head chases the Sun and Moon across the sky, occasionally swallowing them before they re-emerge from his throat.

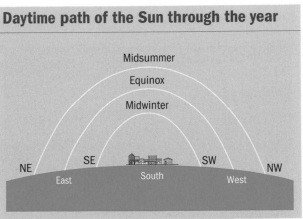

Daytime path of the Sun through the year

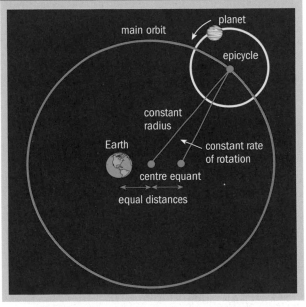

Ptolemy's scheme for planetary motions

Some 3500 years ago the Babylonians noticed that planetary "wanderers" kept to a narrow band of night sky that repeated each year. They divided this band up into 12 zones, each containing a pattern of stars or "constellation" – the origin of the signs of the zodiac. So astrology, the idea that constellations and planets affect human life from birth, is as old as astronomy.

But what moves?

Surely the Earth can't move? Standing in a station, if a train next to yours starts moving, you may think that your train has started. But if the platform moves relative to your train, you know that platforms don't move, so it must be your train.

This is how common-sense thinking about movements seems to work. You never think of the solid ground under your feet as moving because it's what you use to decide whether things are moving or not. So it was with thinking about the universe. The Earth – our human home, the place where we all live – simply couldn't be thought about as moving. It was thus natural to think of objects in the sky as moving around an immovable Earth. It seemed completely obvious. Putting the Earth stationary at the centre of the universe like this is called a geocentric model ("geo" is from the Greek word for Earth). The problem then was how to

describe the motions of objects in the sky.

Greek civilisation was already flourishing 2500 years ago. Applying geometry with considerable skill, Greeks were able to make good estimates of the radius of Earth, and of the size of and distance to the Moon. A number of models were developed to describe simply the observed motions of the Sun, Moon and planets. Viewing the heavens as a sacred and unchanging place, and the circle as a perfect form, it was natural that Greek models were built from uniform, circular motions.

Mediterranean sailors navigated by the position of the Sun, Moon and stars. About AD 120, a Greek mathematician and astronomer called Ptolemy devised a brilliant but complicated geocentric model of the heavens, of wheels within

Key summary: Copernicus' scheme

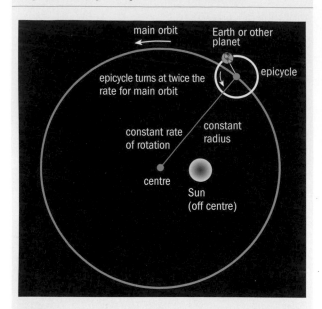

Because the orbits are really ellipses, in his
scheme for planetary motions, Copernicus could
only make the orbits circles by putting the Sun off
centre and by adding epicycles of his own

Key summary: Mars' retrograde motion

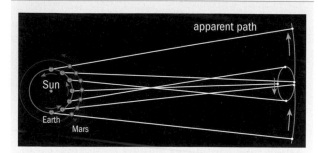

Seen from the moving Earth, Mars seems to make
a backward loop as Earth overtakes on the inside

Copernicus writes candidly to Pope Paul III

"I can well appreciate, Holy Father, that as soon as
certain people realise that in these books which I
have written about the Revolutions of the spheres
of the universe I attribute certain motions to the
globe of the Earth, they will at once clamour for me
to be hooted off the stage with such an opinion...
you expect rather to hear from me how it entered
my mind to dare against the received opinion of
mathematicians and almost against common sense,
to imagine some motion of the Earth."

The Christian church, then a powerful
institution throughout Europe, had adopted a
geocentric model as a picture of God's universe
so Copernicus' view was unpopular, especially
among church people, and eventually his book was
banned. Ever since then the word revolution has
had the connotation of confronting authority.

It is not hard to sympathise with those who
ridiculed Copernicus. He had truly taken the
human race out into space, making Earth a part of
the heavens, just an object somewhere in between
Venus and Mars. Copernicus imagined the Earth
ever travelling – circling annually round the Sun
and spinning daily on its own axis. His model
greatly simplifies the explanation of the retrograde
motion of planets, but it defies common sense:

• if Earth is a moving planet, rotating on an axis,
  why don't objects thrown up in the air get left
  behind?
• if Earth sweeps out an enormous path around
  the Sun, why don't the stars look different as we
  get closer to and farther from them?

Are we alone?

If there really isn't a fundamental difference
between Earth and the heavens, if Earth is just
one planet among many, what is special about life?
Around 1600 the Italian monk Giordano Bruno
got into serious trouble with the Catholic Church
over this question. He had travelled around
Europe, talking with others about the possibility
of life elsewhere in the universe. He believed there
might be an infinite number of inhabited worlds
out in space. Even worse in the eyes of the Church,

wheels. Ptolemy put the spherical Earth near the
centre of the universe. Retrograde motion was
accounted for by imaginary epicycles for each
planet. This model predicted positions with such
accuracy that navigators used it for 14 centuries.

In the 16th century the monk Copernicus, from
Torun in Poland, upset a lot of people with a
book called *On the Revolutions of the Heavenly Spheres*.
Copernicus removed the Earth from the centre of
the solar system to become just another planet.

Five phases of Venus

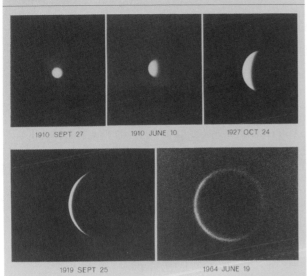

1910 SEPT 27   1910 JUNE 10   1927 OCT 24

1919 SEPT 25   1964 JUNE 19

The heliocentric model can also account for two facts about Venus. It is seen only in the mornings or evenings, never far from the Sun in the sky and, like the Moon, it shows phases.

he suggested that a divine saviour might have lived and died on each of them. This was outrageous heresy, punishable by death. He was handed over to the Inquisition and burned at the stake. Copernicus was right to be cautious.

Celestial clockwork

As a child Tycho Brahe, son of a noble 16th-century Danish family, witnessed an eclipse of the Sun and became fascinated by the power of astronomical prediction. He made regular and precise measurements of the positions of the Sun, Moon and planets for more than 20 years.

German-born Johannes Kepler became convinced that numbers and geometry could express the nature of things; that there must be mathematical reasons behind the relationships found in nature. Kepler first developed a geometrical explanation for the fact that there were just six known planets and for the proportions of their orbital radii. These ideas are now just a curiosity, but they brought him to the attention of Tycho Brahe and Kepler became Brahe's assistant.

Kepler wanted to understand why the planetary system behaves as it does. He wrote a letter in 1605 saying: "My aim is to show that the heavenly machine is not a kind of divine, living being, but a kind of clockwork."

Galileo imagines motion continuing for ever

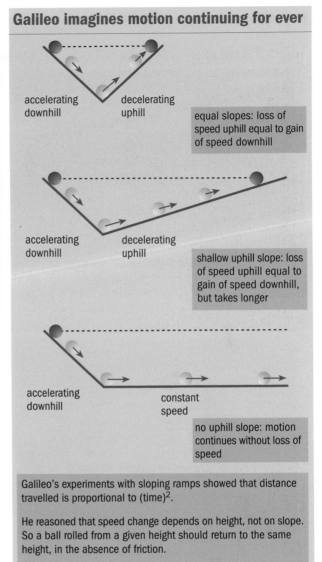

accelerating downhill   decelerating uphill

equal slopes: loss of speed uphill equal to gain of speed downhill

accelerating downhill   decelerating uphill

shallow uphill slope: loss of speed uphill equal to gain of speed downhill, but takes longer

accelerating downhill   constant speed

no uphill slope: motion continues without loss of speed

Galileo's experiments with sloping ramps showed that distance travelled is proportional to $(\text{time})^2$.

He reasoned that speed change depends on height, not on slope. So a ball rolled from a given height should return to the same height, in the absence of friction.

If there is no slope, the ball should roll on forever.

Using Brahe's data for Mars, Kepler found the motion did not match a circular orbit. Although positions differed by only $\frac{8}{60}$ of a degree (your finger's width seen from a bus-length away) this difference was much larger than the uncertainty in Brahe's measurements. Eventually Kepler was able to show that if Mars moved in an elliptical orbit, with the Sun at one focus of the ellipse, everything fitted. He also found how the speed of Mars varied as it went round the Sun.

Kepler wrote one of the first science fiction novels. Called *The Dream*, it told of a journey to the Moon. His fictitious travellers described what it was like to look back at the rotating Earth in the sky above them.

Key summary: Keppler: geometry rules the universe

Law 1: a planet moves in an ellipse with the Sun at one focus

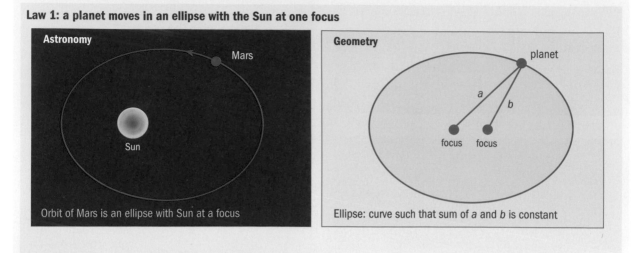

Law 2: the line from the Sun to a planet sweeps out equal areas in equal times

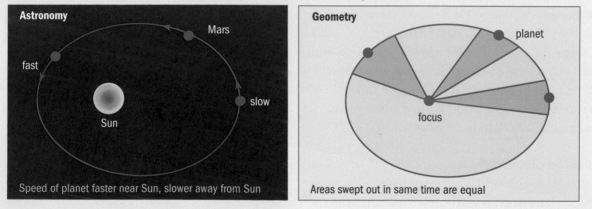

Law 3: square of orbital time is proportional to cube of orbital radius

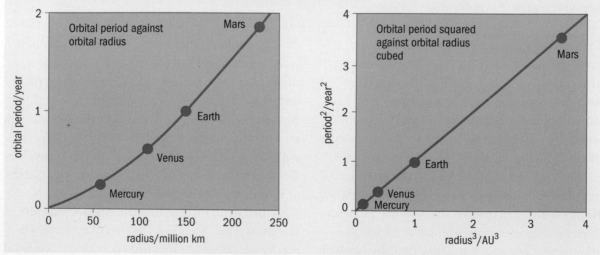

Kepler formulated these three laws governing the motion of planets around the sun

Next on stage is Galileo, a 17th-century teacher at Padua University, also persuaded by this heliocentric view ("helio" is from the Greek word for sun). Galileo was the first astronomer to use the recently invented telescope to aid observations. He made wonderful new discoveries: that the Sun has dark spots and these provide evidence that it too rotates; that Venus exhibits phases and so must orbit the Sun; and that Jupiter has moons. The heavens are not perfect and unchanging.

Galileo also taught the world to think about motion in a new and simpler way. By first imagining an object all by itself, he could then understand how friction and other forces affected its motion. Galileo argued that a body left completely alone continues to move at a constant velocity. This is now known as Newton's First Law.

Going round in circles

Isaac Newton was born in 1642 and grew up in a Lincolnshire village. At the age of 19 he went to Trinity College, Cambridge, to study mathematics. With access to great books, he mostly taught himself. In 1665, to avoid an outbreak of plague, he returned home for two years. During this enforced break, and drawing on earlier thinkers, Newton set out to compile a coherent system of ideas to explain all motions, both on Earth and in the sky.

A major challenge for Newton, before thinking about the planets, was to explain how forces produce a circular path. For example, like everything else on Earth you are going round the Sun at more than 100 000 km an hour. Why do you not notice this speed? It's because everything is travelling together: people, houses, rivers, trees and even the air.

Huygens had experimented with pendulum bobs in clocks moving in circular arcs and had come up with useful ideas. Newton applied them to explaining circular orbits. His conclusion: nothing pushes an object along a circular path. What it takes is a single, external force acting perpendicular to the velocity at every instant.

An object moving along a circular path has an acceleration $\frac{v^2}{r}$ towards the centre of the circle. This means that there must be a force acting on it in that direction. This force will be $F = ma = \frac{mv^2}{r}$, which can also be written $F = m\omega^2 r$ where ω is the

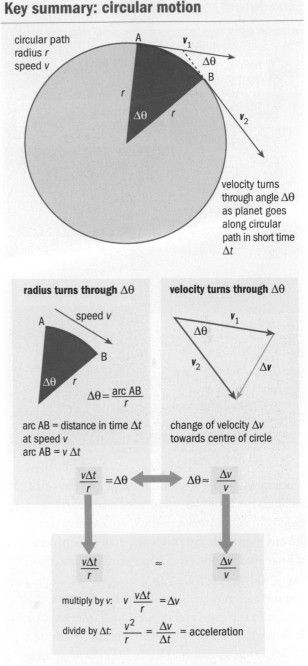

Key summary: circular motion

circular path
radius r
speed v

velocity turns through angle $\Delta\theta$ as planet goes along circular path in short time Δt

radius turns through $\Delta\theta$

speed v

$\Delta\theta = \dfrac{\text{arc AB}}{r}$

arc AB = distance in time Δt at speed v
arc AB = $v\,\Delta t$

velocity turns through $\Delta\theta$

change of velocity Δv towards centre of circle

$\dfrac{v\Delta t}{r} = \Delta\theta \quad\longleftrightarrow\quad \Delta\theta \approx \dfrac{\Delta v}{v}$

$\dfrac{v\Delta t}{r} \quad\approx\quad \dfrac{\Delta v}{v}$

multiply by v: $\quad v\,\dfrac{v\Delta t}{r} = \Delta v$

divide by Δt: $\quad \dfrac{v^2}{r} = \dfrac{\Delta v}{\Delta t} = \text{acceleration}$

The acceleration towards the centre of a circular orbit $= \dfrac{v^2}{r}$

angular velocity in orbit. If the force towards the centre is cut off, the object goes right on ahead in a straight line, towards wherever its velocity vector was pointing when the force stopped. Notice the similarity to projectile motion (chapter 9, *Advancing Physics AS*).

An object in circular motion is moving at constant speed. It gains or loses no energy. The

Speeds and accelerations in the solar system

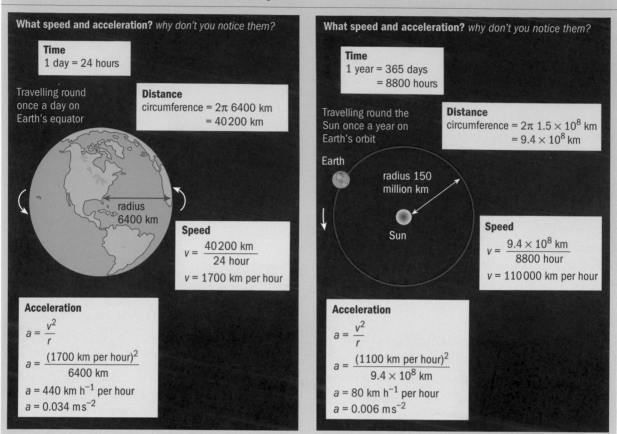

What speed and acceleration? *why don't you notice them?*

Time
1 day = 24 hours

Travelling round once a day on Earth's equator

Distance
circumference = 2π 6400 km
= 40 200 km

radius 6400 km

Speed
$v = \dfrac{40\,200 \text{ km}}{24 \text{ hour}}$
$v = 1700$ km per hour

Acceleration
$a = \dfrac{v^2}{r}$
$a = \dfrac{(1700 \text{ km per hour})^2}{6400 \text{ km}}$
$a = 440$ km h^{-1} per hour
$a = 0.034$ m s^{-2}

What speed and acceleration? *why don't you notice them?*

Time
1 year = 365 days
= 8800 hours

Travelling round the Sun once a year on Earth's orbit

Distance
circumference = 2π 1.5×10^8 km
= 9.4×10^8 km

Earth

radius 150 million km

Sun

Speed
$v = \dfrac{9.4 \times 10^8 \text{ km}}{8800 \text{ hour}}$
$v = 110\,000$ km per hour

Acceleration
$a = \dfrac{v^2}{r}$
$a = \dfrac{(1100 \text{ km per hour})^2}{9.4 \times 10^8 \text{ km}}$
$a = 80$ km h^{-1} per hour
$a = 0.006$ m s^{-2}

accelerating force feeds no energy into or out of the motion. This is because this force is always exactly at right angles to the displacement along the circular path. So the work done, the product of force and displacement in the direction of the force, is always zero.

Spun fast enough, any material will break. The tension in the material needed to accelerate its outer parts towards the centre gets larger than the breaking strength. This is important in large power station generators. Their rotors may be 1 m in diameter and have to rotate 50 times a second. At the rim the speed v is about 150 m s^{-1}. The acceleration $\dfrac{v^2}{r}$ is thus nearly 50 000 m s^{-2}, five thousand times the acceleration of free fall. These rotors must be very strongly built.

Look carefully at "Key summary: circular motion" (p37) to see how the argument works. It depends crucially on the fact that velocity is a vector. A change in velocity can come, not from an increase or decrease in its magnitude, but simply from a change in its direction. As an object travels at uniform speed along a circular path its velocity changes continually. The change in each small step of time Δt is always towards the centre of the circular motion and at right angles to the velocity at that moment. The angle $\Delta\theta$ through which the velocity turns is equal to $\dfrac{\Delta v}{v}$, from the vector diagram adding the change in velocity Δv to the velocity v. Now notice that this angle $\Delta\theta$ can be calculated in a different way. It is just the angle through which the radius turns in time Δt. In this time, a particle travels a distance $v\Delta t$ along an arc of the circular motion, so that the angle $\Delta\theta$ is just equal to the arc divided by the radius r, that is $\Delta\theta = \dfrac{v\Delta t}{r}$. The upshot is that, equating the two ways of calculating $\Delta\theta$:

$$\frac{v\Delta t}{r} = \frac{\Delta v}{v}$$

It follows that the acceleration $a = \dfrac{\Delta v}{\Delta t}$, directed towards the centre of the motion, is given by:

$$a = \frac{v^2}{r}$$

Key summary: forcing things to go in circles

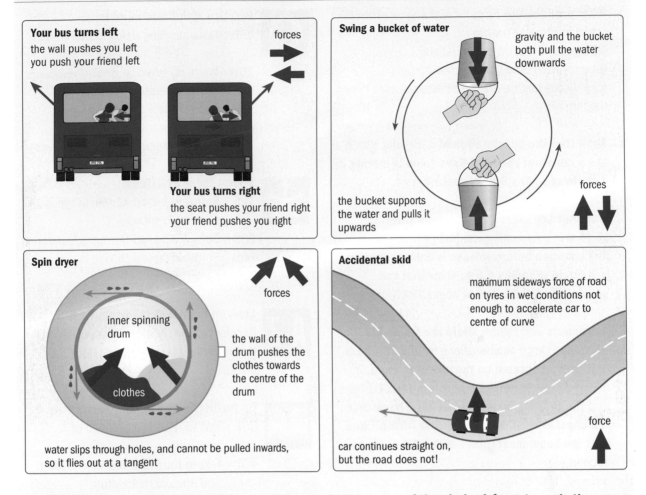

Your bus turns left
the wall pushes you left
you push your friend left

forces

Your bus turns right
the seat pushes your friend right
your friend pushes you right

Swing a bucket of water
gravity and the bucket
both pull the water
downwards

the bucket supports
the water and pulls it
upwards

forces

Spin dryer
inner spinning
drum

clothes

forces

the wall of the
drum pushes the
clothes towards
the centre of the
drum

water slips through holes, and cannot be pulled inwards,
so it flies out at a tangent

Accidental skid
maximum sideways force of road
on tyres in wet conditions not
enough to accelerate car to
centre of curve

car continues straight on,
but the road does not!

force

Anything going in a circle must be accelerated towards the centre of the circle. A force towards the centre of the circle is needed to produce the acceleration.

You may ask why the velocity v is squared in this equation. The reason is that if v increases, this changes the acceleration in two ways at once. One is that the velocity requires a larger change Δv to turn it through the same angle, so increasing the acceleration. Also, at a larger velocity, the radius turns through the same angle in a shorter time Δt, again increasing the acceleration.

This kind of acceleration is called a centripetal acceleration, and the force producing it is called a centripetal force. The word centripetal comes from Latin, meaning "centre seeking".

Because of the fact that in circular motion only the angle of the radius is changing, it is often simplest to consider the angular velocity ω, that

is, the rate $\frac{\Delta\theta}{\Delta t}$ at which the angle changes. With angle measured in radians, the unit of angular velocity ω is radians s^{-1}. The relationship between velocity and angular velocity in this case is very simple. In time Δt, the motion covers an arc of length $v\Delta t = r\Delta\theta$. Thus $\frac{\Delta\theta}{\Delta t} = \omega = \frac{v}{r}$.

The spin cycle of a washing machine is a less dramatic example of a centripetal force than power station generators. A perforated drum carries the wet clothes round and round, while the water from them goes straight ahead into the space outside the drum.

Another example is a centrifuge used, for instance, to separate pieces of DNA in forensic tests. With a high-speed spin, very large accelerations towards the centre can be produced.

Quick check

1. What provides the force to make each of these travel in circular paths:
   (a) a child sitting on a roundabout;
   (b) a toy train on a circular track;
   (c) a discus just before the athlete releases it for the throw?

2. Show that the acceleration of a bicycle, which has a constant speed of $6\,\mathrm{m\,s^{-1}}$ and is moving in a circle of radius $9\,\mathrm{m}$, is about $4\,\mathrm{m\,s^{-2}}$.

3. A cricket ball leaves a fast bowler travelling at $40\,\mathrm{m\,s^{-1}}$. Assuming the last part of the ball's motion before release is uniform motion in a circle of radius $0.8\,\mathrm{m}$, show that the acceleration of the ball is about $2000\,\mathrm{m\,s^{-2}}$.

4. Astronauts and fighter pilots are trained to withstand large accelerations by placing them in a capsule rotated on the end of a boom. If it is required to reach an acceleration of $10\,g$, and the rotating boom is $15\,\mathrm{m}$ long, show that the capsule must travel at almost $40\,\mathrm{m\,s^{-1}}$ and that the boom must rotate about 0.4 turns per second.

5. A child of mass $20\,\mathrm{kg}$ is playing on a swing. Her centre of mass is $4\,\mathrm{m}$ below the pivot supporting the swing when she moves through the bottom of her swing at $5\,\mathrm{m\,s^{-1}}$. Show that a centripetal force of about $125\,\mathrm{N}$ is required with the swing exerting a total force of about $325\,\mathrm{N}$.

6. A satellite circles the Earth with an orbital radius of $6400\,\mathrm{km}$ (just skimming the surface!). Assuming its acceleration is $g = 10\,\mathrm{m\,s^{-2}}$, show that its speed is about $8\,\mathrm{km\,s^{-1}}$ and that its orbital time is about 84 minutes.

Links to the *Advancing Physics* CD-ROM

Practise with these questions:
10D Data handling *Using Kepler's third law*
20W Warm-up exercise *Orbital velocities and acceleration*
30S Short answer *Centripetal force*
40S Short answer *Circular motion – more challenging*

Try out these activities:
10S Software-based *Watching the planets go round*
20S Software-based *Retrograde motion*
60S Software-based *Driving round in a circle*

Look up these key terms in the A–Z:
Circular motion; Kepler's laws; observing the universe; planetary orbit; satellite motion

Go further for interest by looking at:
30T Text to read *The problem of longitude*
40T Text to read *Kepler's second law and angular momentum*

Revise using the revision checklist and:
300 OHT *Kepler's third law*
500 OHT *Geometry rules the universe*
600 OHT *Centripetal acceleration*
700 OHT *Centripetal acceleration is proportional to* v^2

11.2 Newton's gravitational law

Newton had thought about the force keeping the Moon in orbit being gravity, weakened by the extra distance from Earth. Using a geometrical argument, the notion of an inverse square law had been worked out for the way light energy spreads out with distance from a light source. If gravity falls as the square of the distance, it would be reduced to a quarter when the distance doubles, to a ninth when the distance trebles, and so on.

While sitting in an orchard, Newton had the idea that he could test an inverse square law for gravity. He knew how rapidly a falling apple accelerates to the ground. He could calculate the acceleration of the Moon towards Earth using $\frac{v^2}{r}$. Knowing that the radius of Moon's orbit is about 60 times the radius of the Earth, he could check whether the apple's acceleration was 60^2 times the acceleration of the Moon. It was. Newton could now extend gravitational forces out into space.

What Newton had done is startling: movement on Earth and motion in the heavens became all the same thing. The Moon is a falling apple; a falling apple is a moon. Terrestrial and celestial motions were unified.

Newton's *Principia*, or *Mathematical Principles of Natural Philosophy*

By 1687, Newton was able to put together a book explaining the motion not just of the Moon, but of the whole solar system. He had uncovered the mechanism that drives the celestial clockwork. Like many books at the time, it was written in the common language of European scholars, Latin. The word "gravity" simply comes from the Latin word for heaviness.

The centrepiece of *Principia* is the universal gravitational law: all particles in the universe attract all other particles. Expressed in algebraic form, this is

$$F \propto \frac{m_1 m_2}{r^2}$$

Key summary: the acceleration of the Moon and the inverse square law

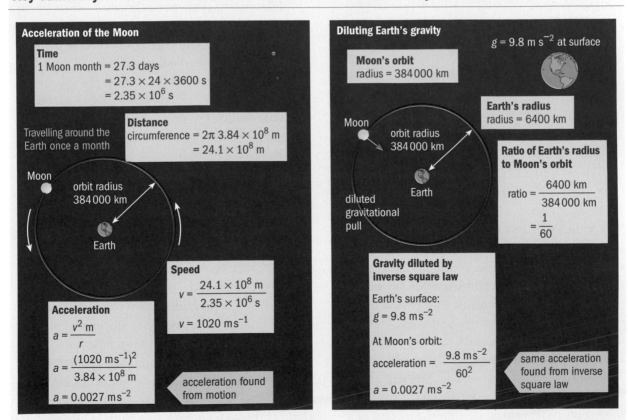

The acceleration of the Moon is diluted Earth gravity. The acceleration measures the gravitational field.

Key summary: Newton's universal law of gravitation

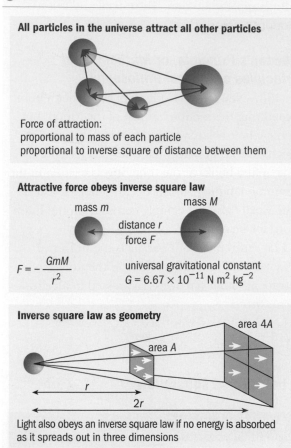

All particles in the universe attract all other particles

Force of attraction:
proportional to mass of each particle
proportional to inverse square of distance between them

Attractive force obeys inverse square law

mass m mass M

distance r
force F

$$F = -\frac{GmM}{r^2}$$

universal gravitational constant
$G = 6.67 \times 10^{-11}$ N m^2 kg^{-2}

Inverse square law as geometry

area $4A$

area A

r

$2r$

Light also obeys an inverse square law if no energy is absorbed as it spreads out in three dimensions

Newton's universal law is $F = -G\dfrac{mM}{r^2}$

In modern form, introducing the gravitational constant G, the radial component of the force is

$$F = -\frac{Gm_1 m_2}{r^2}$$

The minus sign says that the force is always attractive. The force acts between every mass in the universe. If you and a similar 50 kg friend stand one metre apart, the attraction between friends is equivalent to nearly 20 µg – the weight of a speck of dust one or two tenths of a millimetre across. It's small, but it can be measured in the laboratory. The attractive force between your body and the whole Earth is just what you call your weight.

Newton could also prove that the motion of a planet in a closed orbit under this inverse square law force must be an ellipse (of which a circle is a

Inverse square law gives elliptical orbits

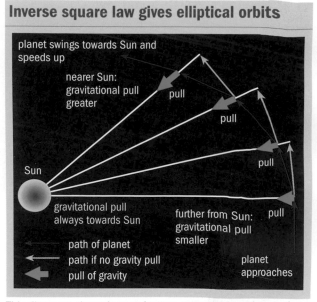

planet swings towards Sun and speeds up

nearer Sun:
gravitational pull
greater   pull

pull

Sun   pull

gravitational pull
always towards Sun   further from Sun: pull
gravitational pull
smaller

path of planet
path if no gravity pull   planet
pull of gravity   approaches

This diagrams shows how an inverse square force towards the sun produces an elliptical orbit.

special case), just as Kepler had seen that it was. In fact Newton's clockwork mechanism could explain all three of Kepler's laws.

To do all this, Newton had to invent a new kind of mathematics: perhaps his greatest achievement of all. You have already seen it at work, because it is in essence the step-by-step method we chose to use in chapters 9 (*Advancing Physics AS*) and 10. Newton's equation $F = ma$ says how the velocity changes in a short period of time. From that you calculate the next position, find the new rate of change of velocity, and repeat. It's a mathematical prediction machine.

This new mathematics, now called calculus, became a tool used by all physicists. Its power also gave the French mathematician Laplace an idea. He imagined how, if a god could have known every motion of every particle in the universe at the start, then the whole future could have been foretold. That grand – or terrifying – vision has established a huge grip on the human imagination over the centuries. It has led to the idea of physics as specialising in making exact predictions. And indeed eclipses are predicted many years ahead.

Newton was unhappy about just assuming that gravitational forces act across a distance. Forces that act through contact are common sense – a push or a pull from someone in a queue, for example. But forces across empty space? That

Newton's thought experiment

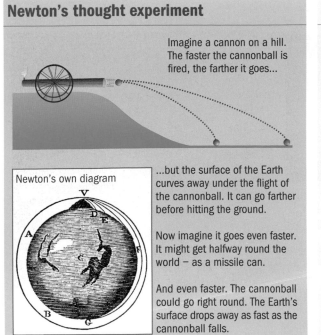

Imagine a cannon on a hill. The faster the cannonball is fired, the farther it goes...

Newton's own diagram

...but the surface of the Earth curves away under the flight of the cannonball. It can go farther before hitting the ground.

Now imagine it goes even faster. It might get halfway round the world – as a missile can.

And even faster. The cannonball could go right round. The Earth's surface drops away as fast as the cannonball falls.

Key summary: pictures of gravitational field

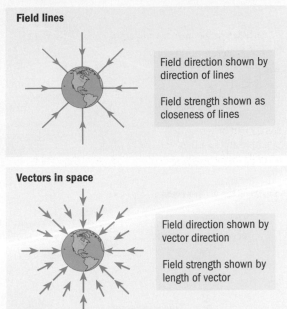

Field lines

Field direction shown by direction of lines

Field strength shown as closeness of lines

Vectors in space

Field direction shown by vector direction

Field strength shown by length of vector

The field has a vector value (magnitude and direction) at every point of space. The field strength at a place is the force per unit mass acting on a small massive body at that place.

was much harder to imagine. However, Newton's gravitational law explained a remarkable number of phenomena. Here are just a few.

- The fact that all particles attract all others accounts for the formation of planets and stars, and for their spherical shape.
- The planets attract each other, diverting them from their smooth orbital paths. Wobbles, or "perturbations", in the orbit of Uranus led to the discovery of Neptune. A big question arises: is the planetary system itself stable?
- Knowing orbital radius and period, the mass of any body that has a satellite can be compared with the mass of Earth. The same idea is used to find the masses of rotating galaxies, even of the supposed black holes at their centres.
- When more than two bodies are involved, the motions become complicated. An example is the study of the rings round Uranus and Saturn.

Gravitational inverse square field

Today, space is imagined as being "filled" with a gravitational field. A mass placed anywhere in the field "feels" a force. You describe the field by specifying at each point the magnitude and direction of the force that would be felt by each kilogram of mass placed there. Chapter 9 of *Advancing Physics AS* introduced this vector quantity,

which is called gravitational field strength; it has the symbol g and units N kg^{-1}. It describes the effect of the gravitating body of mass M regardless of what particular mass m is placed near it. That is

$$\text{radial component of field strength } g = -\frac{GM}{r^2}$$

giving a gravitational force

$$\text{radial component of force} = mg = -\frac{GmM}{r^2}$$

Television by satellite

Newton had long before imagined how a cannon could in principle launch an Earth satellite, linking orbits to free fall. But he could hardly have imagined you sitting at home watching TV programmes relayed from a satellite to a dish on your house pointing up into the sky towards that satellite. The satellite must go round the Earth as rapidly as Earth itself turns, so that it seems to stay in the same place, night and day.

A satellite close to Earth goes round too fast,

Key summary: a geostationary satellite

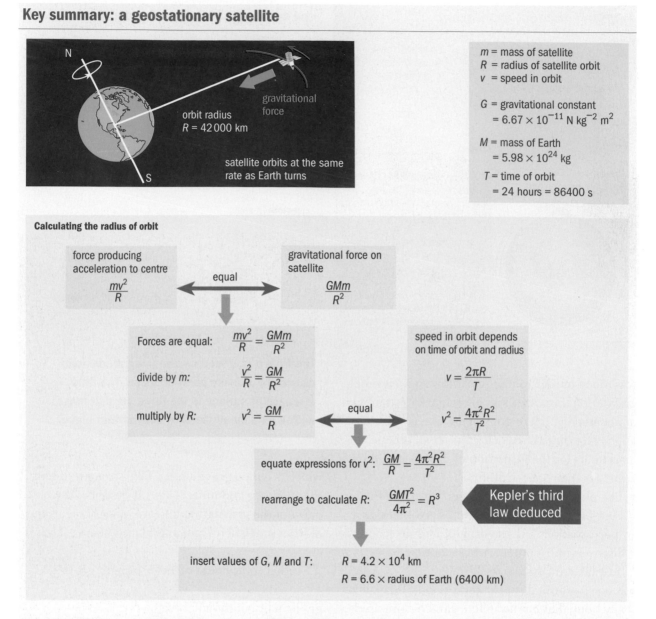

m = mass of satellite
R = radius of satellite orbit
v = speed in orbit

G = gravitational constant
$\quad$ = 6.67×10^{-11} N kg^{-2} m^2

M = mass of Earth
$\quad$ = 5.98×10^{24} kg

T = time of orbit
$\quad$ = 24 hours = 86400 s

orbit radius
R = 42 000 km

gravitational force

satellite orbits at the same rate as Earth turns

Calculating the radius of orbit

force producing acceleration to centre

$$\frac{mv^2}{R}$$

equal

gravitational force on satellite

$$\frac{GMm}{R^2}$$

Forces are equal: $\quad \dfrac{mv^2}{R} = \dfrac{GMm}{R^2}$

divide by m: $\quad \dfrac{v^2}{R} = \dfrac{GM}{R^2}$

multiply by R: $\quad v^2 = \dfrac{GM}{R}$

speed in orbit depends on time of orbit and radius

$$v = \frac{2\pi R}{T}$$

equal

$$v^2 = \frac{4\pi^2 R^2}{T^2}$$

equate expressions for v^2: $\quad \dfrac{GM}{R} = \dfrac{4\pi^2 R^2}{T^2}$

rearrange to calculate R: $\quad \dfrac{GMT^2}{4\pi^2} = R^3$

Kepler's third law deduced

insert values of G, M and T: $\quad R = 4.2 \times 10^4$ km

$R = 6.6 \times$ radius of Earth (6400 km)

Kepler's third law, and the orbit radius of a geostationary satellite, can be deduced from first principles

The satellite TV dishes in this photograph all point the same way – to a satellite that orbits so as to stay above the same part of Earth as it turns.

about every hour and a half. The Moon, farther away, goes round too slowly (once a month). In between must be the right distance. Newton would have known how to calculate the right distance for such a geostationary orbit because he knew how to calculate the time of orbit for a given radius – Kepler's third law (p36).

One small step: the Apollo missions

For many centuries journeys into space were limited to the imagination alone. Finally the combination of Newton's *Principia*, vast engineering

The launch of the *Apollo 11* lunar landing mission on 16 July 1969. Each minute walking on the Moon's surface cost the US $200 000. At its peak the Apollo programme employed 400 000 people. This was politically motivated to regain the lead in the race to space from the Soviet Union.

Astronauts Neil Armstrong, Michael Collins and Edwin Aldrin leave the Kennedy Space Centre during the pre-launch countdown for the *Apollo 11* voyage to the Moon.

Apollo 11 sees Earth rising above the horizon of the Moon, looking across the surface of the Moon.

The *Apollo 11* astronauts, splashing down in the Pacific, were immediately confined to a quarantine capsule. Here they are greeted by the president of the US at that time, Richard Nixon.

effort and the political will to bring it about made it possible for people to get away from Earth.

Between 1969 and 1972 six Apollo missions landed on the Moon – a ball in space 384 000 km away and only 3500 km across. To get them there, NASA engineers had to predict the flight path in detail, using Newton's laws of motion and gravitation. Data from the flight shows how it all works out. While coasting to the Moon, only gravity is acting on *Apollo 11*, slowing it down as it goes away from the Earth. From pairs of recorded velocities at fixed intervals, the rate of change of velocity can be calculated. This deceleration, if in the radial direction of the field, is just the gravitational field at that place. Flight records of the acceleration of a free-floating spacecraft on its voyage are at the same time a record of the gravitational field through which it has passed. So NASA conducted a precision experiment on a huge scale, putting Newtonian calculations to the test. The astronauts travelled safely there and back. Newton did not fail them.

Key summary: *Apollo 11* goes to the Moon

Pairs of observations of speed and distance

Pairs of observations made 600 s apart in time. Distances *r* taken from centre of Earth. *Apollo 11* is coasting with rockets turned off.

	speed/m s^{-1}	distance/10^6 m	mean distance r/10^6 m	mean acceleration/m s^{-2}
pair A	5374 5102	26.3 29.0	27.65	−0.453
pair B	3633 3560	54.4 56.4	55.40	−0.122
pair C	2619 2594	95.7 97.2	96.45	−0.042
pair D	1796 1788	169.9 170.9	170.40	−0.013

Graph of gravitational field against *r*

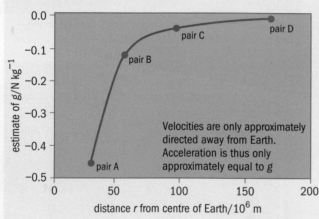

Velocities are only approximately directed away from Earth. Acceleration is thus only approximately equal to *g*

Graph of gravitational field against 1/*r*^{2}

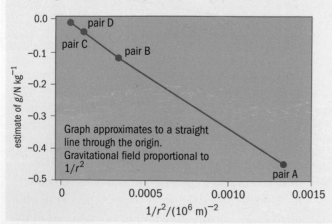

Graph approximates to a straight line through the origin. Gravitational field proportional to 1/*r*^{2}

The path of *Apollo 11*

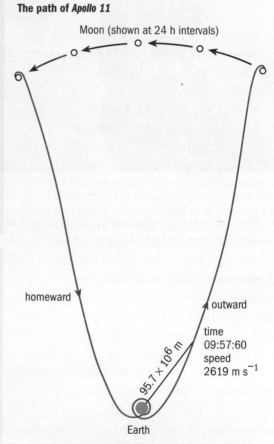

Typical *Apollo* flightpath to the Moon, showing outward journey on the right and homeward journey four days later on the left.

Flight data from *Apollo 11* conforms to Newton's gravitational theory

Quick check

1. Newton correctly estimated the density of the Earth as five to six times that of water. Today we know it is $5500\,\text{kg}\,\text{m}^{-3}$. The volume of a sphere is $\frac{4}{3}\pi r^3$ and the Earth's radius is $6400\,\text{km}$. Show that the mass of the Earth is about $6.0 \times 10^{24}\,\text{kg}$, and use the fact that $g = 10\,\text{N}\,\text{kg}^{-1}$ to show that $G = 6.7 \times 10^{-11}\,\text{N}\,\text{m}^{-2}\,\text{kg}^{-2}$.

2. Show that the gravitational pull of a spherical $5\,\text{kg}$ mass on a spherical $3\,\text{kg}$ mass when their centres are $0.15\,\text{m}$ apart is about $4.4 \times 10^{-8}\,\text{N}$. Compare this with the force of the $3\,\text{kg}$ mass on the $5\,\text{kg}$ mass. $(G = 6.67 \times 10^{-11}\,\text{N}\,\text{m}^2\,\text{kg}^{-2}.)$

3. How would the weight of a body vary on route from the Earth to the Moon? Would its mass change?

4. You are standing on bathroom scales in a lift. Explain what happens to the reading when the lift is (i) stationary, (ii) starting down, (iii) starting up, (iv) moving at constant speed.

5. Another way of thinking about geostationary satellites is that they must have the same angular velocity ω as the aerial that points towards them. Show that their angular velocity ω is about $7.3 \times 10^{-5}\,\text{rad}\,\text{s}^{-1}$.

6. Prove that the kinetic energy per kilogram of a moving mass is $\frac{1}{2}v^2$. Using the *Apollo 11* data, plot a graph to show the kinetic energy per kilogram against distance *r*.

Links to the *Advancing Physics* CD-ROM

Practise with these questions:

10W Warm-up exercise *Testing for an inverse-square law*

80W Warm-up exercise *Newton's gravitational law*

110S Short answer *Finding the mass of a planet with a satellite*

120D Data handling *The gravitational field between the Earth and the Moon*

Try out these activities:

70S Software-based *Variations in gravitational force*

80S Software-based *Gravitational universes*

110S Software-based *Probing a gravitational field*

Look up these key terms in the A–Z:

Gravitational field; inverse square laws; planetary orbit; projectile; weight

Go further for interest by looking at:

70T Text to read *Gravity can pull things apart*

100T Text to read *Supernovae and black holes*

Revise using the revision checklist and:

100O OHT Apollo *goes to the Moon*

110O OHT *A geostationary satellite*

11.3 Arrivals and departures

Collision course

Aboard the space station *Mir* in 1997 the Russian astronaut Vasili Tsibliyev watched the supply ship *Progress*, loaded with essential stores and replacements, creep up to dock with *Mir*. He could see *Progress* through *Mir's* portholes, but he could also see the view of *Mir* taken by a video camera on *Progress* via a television monitor. Each view showed the other craft as moving. The astronaut's job was to make small adjustments to the velocity of *Progress* so that it made as accurate and gentle an approach and contact as possible – always a difficult manoeuvre. Suddenly the video camera aboard *Progress* failed, and the approach had to be aborted.

They tried again. This time, the approach was too fast and there wasn't enough time or braking power to direct the supply ship away from the station. It struck *Mir* hard, damaging a solar panel. *Mir* lost power and went out of control. Only through great effort, courage and skill was a major catastrophe averted.

Don't forget that both *Mir* and *Progress* were hurtling round Earth at 16 000 miles an hour. From the earthbound point of view, two very high velocities had to become almost equal (not quite, or *Progress* would never reach *Mir*). From the point of view aboard *Mir*, *Progress* was approaching at a low relative velocity. But *Progress* has a mass of 7 tonnes (*Mir* much more) and damage is easily done even by a low velocity collision. Think of the damage that can be done to a car with the brake off when it rolls a few metres down a slope into a lamppost.

A law of physics just by thinking?

You wouldn't think that you could find a law of physics just by thinking about things. But Galileo did (p35). Our next example will give you a fresh angle on forces and motion, which proves very helpful in particle physics (chapter 17).

Imagine the collision between *Mir* and *Progress*, but simplified:
● the two craft are identical, with the same mass;
● there is no thrust from the jets.
You are monitoring the event, travelling alongside in a third spacecraft equipped with a video camera.

First, imagine that your observation craft is positioned midway between *Mir* and *Progress*, and that they are going to meet right in front of you. Your camera will show the two craft approaching at equal and opposite velocities $+v$ and $-v$. Their relative velocity is $2v$.

What must happen after they meet and dock? What else can they do except come to a halt together in front of your observation craft? The

Key summary: two crafts approach one another and dock together

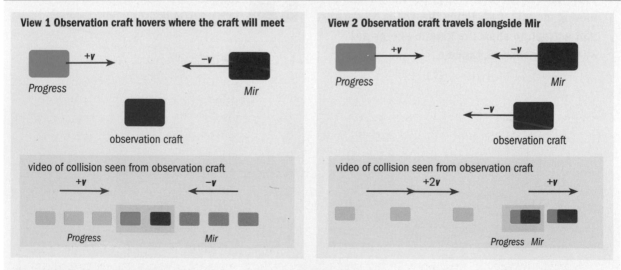

The same event looks different from two different points of view

whole situation is symmetrical: there is no reason for the two linked craft to speed off together in any particular direction. Actually you can predict more than this. If the docking misfired and *Mir* and *Progress* bounced off one another, whatever their rebound velocities were the two velocities would have to be equal and opposite. Symmetry again.

You can try the same thing with two almost friction-free model trucks. If they are pushed so that they come together at equal speeds and lock as one, it is clear that they must come to a stop when they meet. Symmetry says so.

Play it again!

This thought experiment gets more intriguing if you imagine a replay of the docking manoeuvre, but this time with your observation craft and camera flying at the same velocity as *Mir* and just alongside it. What is *Progress* doing, from this point of view? Your video camera, and the astronauts looking out of *Mir's* windows, must see the same thing, since you are travelling along side by side. You will both see *Progress* approach. But at what velocity?

Here's the key point in the thinking. Just changing the point of view of your observation craft cannot possibly affect how fast the other two are coming together. Their relative velocity must still be $2v$. Since *Mir* is now "sitting still" in front of your video camera, the video must show *Progress* coming up at velocity $2v$.

The ships meet and dock. Just by changing your point of view, you can't change the fact that the ships will meet with the same "clunk" as before, join and travel off together. So what your video records is *Progress* bumping into a stationary *Mir* and pushing it into forward motion as they lock together. But your craft isn't involved in the collision, and carries on with no change in velocity. Therefore, as seen on your video, after the collision the locked ships must travel off together at velocity v relative to you.

You can try this with model trucks or air-track gliders too, if you can set up a video camera so that it moves alongside one of the vehicles. As they meet and stop, the camera goes on and the coupled vehicles appear on the video to be carrying on with a velocity opposite to that of the camera.

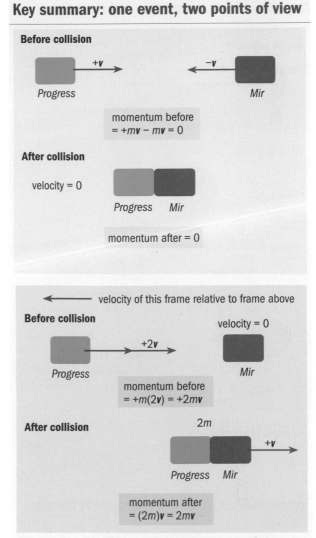

Key summary: one event, two points of view

Before collision

Progress $+v$ $-v$ Mir

momentum before
$= +mv - mv = 0$

After collision

velocity = 0

Progress   Mir

momentum after = 0

← velocity of this frame relative to frame above

Before collision

velocity = 0

Progress $+2v$ Mir

momentum before
$= +m(2v) = +2mv$

After collision $2m$

Progress Mir $+v$

momentum after
$= (2m)v = 2mv$

Momentum is different in the two views of the same event but in each case momentum after = momentum before

Just moving can't change the rules

The difference between the two events was created just by changing how the observer was moving. The two events are the same, so the rule for one must be the same as the rule for the other. The rule must not change if all the velocities involved are changed by the same amount.

Taking the thought experiment out into space, as we chose to do, was a device to help you to see that all points of view moving with different uniform velocities are as good as one another. The physics of an event can't be different just because you walk past it. Named after Galileo, this is sometimes called Galilean invariance.

Key summary: conservation of momentum $p = mv$

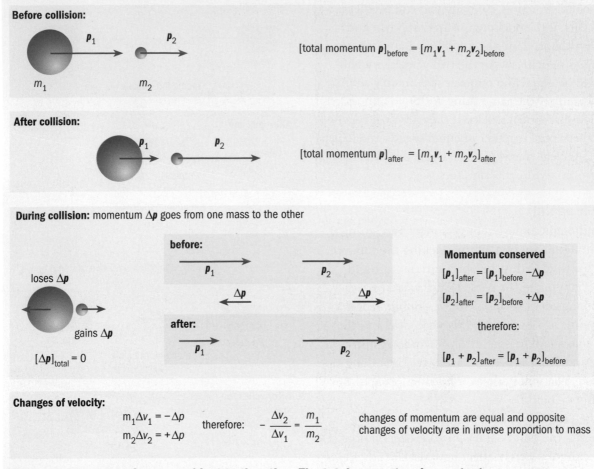

Before collision:

$[\text{total momentum } \boldsymbol{p}]_{\text{before}} = [m_1\boldsymbol{v}_1 + m_2\boldsymbol{v}_2]_{\text{before}}$

After collision:

$[\text{total momentum } \boldsymbol{p}]_{\text{after}} = [m_1\boldsymbol{v}_1 + m_2\boldsymbol{v}_2]_{\text{after}}$

During collision: momentum $\Delta\boldsymbol{p}$ goes from one mass to the other

before:

$\boldsymbol{p}_1$ $\boldsymbol{p}_2$

loses $\Delta\boldsymbol{p}$

$\Delta\boldsymbol{p}$ $\Delta\boldsymbol{p}$

gains $\Delta\boldsymbol{p}$

after:

$\boldsymbol{p}_1$ $\boldsymbol{p}_2$

$[\Delta\boldsymbol{p}]_{\text{total}} = 0$

Momentum conserved

$[\boldsymbol{p}_1]_{\text{after}} = [\boldsymbol{p}_1]_{\text{before}} - \Delta\boldsymbol{p}$

$[\boldsymbol{p}_2]_{\text{after}} = [\boldsymbol{p}_2]_{\text{before}} + \Delta\boldsymbol{p}$

therefore:

$[\boldsymbol{p}_1 + \boldsymbol{p}_2]_{\text{after}} = [\boldsymbol{p}_1 + \boldsymbol{p}_2]_{\text{before}}$

Changes of velocity:

$m_1 \Delta v_1 = -\Delta p$
$m_2 \Delta v_2 = +\Delta p$ therefore: $-\dfrac{\Delta v_2}{\Delta v_1} = \dfrac{m_1}{m_2}$ changes of momentum are equal and opposite
changes of velocity are in inverse proportion to mass

Momentum just goes from one object to the other. The total momentum is constant.

Momentum: a rule

Now a rule can be constructed that the collisions between the two spacecraft must follow. If a mass m with velocity $2\boldsymbol{v}$ collides with a stationary mass m so that they stick together, the combined mass $2m$ will move off at velocity $\boldsymbol{v}$ (all the motions will be in the same straight line). This suggests that the product of mass and velocity must be the same after the event as before. The same rule works just as well, as it must, with the collision seen as two masses approaching one another with velocities $+\boldsymbol{v}$ and $-\boldsymbol{v}$. Before collision, the momentum of one craft is $+m\boldsymbol{v}$ and that of the other is $-m\boldsymbol{v}$, with vector total zero. If they lock together, the mass $2m$ must have zero velocity and zero momentum. Provided that no external forces are acting, in both cases the sum of products of mass and velocity is the same after as before. Similar arguments work just as well for any masses and any velocities.

The product of mass and velocity is called momentum. Interacting bodies pass momentum from one to the other. What one loses the other gains, so the total momentum is unchanged – it is a conserved quantity.

The total momentum before any interaction is equal to the total momentum afterwards. In symbols, this can be written as

$$(m_1\boldsymbol{v}_1 + m_2\boldsymbol{v}_2 + ...)_{\text{before event}} = (m_1\boldsymbol{v}_1 + m_2\boldsymbol{v}_2 + ...)_{\text{after event}}$$

This law of conservation of momentum is very powerful because it applies to all interactions, to every type of collision and explosion, on scales from nuclear to astronomical. So important has momentum become in physics that it is given a special symbol of its own, $\boldsymbol{p}$.

In many sports the skill is to give a ball as much momentum as possible. Here, as the ball is struck, it gains exactly the same amount of forward momentum as the club head loses.

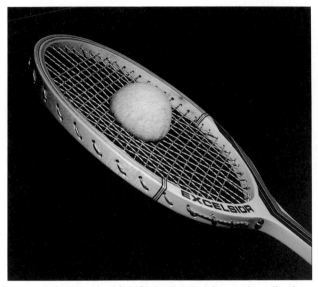

Not until the advent of fast flash photography was it realised how much a tennis ball is squashed out of shape at the moment of impact, as seen in this photograph.

Exchanges of momentum

Skaters collide

velocity *v* velocity −*v*

A golf club hits a ball

velocity *v* velocity zero

You catch a ball

velocity *v* velocity zero

A rocket

small mass ejected at high velocity

rocket changes its velocity

The equivalent of shifting your point of view in the spacecraft example is simply adding the same velocity to every velocity in the equation for momentum conservation. You can see that, although both sides of the equation change, the whole equation stays true. You've just added the same velocity to both sides. The law of momentum conservation holds good from each point of view.

Sporting momentum

Racquets for tennis players, clubs for golfers, bats for cricketers and cues for snooker players all require careful design to exploit momentum changes.

Think about the head of a golf club. As the ball is struck it gains forward momentum and the club head loses exactly the same amount of momentum. But the small ball speeds off at high velocity, while the club head changes its velocity by much less.

Looking at it the other way round, if you are rather slim and you bump into a larger person in the street, your velocity changes by more than theirs does. A child can jump into its mother's arms without knocking her over. In any interaction between two objects, the ratio of their masses

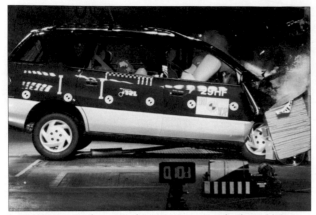

In a collision the forces acting on passengers in the vehicle must be minimised to reduce the risk of injury.

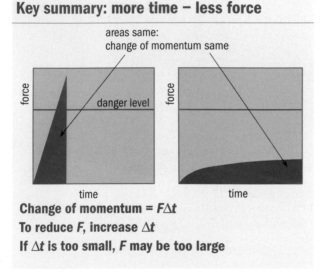

Key summary: more time – less force

areas same:
change of momentum same

danger level

force

time

force

time

Change of momentum = _F_Δ_t_
To reduce _F_, increase Δ_t_
If Δ_t_ is too small, _F_ may be too large

---

determines the ratio of the magnitudes of their changes of velocity:

$$\frac{\Delta v_1}{\Delta v_2} = \frac{m_2}{m_1}$$

because the changes of momentum $m\Delta v$ are equal in magnitude. This result simply says what mass does – the bigger the mass the less the change of velocity. In fact, it defines mass.

Softening the blow

Sports equipment has to be designed not to break under the large forces produced by the impact of bat on ball. The larger the momentum transferred, the bigger is the force on each. But there's another factor: the time for which the impact lasts. The shorter the time, and the faster the rate of change of momentum, the larger the

force. You'll have felt the jar as you jumped down off a chair, or accidentally punched a wall with your fist. To "soften the blow" is to spread out the change in momentum over a longer time, reducing the force.

Parachute instructors give this advice to novices learning to land safely: "Bend your legs and relax, letting yourself fall over." The idea is to make the contact time with the ground as long as possible, so that the momentum with which you hit the ground can be absorbed with the minimum possible force, because otherwise it could easily break a bone.

That the force is just the rate of change of momentum is easy to see:

$$F = \frac{\Delta p}{\Delta t} = m\frac{\Delta v}{\Delta t}$$

Since $\frac{\Delta v}{\Delta t}$ = acceleration a, this is the same as $F = ma$. This defines how to calculate a force. It is Newton's familiar relation $F = ma$, but is obtained by thinking about things from a fresh angle.

Like a parachutist, you are careful to flex your legs when landing after jumping off a wall. From a chest-high wall you fall for half a second and hit the ground at about $5\,\mathrm{m\,s^{-1}}$ – that's just under $20\,\mathrm{km}$ per hour. Your knees can bend enough to spread the deceleration over perhaps $0.5\,\mathrm{m}$, taking about $0.2\,\mathrm{s}$ to make the landing. If your mass is $60\,\mathrm{kg}$ the force in your legs is:

$$\text{time } \Delta t = 0.2\,\mathrm{s}$$

$$\Delta v = 5\,\mathrm{m\,s^{-1}}$$

$$F = m\frac{\Delta v}{\Delta t} = 60\,\mathrm{kg}\,\frac{5\,\mathrm{m\,s^{-1}}}{0.2\,\mathrm{s}} = 1500\,\mathrm{N}$$

If you land with stiff legs, the deceleration time might be ten times less, and the force ten times larger. Real damage to bones and joints could ensue. Gently does it: long slow movements keep the forces small for a given change of momentum. It's what you do when handling eggs or cradling a baby.

Typically in an accident, a vehicle is suddenly brought to a halt. In designing safety systems for trains and cars, engineers will use the strategy of making the impact last longer to reduce the force. For safety, the time Δt must be big enough for the force F to be not too large, given that $F\Delta t$ must

What a sports coach might say

"To give a fast serve…"
"To drive the golf ball as far as possible down the fairway…"
"To hit the cricket ball for six…"
"…swing with your whole body to hit hard, and follow through."

Pairs of forces of interaction at a distance

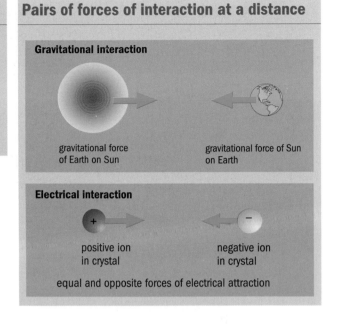

Gravitational interaction

gravitational force of Earth on Sun

gravitational force of Sun on Earth

Electrical interaction

positive ion in crystal

negative ion in crystal

equal and opposite forces of electrical attraction

be equal to $m\Delta\boldsymbol{v}$. The quantity $\boldsymbol{F}\Delta t$ is sometimes called the impulse of the force.

This is what happens when a train goes into the buffers, when a car suddenly stops and the seat belts stretch as they restrain the passengers, or when the crumple zone in a train or car gives way. In every case a structure yields gradually, increasing the time of the interaction. The force acting on passengers is reduced.

Sports coaches use the same principle the other way round, to help increase the momentum given to a ball. The need to hit hard is clear. What the follow-through does is to make the force act on the ball for as long as possible, increasing its momentum as much as possible. The force $\boldsymbol{F}$ is limited by the player's strength but, by increasing Δt, the momentum change $m\Delta\boldsymbol{v} = \boldsymbol{F}\Delta t$ is made bigger.

The crunch in a collision

In a collision, both objects are deformed. The strings of a tennis racquet stretch and the ball is squashed. Each has the same magnitude of change of momentum, and each feels the same magnitude of force acting for the same time. But these forces, like the momentum changes, are opposite in direction:

$$\Delta\boldsymbol{p}_1 = -\Delta\boldsymbol{p}_2$$

so that

$$\boldsymbol{F}_1\Delta t = -\boldsymbol{F}_2\Delta t$$

$$\boldsymbol{F}_1 = -\boldsymbol{F}_2$$

This says that forces always come in pairs, equal and opposite. That's because all forces come from objects interacting – forces don't exist in the abstract. An amount of momentum leaves one

object and goes to the other, so the changes of momentum are equal and opposite and, as a result, the forces are equal and opposite.

Forces come in pairs whether the interacting bodies touch or not. Momentum is conserved in all interactions, so the Earth pulls on the Sun just as the Sun pulls on the Earth.

The fact that the paired forces on interacting masses are equal and opposite is known as Newton's Third Law. We have presented it here as a consequence of the conservation of momentum, and of force being defined as rate of change of momentum.

Up, up and away: imagination into fact

Jules Verne, in his story *From the Earth to the Moon*, imagined a giant cannon used to launch people into space. But, like Newton's gun (p43), the idea of a single shot putting objects into orbit doesn't work. The accelerations needed are enormous: several thousand times the acceleration of free-fall even for a gun a kilometre long. As it is, using rockets, astronauts have to survive accelerations of 10 to 15 g. Beyond this, the forces of acceleration acting on the body's internal organs are so large that they could kill a person.

The first rockets, burning gunpowder, were invented by the Chinese in the 13th century. By the following century, their rocket technology had reached Europe and was used in warfare. Later, as

Key summary: jets and rockets

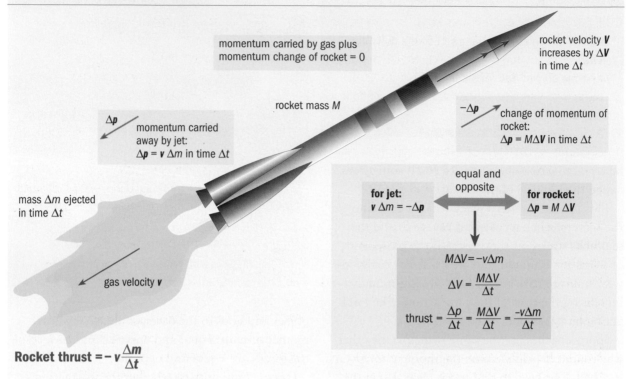

momentum carried by gas plus
momentum change of rocket = 0

rocket velocity **V**
increases by Δ**V**
in time Δt

rocket mass *M*

−Δ**p**   change of momentum of
rocket:
Δ**p** = *M*Δ**V** in time Δt

Δ**p**   momentum carried
away by jet:
Δ**p** = **v** Δ*m* in time Δt

mass Δ*m* ejected
in time Δt

gas velocity **v**

equal and
opposite

for jet:
v Δ*m* = −Δ**p**

for rocket:
Δ**p** = *M* Δ**V**

$$M\Delta V = -v\Delta m$$

$$\Delta V = \frac{M\Delta V}{\Delta t}$$

$$\text{thrust} = \frac{\Delta p}{\Delta t} = \frac{M\Delta V}{\Delta t} = \frac{-v\Delta m}{\Delta t}$$

Rocket thrust $= -\,v\,\dfrac{\Delta m}{\Delta t}$

conventional artillery improved, rockets came to
be used by the military only for signal flares.

Rockets work by throwing away momentum.
The exhaust gases stream out at a high constant
velocity, carrying momentum away from the
rocket in one direction. Because changes in
momentum are equal and opposite, the rocket
gains momentum in the opposite direction. The jet
of gas goes one way and the rocket goes the other.
For the same reason, firemen have to hang on tight
to their fire hoses when the high-velocity water jet
is quenching a fire. How hard do they have to hold
on? The thrust ***F*** created by the jet, whether of a
rocket or a fire hose, is just the rate of change of
momentum:

$$\boldsymbol{F} = \frac{\Delta \boldsymbol{p}}{\Delta t} = -\boldsymbol{v}\,\frac{\Delta m}{\Delta t}$$

The thrust is in the opposite direction to the
jet. In a rocket, as the jet carries mass away
and the rocket mass decreases, the acceleration
increases. Look for this effect next time you are
watching fireworks. And whenever a gust of wind
catches you off balance you have experienced the
momentum carried by a moving stream of matter.

...just by thinking?

The starting point for this section was simply:

- some results of experiments can be predicted just
  by symmetry; and
- looking at an experiment from a different point
  of view can't change anything essential about it.

Out of this we have conjured the law of conservation
of momentum and Newton's laws of motion.

You should be suspicious of getting so much out
of so little, especially without experimental data.
Such seemingly watertight arguments in thought
experiments often turn out to have hidden flaws, so
experimental tests are still needed.

Indeed, the arguments here, which are based
on Galileo's idea of relativity, do turn out to be
not quite correct. Einstein used more thought
experiments to take the ideas further and allow
for the fact that relative velocities cannot exceed
the speed of light. Ideas based on symmetry are
today very important in physics. They were given
a great boost by Amalie (Emmy) Noether, forced
as a Jew to leave Nazi Germany in the 1930s
despite her brilliant theoretical work. She proved
that symmetry and conservation laws always go
together.

Quick check

1. A 10 kg box of tinned beans lands on a conveyor belt that is moving at $2\,\mathrm{m\,s^{-1}}$. Show that the change in the magnitude of its momentum is $20\,\mathrm{kg\,m\,s^{-1}}$ and that the impulsive force that caused this change is $20\,\mathrm{N\,s}$.

2. A car of mass 1000 kg travelling at $13.5\,\mathrm{m\,s^{-1}}$ (30 mph) is stopped in 5 seconds. Show that its change in momentum is $-13\,500\,\mathrm{kg\,m\,s^{-1}}$ and that the brakes exerted a force of 2700 N.

3. A 2 kg bowling ball moving east at $3\,\mathrm{m\,s^{-1}}$ collides with a 1 kg ball travelling in the same direction at $1\,\mathrm{m\,s^{-1}}$. If as a result of the collision the velocity of the 2 kg ball becomes $2\,\mathrm{m\,s^{-1}}$ east, show that the new velocity of the smaller ball is $3\,\mathrm{m\,s^{-1}}$ eastwards.

4. A rocket burns 100 kg of fuel per second and the exhaust speed is $300\,\mathrm{m\,s^{-1}}$. Show that the thrust force created is 30 kN.

5. A golf club of mass 0.20 kg travelling at $50\,\mathrm{m\,s^{-1}}$ hits a golf ball of mass 0.046 kg. If the club speed after impact is $34\,\mathrm{m\,s^{-1}}$, show that the speed of the ball is $70\,\mathrm{m\,s^{-1}}$.

6. A karate expert delivers a blow with his fist (mass 0.7 kg) at $10\,\mathrm{m\,s^{-1}}$. The block of wood he strikes deflects by 2 cm. Assuming a uniform deceleration, show that the impact time is 7.4 ms and that the average force is 1.8 kN.

Links to the *Advancing Physics* CD-ROM

Practise with these questions:
140W Warm-up exercise *Change in momentum as a vector*
150S Short answer *Impulse and momentum in collisions*
160S Short answer *Collisions of spheres*
180S Short answer *Jets and rockets*

Try out these activities:
160S Software-based *Modelling collisions*
180S Software-based *Crunch – gently!*

Look up these key terms in the A–Z:
Momentum; Newton's laws of motion; vectors; vector addition

Go further for interest by looking at:
140P Poster *Momentum, invariance, symmetry*
90T Text to read *"Sling-shotting" spacecraft or "gravity assist"*

Revise using the revision checklist and:
1300 OHT *Two craft collide*
1500 OHT *Conservation of momentum*
1600 OHT *More collisions*
1700 OHT *Jets and rockets*

11.4 Mapping gravity

Since the launch of the Russian *Sputnik 1* satellite in October 1957, hundreds of satellites have been placed into Earth orbit for many purposes:

- mapping Earth and its resources, oceans and atmosphere;
- communications relays from one part of Earth to another;
- military reconnaissance and navigation;
- astronomical observation.

Satellites monitor ice cover in the Arctic to check on global warming; seasonal changes in vegetation and in ocean currents; and the burning of rainforests as land use changes. Near-space is increasingly commercialised.

To get a satellite into orbit around Earth needs a speed of about $8\,\mathrm{km\,s^{-1}}$. This is a kinetic energy of 32 million joules for each kilogram of spacecraft. To make it an artificial planet of the Sun – able to visit other planets – the speed must be more than $11\,\mathrm{km\,s^{-1}}$, with nearly twice as much kinetic energy. To predict the path of a spacecraft, mission controllers need to know in advance how the energy changes when the craft moves in a gravitational field.

If you drew gravitational-field lines above the horizon on the image of the Mediterranean, top right, they would be almost parallel. Close-up, Earth is almost flat and the field is nearly uniform. The gravitational force mg hardly varies with height, so the change $mg\Delta h$ in potential energy is the same for equal increases Δh in height. A map with contour lines of height above sea level is also a good map of the "contours" of constant gravitational potential energy. Find a steep slope with closely spaced lines and you have a place where the energy of falling water might provide a basis for generating power.

It would be useful to know the difference $g\Delta h$ between the top and bottom of the slope, because if you multiplied it by the mass of water likely to be available per second, you would have the power that could be generated (energy per second). That difference $g\Delta h$ is called the difference in gravitational potential. It is simply the difference in gravitational potential energy per kilogram of material. Using the same basic idea, but allowing for changes in the gravitational force with distance

The curvature of the Earth is visible in this satellite view looking across the Mediterranean Sea from the straits of Gibraltar.

This photograph shows the probe *Galileo* bound for Jupiter, in the cargo bay of the Space Shuttle *Atlantis*.

from Earth and knowing a spacecraft's required payload, mission controllers can calculate how much energy will be needed to get it into orbit.

Clearly gravitational potential increases as you go up, or away, from Earth. For earthbound calculations it is common to set the zero value at sea level. But for space calculations it is easier to set the zero as the potential energy per kilogram of a probe that has completely escaped to an infinite distance. This has the advantage of being the same value for probes leaving different planets. As a result, getting away from a planet can be imagined as like climbing out of a deep hole. The larger the planet, the deeper the hole – it's harder to get away from Jupiter than from Earth.

Key summary: gravitational field and gravitational potential energy

Uniform gravitational field

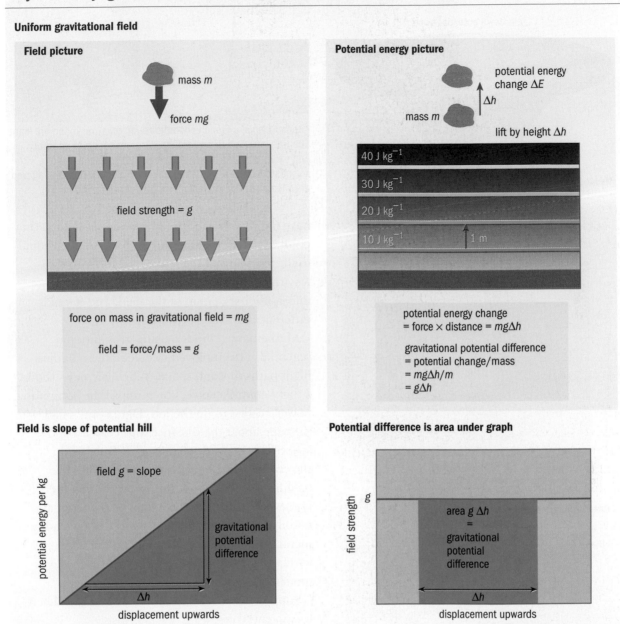

Field picture

mass m

force mg

field strength = g

force on mass in gravitational field = mg

field = force/mass = g

Potential energy picture

potential energy change ΔE

Δh

mass m

lift by height Δh

40 J kg^{-1}

30 J kg^{-1}

20 J kg^{-1}

10 J kg^{-1} 1 m

potential energy change
= force × distance = $mg\Delta h$

gravitational potential difference
= potential change/mass
= $mg\Delta h/m$
= $g\Delta h$

Field is slope of potential hill

field g = slope

potential energy per kg

gravitational potential difference

Δh

displacement upwards

Potential difference is area under graph

field strength

g

area $g\,\Delta h$
=
gravitational potential difference

Δh

displacement upwards

The field is the rate of change of potential with displacement

A French aristocrat and a Nottingham miller

A century after Newton, French mathematicians had greatly developed his ideas. Notably it was the Marquis Pierre Simon de Laplace and Joseph Lagrange who used the idea of mapping the potential energy per kilogram as a useful device for making orbital calculations. English mathematicians were for a time left trailing, until a self-taught Nottingham miller, George Green (chapter 3, *Advancing Physics AS*) showed how to develop the new French thinking further. It was Green who coined the term "potential", seeing its use in electricity as well as in gravitation.

A map of the gravitational potential is a set of contours of constant potential, called equipotentials. Tracking along an equipotential, you can be sure that no gravitational force is acting in that direction because the potential energy is not changing. Equally, look down across the contours to find the steepest slope and you

Key summary: gravitational potential well

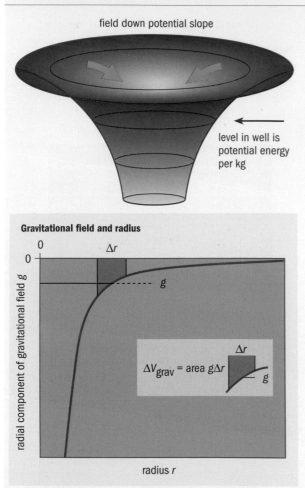

field down potential slope

level in well is potential energy per kg

Gravitational field and radius

ΔV_{grav} = area $g\Delta r$

radial component of gravitational field g

radius r

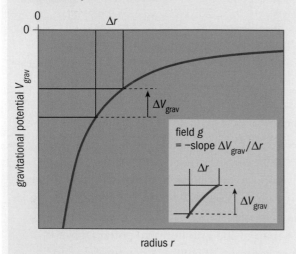

Gravitational potential and radius

field g = −slope $\Delta V_{grav}/\Delta r$

gravitational potential V_{grav}

radius r

The field is the slope of the graph of potential against radius

The difference in potential is the area under the graph of field against radius

Kitsou Dubois, a French choreographer, on a parabolic arc flight. She was intrigued to explore the different sense of internal and external space during weightlessness. She is working with scientists to look at how the body controls movement in altered gravities and is developing new choreographic forms.

have the direction of the gravitational field at that point. Further, the magnitude of the field is equal to the steepness of the "potential hill". In this way the map of the potential contains the same information as a map of the field, and it yields differences in energy directly.

Mathematical models using gravitational potentials are employed to calculate and adjust flight paths through space. When the news shows a flight control centre, you glimpse the computing power needed to do the job. Data come in all the time about the distance and velocity of the spacecraft. Suppose that for the time being the craft is coasting, engines turned off. The data for *Apollo 11* (opposite) show that on the homeward trip from the Moon, engines off and falling towards the Earth, its speed was steadily increasing. This is because the potential energy was falling. Any decrease in potential energy is matched by an increase in kinetic energy.

Suppose *Apollo 11* has total energy E_{total}. Then the potential energy $E_{potential}$ is just

$$E_{potential} = E_{total} - E_{kinetic}$$

Since the kinetic energy is $\frac{1}{2}mv^2$, and the gravitational potential V_{grav} is $\frac{E_{potential}}{m}$, dividing this equation by m gives

$$V_{grav} = \text{constant} - \frac{1}{2}v^2$$

Thus the values of $-\frac{1}{2}v^2$ map out the gravitational potential. If the field varies as $\frac{1}{r^2}$ then the potential varies as $\frac{1}{r}$. The *Apollo 11* data fit a linear variation with $\frac{1}{r}$ very well.

Key summary: gravitational field and gravitational potential

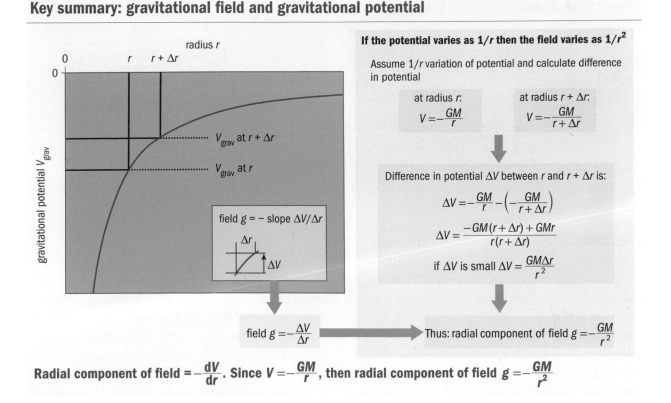

If the potential varies as 1/r then the field varies as $1/r^2$

Assume $1/r$ variation of potential and calculate difference in potential

at radius r:
$$V = -\frac{GM}{r}$$

at radius $r + \Delta r$:
$$V = -\frac{GM}{r + \Delta r}$$

Difference in potential ΔV between r and $r + \Delta r$ is:

$$\Delta V = -\frac{GM}{r} - \left(-\frac{GM}{r + \Delta r}\right)$$

$$\Delta V = \frac{-GM(r + \Delta r) + GMr}{r(r + \Delta r)}$$

if ΔV is small $\Delta V = \frac{GM\Delta r}{r^2}$

field $g = -\frac{\Delta V}{\Delta r}$

Thus: radial component of field $g = -\frac{GM}{r^2}$

Radial component of field $= -\dfrac{dV}{dr}$. Since $V = -\dfrac{GM}{r}$, then radial component of field $g = -\dfrac{GM}{r^2}$

Apollo 11 comes back from the Moon

Pairs of observations of speed and distance

Apollo 11 is coasting home downhill with rockets turned off. Distances r taken from centre of Earth.

distance/10^6 m	speed/m s^{-1}	$-\frac{1}{2}v^2/10^6$ J kg^{-1}	10^8 m/r
241.6	1521	−1.16	0.414
209.7	1676	−1.40	0.477
170.9	1915	−1.83	0.585
96.8	2690	−3.62	1.033
56.4	3626	−6.57	1.774
28.4	5201	−13.53	3.518
13.3	7673	−29.44	7.513

Variation of gravitational potential with distance

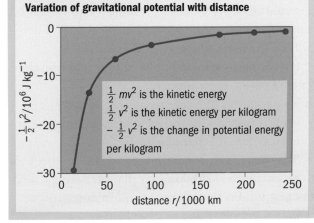

$\frac{1}{2}mv^2$ is the kinetic energy
$\frac{1}{2}v^2$ is the kinetic energy per kilogram
$-\frac{1}{2}v^2$ is the change in potential energy per kilogram

Variation of gravitational potential with 1/r

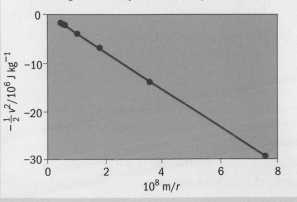

Gravitational potential wells in space

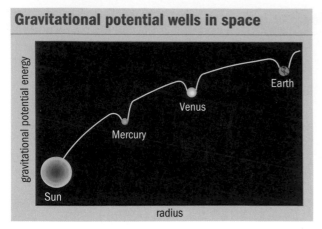

Free float

When a spacecraft is coasting the astronauts feel weightless, just as they do in Earth orbit. They and the spacecraft accelerate together so there is no up or down, and no force pulling them to the floor. Astronauts have to train for this by flying in an aircraft that undertakes parabolic arcs to produce a downward acceleration equal to the acceleration of free fall, causing the passengers to experience weightlessness for short periods.

Hill climbing in space

You are perhaps unaware that you are living at the bottom of a hole, but you are – it is the gravitational potential hollow made by the Earth. Jump in the air and feel yourself drop back to the bottom again. It's exactly like sliding down a hillside.

Picture the gravitational potential in the solar system and beyond. Think of a large flat rubber sheet. A ball on it could roll anywhere once set going, but the Sun makes a deep hollow in the middle and planets must go round it if they are to avoid falling in. Each planet makes its own hollow, in which moons orbit and which makes it hard for the inhabitants to climb out into the space beyond.

To escape from Earth a spacecraft must climb out of the Earth's gravitational potential well. The depth of the well is the potential energy per kilogram needed to take a mass from the Earth's surface to a very long distance away. At the surface – the bottom of the hole – a mass m has a lower potential energy than it would have after escape to any flat part of the sheet at a great distance from Earth. Conventionally, the zero of potential energy is taken as that on the flat part of the sheet (p56).

So the gravitational potential energy for a mass m at the surface of the Earth is

$$E_{grav} = -\frac{GmM_{Earth}}{R_{Earth}}$$

and the gravitational potential is

$$V_{grav} = -\frac{GM_{Earth}}{R_{Earth}}$$

$$= \frac{-6.7 \times 10^{-11} \text{N m kg}^{-2} \times 6.0 \times 10^{24} \text{ kg}}{6.4 \times 10^{6} \text{ m}}$$

$$= -62 \times 10^{6} \text{J kg}^{-1}$$

That is a lot of energy for each kilogram of spacecraft to have to possess. It is needed so that the total energy $E_{kinetic} + E_{grav}$ can be greater than zero, and the spacecraft can coast out of the potential well. That is, in order to escape

$$E_{kinetic} + E_{grav} \geq 0$$

Since $E_{kinetic} = \frac{1}{2}mv^2$, the speed needed to escape is given by

$$\frac{1}{2}mv^2 + mV_{grav} \geq 0$$

$$v^2 \geq 2 \times 62 \times 10^{6} \text{J kg}^{-1}$$

$$v \geq 11 \text{ km s}^{-1}$$

Because of the way rockets achieve their thrust, most of the mass launched continues to have to be fuel. For a spacecraft to journey to the outer planets, not only must it escape Earth's gravitational potential well, but also it must climb up in the Sun's gravitational potential well. The way to the outer planets is still uphill.

Clever methods need to be adopted to find the extra energy. Here are two.

- The rotation of Earth means the launch site itself is moving, so the rocket has kinetic energy even before it is launched.
- The spacecraft can gain kinetic energy through an encounter with another planet, taking energy from that planet's motion, a technique called gravity assist or slingshotting.

Quick check

Useful data: $G = 6.67 \times 10^{-11}\,\text{N}\,\text{m}^2\,\text{kg}^{-2}$. Radius of Earth = 6400 km. Mass of Earth = 6×10^{24} kg.

1. Show that at a height of 2650 km above the Earth's surface its gravitational field has half its value at the surface.

2. Show that the increase in gravitational potential energy when a 20 kg payload is lifted 400 km above the Earth's surface is about 8×10^7 J.

3. Show that the gravitational potential at the Earth's surface is about $-6.3 \times 10^7\,\text{J}\,\text{kg}^{-1}$, increasing to about $-6.2 \times 10^7\,\text{J}\,\text{kg}^{-1}$ at 100 km above the surface.

4. Why should spacecraft be liable to burn up during their descent through the Earth's atmosphere when they did not burn up on their ascent? Hint: compare speeds through the lower atmosphere during ascent and descent.

5. A space probe headed for Mars has escaped Earth. Show that the height of the gravitational potential hill it yet has to climb to reach the orbit of Mars is about $3 \times 10^8\,\text{J}\,\text{kg}^{-1}$. (Mass of Sun = 2×10^{30} kg, radius of Earth's orbit = 150×10^6 km, radius of Mars' orbit = 226×10^6 km.)

6. A coal mine is 1000 m deep. Show that less than $\dfrac{1}{1000}$ of the 40 MJ kg^{-1} that can be obtained by burning the coal is needed to lift it to the surface,

Links to the *Advancing Physics* CD-ROM

Practise with these questions:

200W Warm-up exercise *Pole vaulting*

210S Short answer *Gravitational potential energy and gravitational potential*

220D Data handling *Gravitational potential difference, field strength and potential*

250S Short answer *Summary questions for chapter 11*

Try out these activities:

220S Software-based *Storing energy with gravity*

230S Software-based *Inferring fields*

260S Software-based *Probing gravitational potential*

Look up these key terms in the A–Z:

Gravitational field; gravitational potential; satellite motion

Go further for interest by looking at:

50T Text to read *Training for movement in space*

130T Text to read *The Cassini–Huygens mission to Saturn*

Revise using the revision checklist and:

1800 OHT *Graph showing g against h*

1900 OHT *Field and potential*

2000 OHT *Apollo returns from the Moon*

2100 OHT *Relationship between g and V_g*

2200 OHT *Gravitational equations and graphs*

Summary check-up

Circular motion ✓

- To make an object move in a circular path, a force must act at right angles to its velocity

- A centripetal acceleration $a = \frac{v^2}{r}$. Centripetal force required to cause it is $F = \frac{mv^2}{r}$.

- No work is done when a force on a body acts perpendicular to its motion

Gravitational law ✓

- The radial component of gravitational force between two masses $F = -\frac{Gm_1 m_2}{r^2}$

- Gravitational fields can be represented by field lines or by field vectors
- The radial component of gravitational field strength of a point mass $g = -\frac{GM}{r^2}$

- Gravitational potential energy difference in a uniform field $= mg\Delta h$
- Gravitational potential is gravitational potential energy per unit mass
- Gravitational field strength = −gravitational potential gradient
- Gravitational potential around a point mass $= -\frac{GM}{r}$

- The total energy of a body orbiting in a gravitational field is gravitational potential energy plus kinetic energy

Momentum and collisions ✓

- Momentum $=$ mass $\times$ velocity
- Force is rate of change of momentum
- The thrust of a jet is given by the momentum carried away per second by the jet
- The law of conservation of momentum: the vector sum of momentum is unchanged in an interaction between objects that are not subject to any external forces
- In consequence of conservation of momentum, interaction forces come in equal and opposite pairs

Questions

1. **Wet clothes are spun in a drum of radius 0.4 m so that they experience an acceleration of 16g.**
   (a) Find their speed in m s^{-1}
   (b) Find the number of revolutions of the drum per second.

2. **In Rutherford's 1911 model of the hydrogen atom he imagined the electron orbiting the proton, with an orbital radius of 5.4×10^{-11} m and a speed of 2.2×10^6 m s^{-1}.**
   (a) What force would be needed to keep the electron (mass 9.1×10^{-31} kg) in orbit?
   (b) What could provide such a force?

3. **The Sun, Earth and Jupiter are sometimes positioned so that all three lie on a straight line, with the Earth and Jupiter on the same side of the Sun. When this happens, what is the ratio of Jupiter's gravitational pull on the Earth to the Sun's gravitational pull? (Masses: Sun 2×10^{30} kg, Earth 6×10^{24} kg, Jupiter 1.9×10^{27} kg; orbital radii: Earth 1 astronomical unit (AU), Jupiter 5.2 AU)**

4. **Assume that astrological influences work because of the effect of the force of gravity. Compare the gravitational force of Mars on a newborn baby with that of the midwife who delivered it. What would an astrologer say to these calculations? (Masses: midwife 80 kg, baby 4 kg, Mars 6.4×10^{23} kg; distances: midwife to baby 0.5 m, Earth to Mars 2.2×10^{11} m; $G = 6.67 \times 10^{-11}$ N m^2 kg^{-2})**

5. **Ganymede is Jupiter's largest moon. Its orbit is circular with a radius of 1.07×10^9 m. The orbital period is seven days, three hours and 43 minutes. Use Kepler's third law to find Jupiter's mass.**

6. **The satellites for the Global Positioning System (GPS) orbit Earth twice per day.**
   (a) Use Kepler's third law to calculate the orbital radius for a GPS satellite.
   (b) From this, find the satellite's orbital speed.
   (c) How high is the satellite above the Earth's surface, if it is overhead?
   (d) How long does a radio signal take to reach the satellite, if it is overhead?

7. **The mass of Mars is 0.11 of the Earth's mass and its radius is 0.53 of the Earth's radius.**
   (a) What is the gravitational field strength at the surface of Mars?
   (b) If there were Martian creatures, would they find it easy or difficult to move about on the Earth's surface?

8. **In a movie chase scene, the director wants a 1000 kg car travelling at 15 m s^{-1} to be stopped dead by a head-on collision with a brick wall.**
   (a) What will its change of momentum be?
   (b) What force would the wall exert to stop the car in 0.2 s?

9. **The diagram shows several cars parked in a multi-storey car park. The floors rise in 5 m steps. (Take $g = 10$ N kg^{-1})**

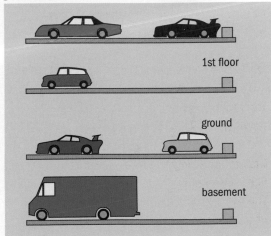

(a) A family car has a mass of 1 tonne. How much potential energy does it gain in moving from the ground floor to: the first floor, the fourth floor, the basement?
(b) What is the gain in energy of each kilogram of the car for each of the moves in part (a)?
(c) What is the gravitational potential difference between the ground floor and the first floor; the fourth floor; the basement?
(d) Write down the gravitational potential difference between the first and the fourth floors; the first floor and the basement.
(e) How much potential energy does a small car of mass 800 kg lose in coming down from the fourth to the first floor?

12 Our place in the universe

In little more than one human lifetime, astronomers using new and powerful types of telescope have dramatically changed ideas about the structure and scale of the universe. We will describe:

● how evidence of the distance, velocity, mass, brightness and composition of astronomical objects is obtained
● how the theory of relativity changes ideas about space and time
● the main evidence that supports current ideas about the origin of the universe

The Andromeda galaxy is just visible to the naked eye on a dark night as a milky patch not far from the "W" shape of Cassiopeia. It is 2.2 million light-years away and is similar in size and shape to the Milky Way galaxy. It is a spiral galaxy, seen partly edge-on. Two smaller satellite galaxies lie near it.

12.1 Observing the universe

On a clear, dark night, what do you see when you look up into the night sky? Hundreds of stars, perhaps a planet or two, the Moon. For thousands of years, the evidence from such naked-eye observation was all that people had to go on. Using records of the position changes of the planets, ancient astronomers put together models of what might be out there, culminating, after the invention of the telescope, with Newton's gravitational theory (chapter 11). The scale of the solar system was known and the sizes and masses of the Sun and its planets could be estimated.

As for the rest, little more was known for more than 100 years; merely that the stars must be a long way off. By the middle of the 19th century the philosopher Auguste Compte was suggesting that astronomy as a science is a waste of time because nobody can reach out to the stars to tell what they are made of or how they work.

Besides stars, the sky contains numerous fuzzy patches called nebulae. The new tool, photography, showed some to be spiral, others elliptical. Nobody knew what they were, or how far away they were. As it turned out, the nebulae were the key to the modern picture of the expanding universe. But it took until the 1920s to know that the nearest large galaxy of this kind, M31

Stars collect together to form galaxies, and galaxies form clusters. This image shows part of the Virgo cluster, which contains both spiral and elliptical galaxies.

The galaxy M100 happens to lie face-on to us revealing its spiral structure, which is thought to be similar to that of our own galaxy, the Milky Way. It is a member of the Virgo cluster, which is about 50 million light-years away.

Key summary: Distances in light travel time

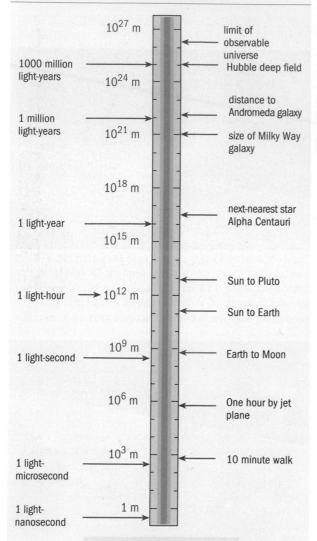

10^{27} m	limit of observable universe
1000 million light-years	Hubble deep field
10^{24} m	distance to Andromeda galaxy
1 million light-years	
10^{21} m	size of Milky Way galaxy
10^{18} m	
	next-nearest star Alpha Centauri
1 light-year	
10^{15} m	
	Sun to Pluto
1 light-hour → 10^{12} m	Sun to Earth
10^{9} m	
1 light-second	Earth to Moon
10^{6} m	One hour by jet plane
10^{3} m	
1 light-microsecond	10 minute walk
1 light-nanosecond	1 m

Speed of light
300 000 kilometres per second
300 km per millisecond
300 m per microsecond
300 mm per nanosecond

Light travel-time
1 light-second = 300 000 km
1 light-millisecond = 300 km
1 light-microsecond = 300 m
1 light-nanosecond = 300 mm

Time
1 year = 31.5 million seconds

Light-year
1 light-year = 9.4×10^{15} m
= 10^{16} m approximately

Distances can be expressed in units of light travel time

This is an image of galaxy NGC 891 in Andromeda. It lies edge-on, showing the thin disc and central bulge that is characteristic of a spiral galaxy. A dark dust lane is visible along its edge.

in Andromeda, is right outside the Milky Way. M31 is the most distant object you can see with the naked eye. Its distance is now estimated at 2.2 million light-years. The Sun is now thought of as a very ordinary star in one of the outer spirals of the Milky Way galaxy.

Today, astronomers think they know the distances, energy output, masses and composition of astronomical objects that are thousands of millions of light-years away. They have estimates (currently about 13.7 thousand million ± 1 thousand million years) for the "age scale" of the universe. They also have theories to account for the origin, structure and expansion of the universe.

How far, how fast?

The modern way to measure distances in the solar system is to use radar, just as air-traffic controllers or traffic police do. Venus was successfully targeted for the first time in 1961; a pulse of radio waves was sent out from Earth, and the reflected waves detected returning after a delay. The distance to the planet can be calculated from the delay and the known speed of the waves. With a distance to Venus of the order of 100 million kilometres, the out-and-back delay is several minutes and the reflected signal is very weak. There is no hope of measuring the distance to even the nearest star

This Second World War German JU88G fighter has its radar aerial clearly visible on its nose. Radar (**ra**dio **d**etection **a**nd **r**anging) systems were used and developed extensively by both sides during the war, following development work during the 1930s in Britain, France, Germany and the US. The first patent for a radar type system went to a German, Christian Hulsmeyer, in 1904 but much of the early development work was done by the Scot Robert Alexander Watson-Watt, resulting in operational radar stations being in place in Britain at the start of the war. The development of radar provides an example of technological development being driven by the needs of warfare.

In the 1990s the *Magellan* spacecraft was put into orbit around Venus. It mapped the planet's surface by beaming down radio waves and detecting the reflected waves. A bat images its own surroundings in much the same way, but uses pulses of ultrasound waves. The reflected waves tell it how far away the reflecting surface is, how fast it is moving and its texture.

Key summary: radar ranging

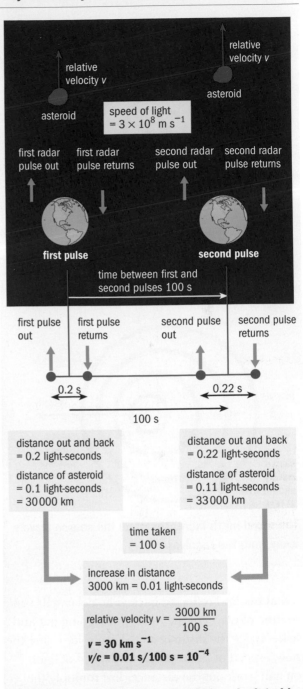

The relative velocity of an asteroid is calculated by measuring two out-and-back radar pulses

(not counting the Sun), which is 4.2 light-years away, in this way.

Imagine that an asteroid is reported as passing close to Earth. Urgent question: how far away is it and how fast is it approaching or receding from Earth? To get a range, you need a radar pulse to

Key summary: Doppler shift

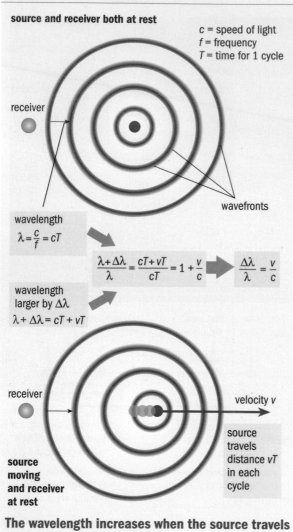

source and receiver both at rest

c = speed of light
f = frequency
T = time for 1 cycle

receiver

wavefronts

wavelength
$\lambda = \dfrac{c}{f} = cT$

$$\dfrac{\lambda + \Delta\lambda}{\lambda} = \dfrac{cT + vT}{cT} = 1 + \dfrac{v}{c} \qquad \dfrac{\Delta\lambda}{\lambda} = \dfrac{v}{c}$$

wavelength
larger by $\Delta\lambda$
$\lambda + \Delta\lambda = cT + vT$

receiver

velocity v

source
travels
distance vT
in each
cycle

**source
moving
and receiver
at rest**

The wavelength increases when the source travels away from the receiver

Johann Christian Doppler predicted the change in frequency of sound when the source and observer are in relative motion. In Utrecht in 1845 an open railway carriage was used to carry a group of trumpeters past musicians with a good enough ear for music to detect the variation in pitch of the note the trumpeters played.

We have tried to make the calculation of distance and speed from radar measurements look as obvious and simple as possible. But actually we hid some assumptions. We assumed that the speed of the signal was the same both ways, and that the moment of reflection was just halfway through the time delay of the pulse as observed on Earth. By contrast, a spacecraft sent there and back could change speed, and could pick up or lose speed in the encounter with the asteroid or planet.

Both of these assumptions require that the speed of light is constant no matter how you are moving relative to other objects. That is the basis of the theory of relativity, and this calculation of distance and relative velocity is correct in that theory. Some parts of relativity are quite easy, after all!

The Doppler shift: measuring velocities

It isn't possible to use radar to measure the velocities of distant stars and galaxies. The signal would take far too long to return and would be hopelessly weak. But it is still possible to measure the velocities of even the most distant objects. You use the fact that atoms in the distant star emit light at particular wavelengths in their spectrum. These wavelengths are accurately known from laboratory measurements done here on Earth.

When the light from the star reaches Earth the wavelengths of all the lines in its spectrum are changed. The wavelength increases if the star is moving away from you, and decreases if the star is coming towards you. These changes in wavelength are called Doppler shifts. Johann Christian Doppler discovered the same effect in the sound from a moving source. You'll have heard the fall in pitch of the siren of a police car or ambulance

be sent out and reflected back. The delay tells you the time of travel of the radar pulse going out and back – twice the distance to the asteroid. Using the known speed of radar pulses (the speed of light) you can work out the distance. Just multiply the speed of light by half the out-and-back time.

To get the relative velocity of the asteroid and Earth you need to measure the distance again. If the out-and-back time increases, the asteroid is moving away from Earth and all is well. If the out-and-back time decreases, the asteroid is getting closer and there might be danger of a collision. The component of velocity along the line of sight is just the change in distance divided by the time between the two measurements.

when a vehicle coming towards you goes past and then travels away from you.

It's easy to understand how the Doppler effect arises. If the source is not moving relative to you, it emits waves whose peaks are a distance λ (the wavelength) apart, emitted at equal intervals of time T, with $\lambda = cT$. If the source is moving away from you at speed v, the wavelength is stretched out. This is because between the emission of any two peaks, the source has moved an extra distance vT away from you. So the wavelength has increased from:

$$\lambda = cT$$

by an amount

$$\Delta\lambda = vT$$

The fractional increase in wavelength is thus:

$$\frac{\Delta\lambda}{\lambda} = \frac{v}{c}$$

To get an idea of how big the change in wavelength might be, recall that the Earth's orbital speed round the Sun is about $10^{-4}c$, so the fractional change in wavelength for a speed of this magnitude would be one part in $10\,000$.

For speeds v much less than the speed of light c, the equation $\frac{\Delta\lambda}{\lambda} = \frac{v}{c}$ is accurate. However, it is not the full story. In section 12.2 we will explain how Einstein's special theory of relativity modifies the calculation. The difference made by relativity is fundamental, affecting ideas of both space and time. But even for the large speed of the Earth round the Sun, the correction to the equation above is only 1 part in 100 million. In practice then, the approximate equation is remarkably accurate for all moderate speeds.

The Doppler effect led to one of the most important steps in our understanding of the universe. In 1912 Vesto Slipher took spectra of light from the Andromeda galaxy, M31 (p65), and found that the wavelengths of the light were slightly shrunk. This meant that this galaxy, the nearest one to us outside the Milky Way, is moving towards the Milky Way at $300\,\mathrm{km\,s^{-1}}$. Soon other galaxies were studied and most were found to

The position of a nearby star relative to a distant star changes as the Earth orbits the Sun. In the 1720s, English astronomer James Bradley built this telescope at Greenwich in an attempt to measure the "annual parallax" of stars that appear at the zenith (directly overhead). Bradley failed, but his measurements did provide unarguable proof that the Earth goes round the Sun. Annual parallax was eventually measured for the first time over a century later by the German Friedrich Bessel (1784–1846), who detected the parallax of the star 61 Cygni in 1838. Just afterwards, the 4.2 light-year distance to the nearest star, Alpha Centauri, was measured in the same way.

be travelling away from us, not approaching. This was the first hint of the expansion of the universe. We will complete this story in section 12.3.

How far is it to the stars and galaxies?

The velocity of a distant star or galaxy is much easier to measure – using the Doppler effect – than its distance. Uncertainties in astronomical distances have led to frequently changing estimates of the age of the universe (p85).

The starting point has to be the nearer stars, with the Earth's orbit as the baseline of measurement. Slight shifts in the positions of the stars during the course of a year can be observed. The effect is called parallax. Look around you and alternately shut your left and right eyes. You'll

Henrietta Swan Leavitt was a leader in the then-new science of photographic measurement, setting standards that were adopted worldwide. She discovered a total of 2400 new variable stars as well as four novae. She began work in a voluntary capacity at Harvard College Observatory and ultimately became head of the department of photographic stellar photometry. She died of cancer in 1921 aged 53. The Cepheid variables that Leavitt observed were all in the Small Magellanic Cloud, a satellite galaxy of the Milky Way. Because they are far away compared to the size of the galaxy, differences in their apparent brightness could be assumed to be due to differences in their absolute brightness.

see nearby objects, like this book, seem to shift position against the background. The distance between your eyes is the baseline and the nearer an object is to you, the farther it seems to move sideways as you switch eyes. To see stars shift, however, you have to wait six months between "blinks" with a telescope.

The apparent angular shift (in radians) is just the ratio of the baseline to the distance. The nearest stars are more than a hundred thousand times the diameter of the Earth's orbit away, so the angles to be measured are tiny, and the parallax method only works for a few thousand stars. In recent years satellites designed for the job have made much more precise parallax measurements and have extended them to larger distances.

Estimates of greater distances often involve comparing how bright a star or galaxy appears to be with how bright it actually is. From the inverse square law for intensity of light, an object twice as far away will look a quarter as bright; 10 times as far away and it will look one-hundredth of the actual brightness. But how do you know the actual brightness (astronomers call it the absolute magnitude) of a star or galaxy? Astronomers need to identify what they call standard candles – stars and galaxies whose true brightness can be estimated – starting with the few stars whose distance is known, for which the inverse square law will tell you the brightness. Many methods have been tried, at successively greater and greater distances, including:

- using the fact that blue stars of a given type are

hotter and brighter than red stars of the same type. The brightness is estimated from the colour;
- using the fact, discovered in 1912 by Henrietta Swan Leavitt, that the brightness of certain very bright variable stars (Cepheids) is related to the time period of their variation. Measure how the light varies and you know the brightness;
- assuming that the brightest stars in any galaxy (blue supergiants) are all about the same brightness;
- assuming that the brightest galaxies in one cluster are as bright as the brightest in another cluster;
- using the fact, discovered by Tully and Fisher in 1977, that the speed of rotation of a galaxy (measured by Doppler shifts) is related to its brightness;
- and using the fact that a certain type of exploding star (Type Ia supernovae) all have the same brightness, because the stars explode after reaching the same critical mass achieved by attracting material from a nearby star.

It was Leavitt's work that enabled Edwin Hubble – detecting Cepheid variables in the Andromeda nebula with the then new 100 inch Mount Wilson telescope – to show for the first time that the nearest galaxy is well outside the Milky Way. Up until then, the Milky Way had been thought by many to be the whole universe; now it was seen to be just one galaxy among many.

Getting a good distance scale has proved very difficult, and even now the scale is not certain. Between the 1920s and 1980s it was revised several times, changing by a factor of 10 (see p85). In the late 1970s a bitter argument broke out between two groups, one led by Allan Sandage and the other by Gerard de Vaucouleurs. Their calculated distance scales differed by as much as a factor of two. Their arguments were all about how to build one reliable ladder of distance measurement from several whose tops and bottoms overlap. Each side thought the other was making unjustifiable guesses and assumptions; many harsh words were spoken. For a decade or so, astronomers had two competing distance scales. Only very recently has there been a better degree of agreement.

Key summary: measuring the mass of a black hole

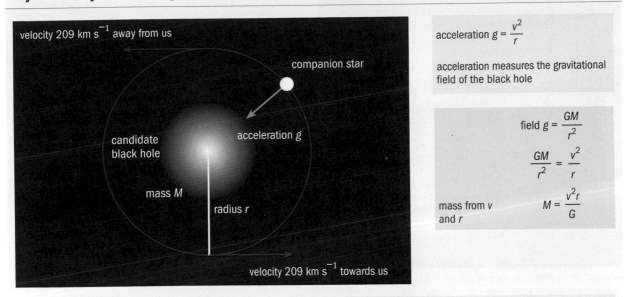

velocity 209 km s^{-1} away from us

companion star

acceleration g

candidate black hole

mass M

radius r

velocity 209 km s^{-1} towards us

acceleration $g = \dfrac{v^2}{r}$

acceleration measures the gravitational field of the black hole

field $g = \dfrac{GM}{r^2}$

$$\dfrac{GM}{r^2} = \dfrac{v^2}{r}$$

mass from v and r $M = \dfrac{v^2 r}{G}$

Example: mass of a candidate black hole V404 Cygnus

V404 Cygnus emits X-rays perhaps due to matter from an ordinary star falling into a massive black hole companion which it orbits. Doppler shifts of light from the ordinary star show its velocity varying by plus or minus 209 km s^{-1} over a period of 6.5 days.

speed v in orbit	from Doppler measurements, assuming orbital plane is in the line of sight	$v = 209$ km s^{-1}
radius r of orbit	from time of orbit and speed of star: star travels $2\pi r$ in 6.5 days at 209 km s^{-1}	$r = 18.7 \times 10^9$ m
acceleration a towards black hole	acceleration $= \dfrac{v^2}{r} = \dfrac{(209 \times 10^3 \text{ ms}^{-1})^2}{18.7 \times 10^9 \text{ m}}$	$a = 2.34$ m s^{-2}
gravitational field g of black hole	gravitational field = acceleration	same quantity $g = 2.34$ N kg^{-1}
mass M of black hole	from gravitational inverse square law, field $g = \dfrac{GM}{r^2}$ $$M = \dfrac{gr^2}{G} = \dfrac{2.34 \text{ N kg}^{-1} \times (18.7 \times 10^9 \text{ m})^2}{6.67 \times 10^{-11} \text{ N m}^2 \text{ kg}^{-2}}$$	$M = 1.2 \times 10^{31}$ kg

mass of Sun $= 2 \times 10^{30}$ kg   mass $M = 6$ times mass of Sun

The mass of a black hole can be calculated using the velocity of a star orbiting it

Weighing the universe

Armed with Doppler-shift measurements of velocities it becomes possible to determine the masses of stars, black holes and even galaxies. Here's an example.

Ever since the idea of gravitational black holes was suggested, astronomers have looked for possible traces of them, even though a black hole

is itself invisible. However, the presence of a black hole can be inferred from the gravitational effects of its mass. So the hunt has been on to find the masses of candidate black holes.

One rather good candidate is the star system V404 Cygnus, suspected of being a normal star in orbit round a massive black hole. X-ray telescopes show it emitting intense and rapidly

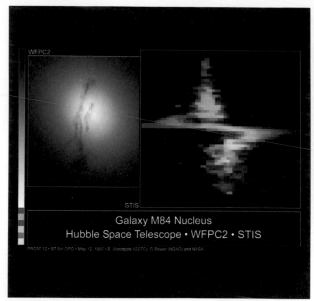

Doppler shifts across a slice of galaxy M84. The spectroscopic image is plotted to show how the velocities obtained from Doppler shifts vary across the centre of the galaxy. Velocities near the centre reach 400 km s^{-1} suggesting a mass of perhaps 300 million solar masses in the centre.

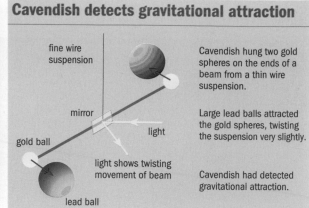

Cavendish detects gravitational attraction

fine wire suspension

mirror

gold ball

light

light shows twisting movement of beam

lead ball

Cavendish hung two gold spheres on the ends of a beam from a thin wire suspension.

Large lead balls attracted the gold spheres, twisting the suspension very slightly.

Cavendish had detected gravitational attraction.

varying amounts of X-rays, which could be caused by matter from the normal star falling into the black hole. Doppler shifts can track the velocity of the star in its orbit. In 1990 Prof. Philip Charles showed by careful spectroscopy that the wavelength of light from the star shifted alternately to the blue and to the red, with a period of 6.5 days. The velocity in the line of sight (calculated from $\frac{\Delta\lambda}{\lambda} = \frac{v}{c}$) went from 209 km s^{-1} towards us to 209 km s^{-1} away from us and back again. Surprisingly, this is enough information to estimate the mass of the candidate black hole.

From the speed and time of the star's orbit you can find the radius of that orbit. Knowing the speed and the radius you can find the acceleration of the star, from $\frac{v^2}{r}$ (see chapter 11). But this is just the gravitational field of the possible black hole near the star. Knowing the field, the gravitational inverse square law $g = \frac{GM}{r^2}$ can be used to calculate the mass of the black hole.

These calculations put the star much nearer to the black hole than Earth is to the Sun: one-eighth of the distance. The black hole's gravitational field is 400 times as strong as that of the Sun's on Earth. The mass to give this field at that distance is big: six times the mass of the Sun. This makes the object at the centre of V404 Cygnus a good

candidate for being a black hole – no dark object of that mass could fail to collapse. You may be able to prove that if the orbit is in fact tilted, the estimated mass is larger still.

We think that this is a remarkable example of the power of theory in physics. The mass of a mysterious distant object is found just from observations of how rapidly and by how much the wavelength of light from a star near it is shifted. It's interesting that the measurement uses Newton's law of gravitation to find the mass of an object for which that law actually breaks down. Inside the black hole, Newton's law has to be replaced by Einstein's theory of general relativity. But outside, far enough away, Newton's law is still useful. Classical mechanics, including $F = \frac{mv^2}{r}$, provides a powerful tool for astronomers.

It all only works if you actually know the gravitational constant G, and you can't find that from astronomical observation. You have to measure the force between two lumps of matter of known mass in the laboratory. This force is so tiny that it is hard to detect at all, and very hard to measure accurately. The first person to attempt it was Henry Cavendish in 1798, measuring the force between balls made of gold and of lead. From these measurements, Cavendish was the first to be able to estimate the mass of the Earth.

Others, including Charles Vernon Boys and Lorand Eötvös, improved the measurement. But the gravitational constant G is so difficult to measure with precision that it remains the least accurately known of all the fundamental constants. Even the most recent measurements only give it to a precision of about one part in 10 000.

These dishes looking out into space form part of the Very Large Array of radio telescopes in New Mexico in the US.

More windows on the universe: astronomy at many wavelengths

Only after the Second World War, using converted radar equipment, did astronomers look at the universe through a wider window than the narrow band of visible photons. An important early achievement was to pick up the 21 cm wavelength line in the radio spectrum of neutral hydrogen gas. Radio astronomers began to map the cool hydrogen clouds in our galaxy. Where optical astronomers could see only masses of stars and dust obscuring the farther reaches of the galaxy, radio astronomers could see right through this dust because it is transparent to such radio waves.

It was frustratingly difficult to identify the objects that radio astronomers started picking up. Early on, this was because the positions of radio sources were not precisely located. Later, it became clear that another reason was that the radio astronomers were seeing farther than ever before. They were detecting powerful sources so distant that optical telescopes had trouble seeing them at all. Among these were the so-called quasars – strong radio sources that looked like tiny stars in the optical telescope. They turned out to be the active cores of enormously distant galaxies, emitting radiation from when the universe was much younger than it is now.

Martin Ryle, a radio astronomer at Cambridge, started a huge row when he announced that there were too many distant radio sources. That is, if he plotted the density of sources in space, he found that the more distant, younger ones were more closely packed. To Ryle, this showed that the universe was more dense in the past than it is now. Others, including Fred Hoyle, held a different theory. It said that the universe should be uniform

This map of the spiral arms of the Milky Way galaxy was drawn from measurements of emission and absorption of radio waves at 21 cm wavelength. Doppler shifts provide further information about the velocities of movement of the spiral arms. A small part of the galaxy is obscured by its dense central region.

in time as well as in space. The argument raged, all dependent on whether Ryle's graphs bent a little from a straight line or not. In the end, with more data, Ryle's ideas won out.

Detecting the spectral signature of different atoms and molecules has given astronomers a good picture of what the visible matter in the universe is made of. The answer is mostly hydrogen, about 10% helium particles, a little lithium and traces of heavier elements.

With the coming of rockets and satellites, astronomers made further use of military technology. The Earth's atmosphere absorbs quite strongly in the infrared and very strongly in the ultraviolet, X-ray and gamma-ray regions. So telescopes for all these wavelengths were built, and sent high above the atmosphere to look at the universe through these new windows.

Infrared telescopes can see through dust clouds and watch stars being formed, and from their spectra find the elements going into their making. They have detected dust clouds around stars, which are thought to be planetary systems being formed. They have also helped to push observations nearer to the limits of the observable universe, seeing it as it was a thousand million years ago. X-ray and gamma-ray telescopes keep springing surprises, particularly enormously energetic gamma-ray bursts, now thought to result from collisions between neutron stars.

Quick check

Useful data: The speed of light in free space is $3 \times 10^8 \, \text{m} \, \text{s}^{-1}$. One light-year is $10^{16} \, \text{m}$. The mass of the Sun is $2 \times 10^{30} \, \text{kg}$. The gravitational constant G is $6.67 \times 10^{-11} \, \text{N} \, \text{m}^2 \, \text{kg}^{-2}$.

1. Show that a tree 600 m away is 2 µs distant in light travel-time.

2. Radar signals reflected from Mars return 30 minutes after they were sent. Show that Mars is $2.7 \times 10^{11} \, \text{m}$, or 900 light-seconds away.

3. The star system Alpha Centauri is observed to shift in direction by 4.2×10^{-4} degrees, when observed from opposite sides of the Earth's orbit (diameter $3 \times 10^{11} \, \text{m}$). Show that Alpha Centauri is about $4 \times 10^{16} \, \text{m}$, or about 4 light-years away.

4. A Cepheid variable of period 100 days is 50 times brighter than one with a period of one day. Show that if the two appear equally bright, the brighter one is about seven times as distant as the fainter one.

5. 21 cm radio waves from a hydrogen gas cloud in an arm of the Milky Way are observed to be shifted in wavelength by 0.1 mm. Show that the gas cloud is moving relative to us at a speed of about $140 \, \text{km} \, \text{s}^{-1}$.

6. Part of the arm of a galaxy rotates around the galactic centre at a speed of $100 \, \text{km} \, \text{s}^{-1}$ and at a distance of 20 000 light-years from the centre. Show that:
   (a) the acceleration towards the centre is about $5 \times 10^{-11} \, \text{m} \, \text{s}^{-2}$;
   (b) the gravitational field at this radius is about $5 \times 10^{-11} \, \text{N} \, \text{kg}^{-1}$;
   (c) if the mass of the galaxy inside this radius acts as if it were at the centre, this mass is more than 10^{10} times that of the Sun.

Links to the *Advancing Physics* CD-ROM

Practise with these questions:
20W   Warm-up exercise *Using time to measure distance*
30W   Warm-up exercise *Units for distance measurement*
50S   Short answer *Trip times tell distances*
55S   Short answer *Doppler shifts in astronomy*
70D   Data handling *Using orbital data to calculate masses*

Try out these activities:
90S   Software-based *The space police*
60H   Home experiment *Two million year old light: Seeing the Andromeda nebula*

Look up these key terms in the A–Z:
Doppler effect; galaxy; gravitational field; inverse square laws; observing the universe

Go further for interest by looking at:
20T   Text to read *The ladder of astronomical distances*

Revise using the student's checklist and:
100   OHT *Distances in light travel time*
900   OHT *Velocities from radar ranging*
950   OHT *Non-relativistic Doppler shift*
1900 OHT *Measuring black holes*

12.2 Special relativity

In everyday life we think we are clear what we mean by being at rest. It means not moving relative to all the things around you that you regard as fixed in place: houses, physics laboratories, trees, streets and so on. So in the limerick (below), Einstein's idea that Madrid was moving towards him while he sat in a fixed train seems at best fanciful. Even so, when you sit still in your seat in an aeroplane and watch the ground below slipping past, it is slightly less crazy to wonder when your destination will glide into sight.

However, think about observing speeds of objects in the universe. You know that you sit on a moving platform (the Earth) that goes round the Sun, which moves round the Milky Way galaxy, which moves relative to other galaxies, and so on. Nothing is fixed. All velocities are relative. From this point of view, Einstein's question is not so absurd. In fact, it becomes fundamental.

What being at rest now means is simply "moving with me". A clock at rest is just one that you carry with you. We will call the time that such a clock records "wristwatch time". Notice that one observer's state of rest may not be the same as another's. They differ by their relative velocity.

You may say: "So what?" and argue that this is a totally unimportant difference. The two observers will see things a bit differently, but that can't change anything about how the world is. And you would be right. Indeed, this is the first fundamental assumption or postulate of the theory of relativity, see "Postulate 1".

Less formally this says that there is nothing special about uniform velocity. Travelling at uniform velocity relative to another object needs no explanation and changes nothing about physics.

We used this idea in chapter 11, when thinking about changes of momentum as seen from differently moving platforms. In that case, the physical law that momentum is always conserved took the same form whatever the motion of the platform. This idea goes as far back as Galileo, but it took Einstein to derive yet more important consequences from it.

Constant speed of light

In 1983, the community of physicists worldwide made a momentous decision. They decided that from then on, the speed of light c would become a defined constant:

$$c = 299\,792\,458\,\mathrm{m\,s^{-1}}\ (\text{exact})$$

More precisely, they decided that the metre, instead of being the distance between two marks on a special bar of metal, would be the distance travelled by light in a vacuum in a time of $1/299\,792\,458$ of a second.

As a result, all speeds became fractions of the speed of light. That is, since

$$v = \frac{\text{distance moved by object}}{\text{trip time for object}}$$

and because now

$$c = \frac{\text{distance moved by light}}{\text{trip time for light}}$$

then it follows that

$$\frac{v}{c} = \frac{\text{trip time for light}}{\text{trip time for object}}$$

Einstein Postulates

Postulate 1
Physical behaviour cannot depend on any "absolute velocity". Physical laws must take the same form for all observers, no matter what their state of uniform motion in a straight line.

Postulate 2
The speed of light c is a universal constant. It has the same value, regardless of the motion of the platform from which it is observed. In effect, the translation between distance and time units is the same for everybody.

When Einstein was travelling to Spain,
He drove the conductor insane:
"It may be a while,"
He would ask with a smile,
"But when does Madrid reach this train?"

Schematic of the Diamond facility

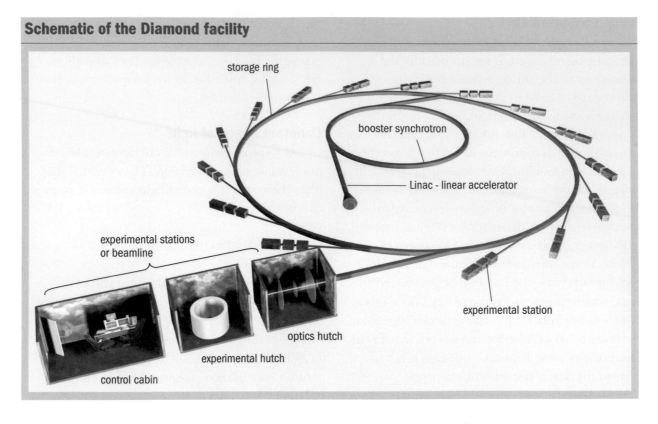

Michelson–Morley experiment

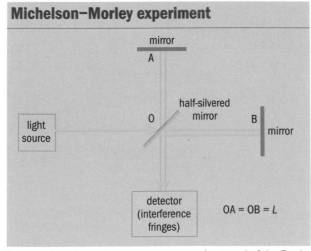

The experiment was designed to measure the speed of the Earth through space. Light from a source was split at a half-silvered mirror, O. Half the light travelled a distance OA and back at right angles to the original direction. The other half travelled an equal distance OB and back along the original direction. Calculations show that if the Earth is moving along one of these directions, the travel times will differ. Differences in travel time were to be detected by looking the interference fringes formed when the two beams recombined. By slowly rotating the whole apparatus, Michelson and Morley could ensure that at some point the two beams would point along and across the Earth's direction of travel. The fringes were expected to shift as the apparatus turned. In fact, the interference fringes obstinately stayed put. No difference in travel time of the two beams was observable. Either the Earth was not moving in space, or it was but the light was unaffected by any such movement.

Note that the ratio $\frac{v}{c}$ has no units. The speed of an object is judged by comparing the time taken for light to cover the distance moved by the object to the time taken by the object to cover that distance.

The decision to define the speed of light as constant was taken for technical reasons to do with precision of measurement. At the same time it agreed with a fundamental shift in thinking, introduced into physics by Einstein. This was to regard the speed of light as a fixed conversion of units of distance and time. The idea can be stated as the second postulate of the theory of relativity. See "Postulate 2" (p75).

There is good experimental evidence that the speed of light is constant, as well as firm theoretical arguments. Some years before Einstein formulated his ideas, Albert Michelson and Edward Morley had tried – and failed – to find any effect on the speed of light due to the Earth's movement in space.

Michelson and Morley's idea was simple: split a light beam into two and send the two beams along paths of the same length but at right angles to one another. Find a way to time the beams along the two paths and, if one path lies along the direction of the Earth's motion and the other across it, the two beams should take different times to travel the two paths. The time difference was to be detected by looking at the dark and bright interference fringes where the two beams were recombined. If the whole apparatus is rotated, the quicker path becomes the slower path, and vice versa, and the fringes should shift in position.

No such fringe shift was detectable, even though the effect of a speed equal to that of the Earth in orbit round the Sun should have shown up.

Nowadays, it is easy to get even more direct evidence of the constancy of the speed of light. For example, in the modern research facility Diamond, light is generated from electrons accelerated to very close to the speed of light. The light comes out along the direction of motion of the electrons. You might expect it to travel with a speed of almost $2c$, but no – the measured speed is exactly c, just as if the electrons emitting it were not moving.

Einstein's second postulate turns this experimentally discovered fact into a fundamental principle. You could even see the second postulate as simply an extension of the first. If the speed of light is a physical property of electromagnetic radiation, then its value ought not to depend on the motion of the observer.

But this creates a problem: what makes the speed of light so special? The answer lies in thinking harder about how speeds have to be measured. It then turns out that these two postulates are enough to generate the whole theory of special relativity with all its consequences. It is this deep simplicity that makes relativity seem so compelling, even obvious, at least to the modern mind.

Measuring distance and speed by radar

In section 12.1 we gave a numerical example of the radar measurement of the distance and speed of an asteroid (p67). No measuring rod sticking out into space was or could have been used. Everything depended on the times of travel of radar pulses, travelling at the constant speed of light, out to the asteroid and back.

We will now look at this kind of measurement again and discuss some less obvious consequences. One of the most startling is that identical clocks moving relative to one another do not run at the same rate – the relativistic time dilation.

"Key summary: space–time diagrams" (p78) introduces a way of picturing objects moving relative to one another. Space–time diagrams have the following features:

- time is conventionally shown on the y-axis;
- distance is shown on the x-axis, in units of $\frac{x}{c}$. For simplicity, only one space dimension is shown;
- every diagram is drawn from the point of view of a given platform. This platform (for example a place on Earth) moves up the time axis as time passes. Observers, clocks etc. are all carried with the platform. Other platforms are taken to move relative to this one;
- any other object at rest relative to the platform also moves vertically up the time axis as time passes, staying a constant distance away;
- lines representing paths of objects through space and time are called worldlines;
- an object that moves relative to the platform at constant velocity moves along a sloping worldline across the diagram, changing its distance from the observer;
- a light pulse travels at 1 light-second per second, so its worldline travels at 45° across the diagram. Because the speed of light is constant, independent of the observer's motion, this is true for every space–time diagram.

A vertical line on the diagram shows something moving only in time, as you do when sitting still in a chair. The time shown is the wristwatch time on a clock moving with you. Worldlines of objects moving relative to you slope more steeply (nearer the vertical) than the worldline of a light pulse because they always go more slowly than light.

Measuring speed

To measure the speed of a distant object, you have to send two radar pulses, one after the other. This lets you measure the distance at two times and calculate the speed. If the distance at the second time is greater than at the first, the object is moving away from you.

"Key summary: two-way radar speed measurement" (p79) shows how radar can be used to measure relative speeds. Everything is measured from an observer moving in time up the y-axis. To keep the algebra as simple as possible we have chosen a special case in which

Key summary: space–time diagrams

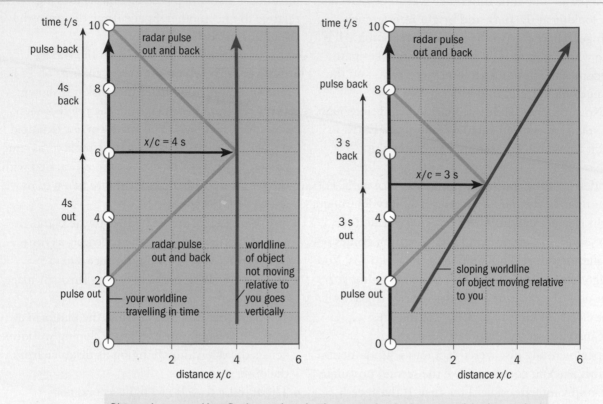

Distance is measured by reflecting a radar pulse then measuring the time sent and the time received back. The clock used travels with you and the radar equipment (at rest).

Assumptions:
  speed c is constant so reflection occurs halfway through the out-and-back time
  speed c is not affected by the motion of the distant object

Radar measures distance using light travel times. The constant speed of light is assumed.

the object moving relative to the observer passes close by at $t=0$. We suppose that a first radar pulse is sent out at that instant and comes back in negligible time. Then a second pulse can be used to find out how far the object travels away from the observer in a given time.

If the second pulse is sent out at time t_1, and returns at t_2, then the pulse travelled out and back in time t_2-t_1. Taking the speed of light c to be constant, the outward trip must have taken the same time as the return trip, that is $\frac{1}{2}(t_2-t_1)$. So the distance at which the reflection took place was:

$$\text{distance}=c\frac{1}{2}(t_2-t_1)$$

But you also know that this is the distance travelled at the speed v of the moving object since it passed

the observer. Again taking the speed of light to be constant, the observer has to assume that the reflection took place halfway between t_1 and t_2, that is at time $\frac{1}{2}(t_2+t_1)$. So the distance travelled is also given by

$$\text{distance}=v\frac{1}{2}(t_2+t_1)$$

Equating the two expressions for the distance gives the speed v as a fraction of the speed c of the pulse:

$$\frac{v}{c}=\frac{t_2-t_1}{t_2+t_1}$$

Notice how everything is measured from a single consistent relativistic point of view. The measurements (times of sending and receiving a pulse) are all made in the wristwatch time of

Key summary: two-way radar speed measurement (simplified)

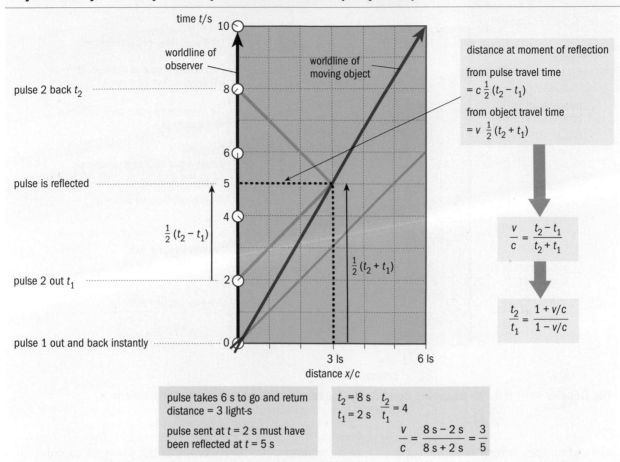

time t/s

worldline of observer

worldline of moving object

pulse 2 back t_2 8

distance at moment of reflection

from pulse travel time
$= c \frac{1}{2}(t_2 - t_1)$

from object travel time
$= v \frac{1}{2}(t_2 + t_1)$

pulse is reflected 5

$\frac{1}{2}(t_2 - t_1)$

$\frac{1}{2}(t_2 + t_1)$

pulse 2 out t_1 2

pulse 1 out and back instantly 0

$$\frac{v}{c} = \frac{t_2 - t_1}{t_2 + t_1}$$

$$\frac{t_2}{t_1} = \frac{1 + v/c}{1 - v/c}$$

3 ls   6 ls

distance x/c

pulse takes 6 s to go and return
distance = 3 light-s

pulse sent at $t = 2$ s must have
been reflected at $t = 5$ s

$t_2 = 8$ s $\frac{t_2}{t_1} = 4$
$t_1 = 2$ s

$\frac{v}{c} = \frac{8\,s - 2\,s}{8\,s + 2\,s} = \frac{3}{5}$

Speed is measured by comparing the interval between returning pulses with the interval at which they were sent

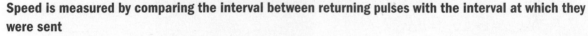

a clock that travels with you. The constancy of the speed of light is built-in by supposing that the out-and-back pulses travel at the same speed, regardless of any relative motion. The time of the reflection is calculated (not measured directly) making this assumption.

The Doppler shift

The ratio of times t_1 and t_2 in the above calculation

$$\frac{t_2}{t_1} = \frac{1 + v/c}{1 - v/c}$$

is related to the Doppler shift (p68). Time t_1 can be thought of as the interval between sending the first pulse (at $t = 0$) and sending the second pulse. Time t_2 can be thought of as the time between receiving the first pulse back (also at $t = 0$) and receiving the second pulse back. Thus, more generally, this ratio

is the ratio of intervals between receiving a pair of pulses back, to the interval between sending them out.

$$\frac{\text{interval between pulses coming back}}{\text{interval between pulses sent out}} = \frac{\Delta t_{\text{back}}}{\Delta t_{\text{out}}}$$

$$= \frac{1 + v/c}{1 - v/c}$$

This stretching of the interval between a pair of pulses is just the Doppler shift, twice over. First, pulses sent at an interval Δt_{out} will be farther apart in time when they arrive at the moving object, because of the relative motion. Suppose this Doppler stretching is by a factor k. Then the pulses arrive at and start back from the moving object with an interval between them. Now they return to the observer, and the Doppler shift works again, with a further factor k. So the interval between

Key summary: Doppler shift – two-way and one-way

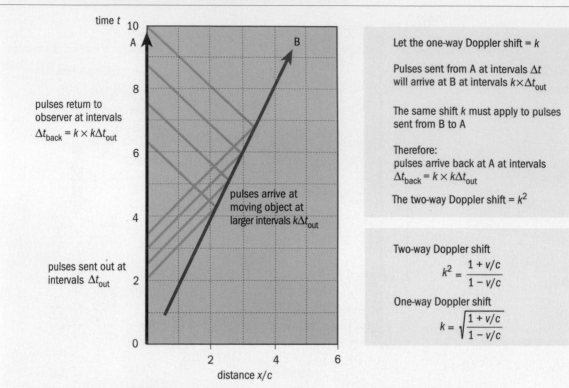

Let the one-way Doppler shift = k

Pulses sent from A at intervals Δt will arrive at B at intervals $k \times \Delta t_{out}$

The same shift k must apply to pulses sent from B to A

Therefore:
pulses arrive back at A at intervals $\Delta t_{back} = k \times k \Delta t_{out}$

The two-way Doppler shift = k^2

Two-way Doppler shift
$$k^2 = \frac{1 + v/c}{1 - v/c}$$

One-way Doppler shift
$$k = \sqrt{\frac{1 + v/c}{1 - v/c}}$$

The Doppler shift k is the observed quantity that measures the speed of a remote object

pulses when they return is

$$\Delta t_{back} = k \times k \Delta t_{out} = k^2 \Delta t_{out}$$

This result tells you at once how the Doppler shift k is related to the ratio $\frac{v}{c}$:

$$\frac{\Delta t_{back}}{\Delta t_{out}} = k^2 = \frac{1 + v/c}{1 - v/c}$$

The fact that the two Doppler shifts must have the same value of k follows from the first postulate. The motion is purely relative, so it can't matter who is the sender and who is the receiver.

The two-way radar shift is equal to k^2. The one-way Doppler shift k is the stretching of time between pulses on a one-way trip

$$k = \sqrt{\frac{1 + v/c}{1 - v/c}}$$

k is also the stretching factor for wavelengths of light received by an observer when emitted by a source moving relative to the observer, since the wavelength λ is proportional to the time for one

cycle. Relativity replaces the previous calculation of the Doppler shift (p68) by the relationship

$$\frac{\lambda_{received}}{\lambda_{sent}} = \frac{\lambda + \Delta\lambda}{\lambda} = \sqrt{\frac{1 + v/c}{1 - v/c}}$$

The relativistic result looks very different from the previous not-quite-correct expression

$$\frac{\lambda + \Delta\lambda}{\lambda} = 1 + v/c$$

However, it is not so different. "Key summary: relativistic effects" (opposite) shows that the relativistic result can be written as

$$\frac{\lambda + \Delta\lambda}{\lambda} = \gamma(1 + v/c)$$

with the relativistic factor γ given by

$$\gamma = \frac{1}{\sqrt{1 - v^2/c^2}}$$

Even for a speed v equal to 0.1 of c, the factor γ is only 1.005, barely different from 1. So the simpler expression for the Doppler shift is good for all

Key summary: relativistic effects

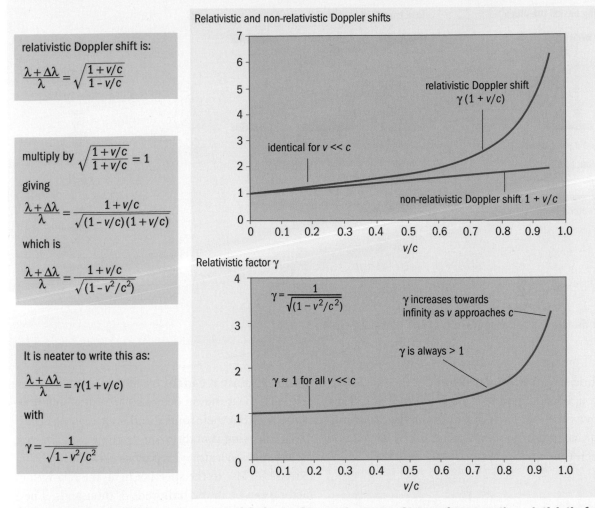

relativistic Doppler shift is:

$$\frac{\lambda + \Delta\lambda}{\lambda} = \sqrt{\frac{1 + v/c}{1 - v/c}}$$

multiply by $\sqrt{\dfrac{1 + v/c}{1 + v/c}} = 1$

giving

$$\frac{\lambda + \Delta\lambda}{\lambda} = \frac{1 + v/c}{\sqrt{(1 - v/c)(1 + v/c)}}$$

which is

$$\frac{\lambda + \Delta\lambda}{\lambda} = \frac{1 + v/c}{\sqrt{(1 - v^2/c^2)}}$$

It is neater to write this as:

$$\frac{\lambda + \Delta\lambda}{\lambda} = \gamma(1 + v/c)$$

with

$$\gamma = \frac{1}{\sqrt{1 - v^2/c^2}}$$

Relativistic and non-relativistic Doppler shifts

identical for $v \ll c$

relativistic Doppler shift $\gamma(1 + v/c)$

non-relativistic Doppler shift $1 + v/c$

v/c

Relativistic factor γ

$$\gamma = \frac{1}{\sqrt{(1 - v^2/c^2)}}$$

γ increases towards infinity as v approaches c

γ is always > 1

$\gamma \approx 1$ for all $v \ll c$

v/c

At low speeds relativistic and non-relativistic results are the same. At speeds near c the relativistic factor γ increases rapidly.

speeds that are slower than this.

Time dilation and the relativistic factor γ

The Doppler shift is what you see happening to light from a source in relative motion towards or away from you. The term $1 + \frac{v}{c}$ is just the result of the source moving away from you, stretching out the light waves. The term γ in $\gamma(1 + \frac{v}{c})$ is less obvious. It arises because clocks in systems moving relative to one another cannot agree. This effect makes a clock moving relative to you run slower by a factor γ than a clock moving with you showing your wristwatch time.

To see that there must be this time-dilation effect it is easiest to think about a clock moving sideways

past you. Then there is no classic Doppler shift, but only the effect of time dilation.

The simplest clock to think about is a light-clock: a pair of mirrors between which pulses of light bounce back and forth. The first postulate says that the wristwatch time τ between ticks of such a clock must be the same for anyone travelling with the clock, no matter how it moves relative to other things. In other words: the one time that everyone can trust is their own wristwatch time.

But this does not mean that the clock will tick at this rate if seen from a platform relative to which the clock is moving. Seen from such a platform the clock must be running slow. This is a consequence of the second postulate: that the speed of light is

Key summary: the light clock

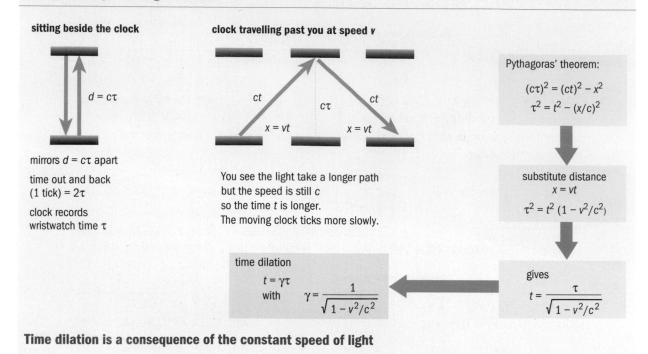

Time dilation is a consequence of the constant speed of light

the same seen from any platform.

The upshot is that the clock going past you takes a longer time t for each tick than the time τ from a similar clock ticking wristwatch time. The relation between the two is simply:

$$t = \gamma\tau$$

with

$$\gamma = \frac{1}{\sqrt{1 - v^2/c^2}}$$

This effect of time dilation has been confirmed experimentally in many ways. Modern technology actually takes it for granted. For example, a global positioning system has to allow for the effects of time dilation on the signals from its satellites to tell you where you are to within a metre or less.

Time dilation is also seen at work every day in high-energy physics laboratories. For instance, unstable particles called π-mesons are routinely produced from particle collisions. The mean lifetime of a π-meson at low speeds is well known: 2.6×10^{-8} s. Multiply this by the speed of light, $3 \times 10^8 \, \mathrm{m \, s^{-1}}$, and you would expect it to decay in a distance of less than 7 or 8 m from where it was

produced, since it cannot travel farther than that.

In fact, the π-meson detectors are often built tens of metres from their source and no π-mesons decay while travelling that distance. In these experiments, at very high energies with particles travelling extremely close to the speed of light, the relativistic factor γ can be in the hundreds or thousands. The "inner clock" telling the π-mesons how long they have to decay has slowed down enormously.

Finally, one especially startling consequence of time dilation is that photons never grow old. Their wristwatch time is given by

$$\tau = \frac{t}{\gamma}$$

and if $v = c$ the factor γ becomes infinite. So whatever the value of t between a photon being emitted and absorbed, the photon sees no time pass by. More carefully stated, it is impossible to travel alongside a photon. This is what started Einstein off thinking about relativity. He realised that if you imagine travelling along with an electromagnetic wave, there can't be any wave, because the electric and magnetic fields would be static. This is what makes light special: it simply has to have the one particular speed that it has.

Quick check

Useful data: $c = 3 \times 10^8 \, \text{m s}^{-1}$

relativistic factor $\gamma = \dfrac{1}{\sqrt{1 - v^2/c^2}}$. If $v \ll c$ then

$\gamma = 1 + \dfrac{1}{2} \dfrac{v^2}{c^2}$.

1. A space probe is travelling away from Earth at $30 \, \text{km s}^{-1}$. Show that the Doppler shift of the wavelength of 1 GHz signals sent from the probe to Earth is about 0.03 mm.

2. Show that for $\dfrac{v}{c} = \dfrac{3}{5}$ the time dilation factor $\gamma = \dfrac{5}{4}$.

3. Show that for a time dilation factor $\gamma = 2$ the ratio $\dfrac{v}{c} = 0.866$.

4. Show that for $\dfrac{v}{c} = \dfrac{1}{10}$, the time dilation factor γ is only about 0.5% larger than one.

5. Atomic clocks in the global positioning system can measure time differences to one part in 10^{12}. Show that for this precision, corrections for relativistic time dilation start to become needed at speeds of the order of $300 \, \text{m s}^{-1}$.

6. A satellite circles Earth, radius 6400 km, every 90 minutes at low altitude. Show that the relativistic time dilation factor $\gamma = 1 + 3 \times 10^{-10}$ approximately.

Links to the *Advancing Physics* CD-ROM

Practise with these questions:

50C   Comprehension *Thinking relatively*

50W   Warm-up exercise *When does the speed of light matter?*

60W   Warm-up exercise *The relativistic time dilation factor* γ

70S   Short answer *Practice with the relativistic Doppler shift equation*

80S   Short answer *Practice with the relativistic time dilation equation*

Try out these activities:

50S   Software-based *Investigating the relativistic Doppler shift*

60S   Software-based *Investigating the time dilation factor* γ

100S   Software-based *The relativistic Doppler effect*

110S   Software-based *The light clock*

Look up these key terms in the A–Z:

Doppler effect; invariance of the speed of light; theory of special relativity

Go further for interest by looking at:

100T   Text to read *Why we believe in special relativity: Experimental support for Einstein's theory*

Revise using the student's checklist and:

1000   OHT *Space–time diagrams*

1100   OHT *Two-way radar speed measurement*

1200   OHT *Doppler shift: Two-way and one-way*

1250   OHT *Relativistic effects*

1300   OHT *The light clock*

12.3 Was there a Big Bang?

Between the 1920s and the 1950s an idea developed that the universe originated in a very hot, very dense state from which it has expanded and cooled. The astronomer Fred Hoyle, who was opposed to the idea, rudely dubbed it the Big Bang. The name stuck, but is misleading. The beginning of the universe was not, according to present theories, an explosion in an existing empty space, but the very creation of an expanding region of space and time. Everything exists only inside the Big Bang; there is no outside.

In this section we will tell the story of how evidence for this picture built up, and describe some of the problems that remain to be solved. One of them, the puzzle that most of the matter in the universe seems to be missing, is taken up in the final chapter "Advances in Physics".

The universe is expanding

In 1929, of the millions of galaxies in the universe, just 24 of them gave Edwin Hubble publishable evidence of a general expansion of the universe. His colleague Vesto Slipher had been collecting spectra of the light from galaxies since 1914 and with the exception of the two nearest (M31 and M33) the light was red shifted. If this was the Doppler effect, the galaxies were receding. But nobody knew how far away they were, or how big.

Then in 1925 Hubble used Cepheid variables to show that M31, the Andromeda nebula, was well outside the Milky Way galaxy (p69–70). He could not detect Cepheid variables in smaller, fainter and presumably more distant galaxies but he could look at how bright and how big these galaxies appeared to be in the telescope and plot how many there were in different directions. This gave Hubble at least an idea of the relative distances. Having Slipher's (largely unpublished) red-shift data to hand, Hubble plotted the speeds of recession of 24 galaxies against his estimates of their distance – in effect, against apparent brightness. The points fell on a more or less straight line: the farther away, the faster the recession.

This sparse evidence had a big impact, because theoreticians including Einstein, de Sitter, Lemaitre and Friedman had concluded that Einstein's theory of gravitation made it impossible

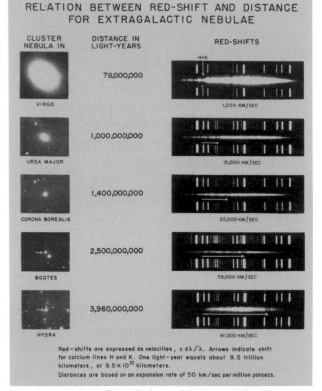

This diagram shows five galaxies and their red shifts. The spectra appear as smears, crossed by a pair of prominent absorption lines.

for the universe to stand still. It had to expand or contract. Attempts to "fix" the theory to get the universe to be static were ugly so Hubble's news was welcome.

Throughout the 1930s, Hubble, with Humason taking the spectra, pursued galaxies to greater and greater depths in space. Despite the problems of establishing distances, Hubble's law continued to stand. With the red shift interpreted as a recession velocity:

$$\text{recession velocity} = \text{constant} \times \text{distance}$$

$$v = H_0 r$$

The constant H_0 became called the Hubble constant.

The bigger the Hubble constant, the faster the universe expands, and the younger it must be to have got to its present size. So the Hubble constant gives you a time scale for the universe. This time is just its reciprocal $\frac{1}{H_0}$. The universe must be younger than the age calculated from this time,

Key summary: Hubble's law and the age of the universe

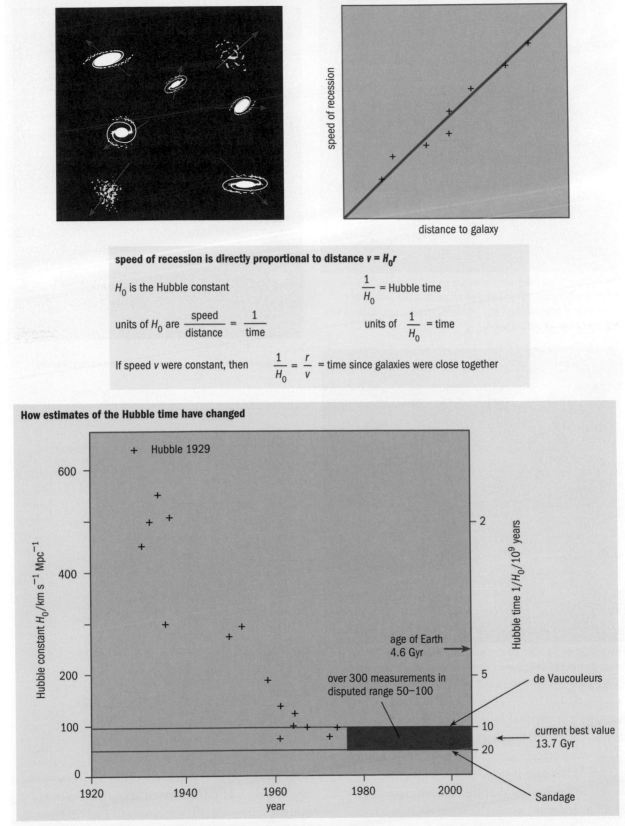

speed of recession is directly proportional to distance $v = H_0 r$

H_0 is the Hubble constant

$\dfrac{1}{H_0}$ = Hubble time

units of H_0 are $\dfrac{\text{speed}}{\text{distance}} = \dfrac{1}{\text{time}}$

units of $\dfrac{1}{H_0}$ = time

If speed v were constant, then $\dfrac{1}{H_0} = \dfrac{r}{v}$ = time since galaxies were close together

How estimates of the Hubble time have changed

The smaller the value of H_0 the older the universe

The Hubble deep field: seeing as far as you can see. This image combines hundreds of exposures to identify thousands of galaxies in a tiny patch of sky, reaching deeper into space than had ever been done before.

This image of gigantic dust clouds in the Eagle nebula was taken by the Hubble space telescope. New stars form as a mass of gas and dust collapses under its own gravitational pull.

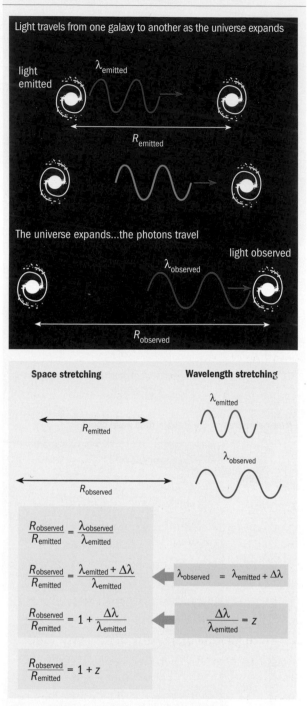

Key summary: cosmological red shift

Light travels from one galaxy to another as the universe expands

The universe expands...the photons travel

The cosmological red shift measures the stretching of space

because gravity will have slowed the expansion down, the rate being higher nearer the beginning of the expansion.

This fact made it very embarrassing for astronomers to find that the first estimates of the Hubble constant made the universe much too young. Up until the 1960s the universe seemed to be younger than Earth! As the distance scale was revised, so the estimated age of the universe rose. Settling on a Hubble time of about 14 thousand million years, it's still an awkward fact (in late 2007) that there are star clusters that seem to be a bit older than this.

Expansion of the universe

Imagine a balloon with dots drawn on it to represent galaxies being slowly blown up.

As the balloon grows, the space between the galaxies grows. On the balloon, the dots also grow. In the universe, however, the galaxies stay the same size.

The universe has evolved

The best optical telescopes can peer out to about 1000 million light-years, and so back in time 1000 million years. Radio and infrared telescopes beat them and see even farther. At these distances, red shifts get really large. They have to be thought of not as signalling velocities through space but the expanding of space itself.

At large distances, the red shift is not best thought of as a velocity of recession. Rather, the red shift is simply the waves stretching as the space stretches. If the red shift

$$z = \frac{\text{change in wavelength}}{\text{wavelength}}$$

then the space has stretched by a factor $1 + z$. Observing large red shifts is equivalent to looking at a younger smaller universe. Only for red shifts much less than one is it possible to interpret the red shift as a relative velocity.

The first objects with startlingly large red shifts were quasars. Picked up initially as radio sources, some were identified optically in the early 1960s as faint point-like stars, often with a visible "jet" extending to the side. Allan Sandage, Hubble's successor in the work on red shifts and an expert on stellar spectra, could make no sense of their spectra. There were lines where no lines ought to be. One quasar in particular, 3C48, kept puzzling him. Then in 1962 Maarten Schmidt realised that he could after all make sense of the spectrum of a quasar called 3C273. All he had to do was to shift its lines by 16%, and he was looking at the spectrum of hydrogen. When 3C48 was looked at like this, its red shift came out at a huge 37%.

Sandage had just missed realising that he was seeing an object from a time when the universe was smaller than it is now by a factor of 1.37.

Today, helped by infrared and other types of telescope, red shifts of up to 10 have been detected. The universe at a tenth of its present size is coming into view. Things now are not what they used to be. There were many more quasars then than there are now. Sources when counted show denser packing, and the galaxies you can see look more like prototype galaxies in formation, or pairs of galaxies in collision, than most of those at smaller distances. Looking far back in time has shown that the universe really did evolve; that it has not been the same forever.

A hot, dense early universe

Stars generally seem to have the same ratio of helium to hydrogen, about 10% of helium particles. But in old stars this is all there is, while in younger ones there are traces of heavy elements as well, which is more evidence that the universe has evolved. The heavy elements are thought to have been made in the stellar explosions, called supernovae, at the end of a star's life. The Earth and your body are made of the ash of exploding stars. But where did the ratio of helium to hydrogen come from?

George Gamow and his students Ralph Alpher and Robert Herman suggested in the 1940s that the universe might once have been much smaller and hotter than it is now. They could show that nuclear fusion of hydrogen would lead to the observed proportions of helium to hydrogen. They suggested that their theory could be tested by

Cosmic microwave background radiation

In the beginning...

there was the Big Bang...

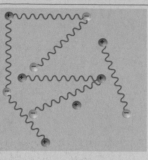

...the universe is filled with a plasma of elementary particles exchanging energy with photons of electromagnetic radiation.

300 000 years after the Big Bang

temperature 3000 K
typical wavelength of radiation: 1 μm

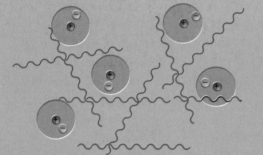

As the temperature falls, atoms form as electrons are held in orbit around nuclei of protons and neutrons.

The universe becomes transparent to photons which no longer interact so easily with atoms and so travel unaffected through the universe.

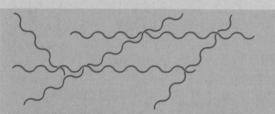

Interstellar space is filled with a photon "gas" (and some atoms). The temperature of this gas is proportional to the energy of the photons.

The energy of a photon is proportional to its frequency. Therefore the temperature of the photon gas is proportional to the frequency of the radiation.

...14 billion years after the Big Bang

temperature 2.7 K
typical wavelength of radiation: 1 mm

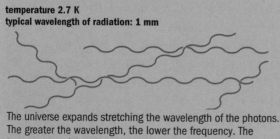

The universe expands stretching the wavelength of the photons. The greater the wavelength, the lower the frequency. The temperature of the photon gas falls.

Arno Penzias and Robert Wilson used this giant horn to collect microwave radiation. They were surprised to find that they detected microwaves no matter which direction they pointed the antenna. Other physicists realised that this was the predicted cosmic background radiation that was left over from the hot beginnings of the universe.

looking for the leftover radiation.

Their work remained largely forgotten for many years. The radiation wasn't discovered until 1965 when, at Bell Laboratories, Arno Penzias and Robert Wilson, meticulously calibrating an antenna designed for use with new communications satellites, noticed something odd in their results. There was always just a little too much noise in the signal. And the noise seemed to come equally from every direction. Nearby in Princeton, a different group led by Robert Dicke and including Jim Peebles were planning to build an antenna to look for the same thing. But they knew what they were looking for: the cooled down radiation from the Big Bang. Finally the two groups each published a paper in the same journal: one describing the unexpected extra noise signal and the other explaining its significance as the first direct evidence that the universe was once very hot.

The cosmic microwave background has the biggest cosmological red shift known. It was produced when the universe became just cool enough for electrons and ions to combine into neutral atoms, emitting photons. That happens at around 3000 K, when the typical wavelength of the photons is around 1 μm. Today these photons are seen, stretched in wavelength, as microwaves with a wavelength of the order of 1 mm – 1000 times as great. The temperature, too, has fallen by a factor of 1000, to just below 3 K. The cosmological expansion $z+1$ is the same ratio:

$$z + 1 = \frac{\text{radius of the universe now}}{\text{radius of the universe then}}$$

By itself, the expansion of the universe is not strong evidence for it having evolved from an early very hot, very dense state.

An alternative explanation, promoted by Fred Hoyle, Hermann Bondi and Thomas Gold, was that maybe matter is being continually created in empty space, so that the universe expands without really changing. That idea was challenged by the discovery that distant radio sources seem to be more closely packed than they are now (p73). But the clinching evidence was the detection of cosmic microwave background radiation. A hot, dense early universe seems to be the only explanation and the expansion of the universe becomes just what should be expected.

...and some problems

There are plenty of unsolved problems about the nature of the universe. Here are some of them.

- **Dark matter** The way the stars in a galaxy move shows that galaxies are surrounded by large amounts of unseen matter. There is similar evidence for dark matter in clusters of galaxies. The problem is that the mass of this unseen part of the universe seems to be at least 10 times that of the visible matter. There are competing ideas about what such dark matter could be made of. Thre is more about this problem in the chapter "Advances in Physics".

- **The same everywhere** The fluctuations in cosmic microwave background (opposite), although looking impressive in images to display them, are in fact extremely small, around one part in 10^4. The universe is remarkably isotropic, that is, it looks the same in all directions. Because gravity tends to accentuate any fluctuations, this means that the universe must have been born even "smoother", indeed to a remarkable extent. Yet there must have been some variations in density, or matter would never have clustered together in galaxies. Accounting for the amount of variation

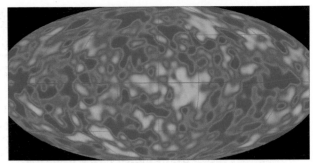

This image of the whole universe is taken from the Cosmic Background Explorer satellite. The satellite measured the background temperature in different directions within the universe. The pink areas are very slightly warmer than the average and the blue areas are slightly colder. This shows that the universe wasn't quite uniform in its early stages, and it was from these non-uniformities that stars and galaxies are thought to have subsequently formed.

found is a live problem.

- **The age problem** How the universe evolves depends on the density of matter in it. Too much matter and gravitational attraction would have re-collapsed it long ago. Too little, and galaxies and stars wouldn't have formed at all. In either case, you wouldn't be here to think about it. In between is a critical density, at which the universe is said to be flat and there is time for it to evolve as it seems to have done. Present evidence is that the actual density is close to the critical density. But that could easily change.

- **Evolution of galaxies** The universe is certainly not the same now as it used to be. For example, at red shift $z=2$ to 3, the number of quasars seems to be maybe 30 times as great as it is today. But at even earlier epochs, there were few. So there is evidence of evolution. The Hubble space telescope and infrared telescopes have just begun to identify what look like protogalaxies – galaxies in formation. Our knowledge of how galaxies are formed and evolve is steadily increasing.

- **Dark energy** Studies of distance, brightness and speed of supernovae in very distant galaxies have recently indicated that the expansion of the universe is accelerating with time. To explain this, a mysterious form of energy, dubbed dark energy, is inserted into gravitational equations. Despite dark energy accounting for up to 70% of the contents of the universe, nobody knows what it is, if indeed it exists.

Quick check

Useful data: The speed of light $c = 300\,000\,\text{km s}^{-1}$. The distance 1 megaparsec (Mpc) is approximately 3 million light-years, or $3 \times 10^{19}\,\text{km}$. There are approximately 30 million seconds in a year.

1. The radio galaxy 3C324 has a red shift $z = 1.12$. Show that the 21 cm hydrogen line is observed at a wavelength of about 44 cm.

2. A galaxy in the constellation Hydra has a red shift $z = 0.203$ (p84). Show that the universe has expanded by about 20% since the light from it was emitted.

3. Show that ultraviolet radiation, wavelength 100 nm, originally part of what is now the cosmic microwave background, is now detected at a wavelength of about 0.1 mm.

4. A nearby galaxy has a small red shift $z = 0.003$. Show that its speed of recession is about $1000\,\text{km s}^{-1}$ (you may approximate $z = \frac{v}{c}$).

5. Converting distances in megaparsec to kilometres, show that a Hubble constant of $100\,\text{km s}^{-1}\,\text{Mpc}^{-1}$ corresponds to a timescale for the universe of roughly 10^{10} years.

6. Explain why a small estimate of the Hubble constant corresponds to a long timescale for the universe.

Links to the *Advancing Physics* CD-ROM

Practise with these questions:

40W Warm-up exercise *Cosmological expansion*

80D Data handling *Astronomical distances*

90D Data handling *Calculating the age of the universe*

100S Short answer *Cosmic microwave background radiation*

110C Comprehension *Evidence for a hot early universe*

Try out this activity:

130S Software-based *The cosmological red-shift*

Look up these key terms in the A–Z:

Expansion of the universe; microwave background radiation

Go further for interest by looking at:

60T Text to read *The sky is dark at night: A reason to think that the universe has evolved*

Revise using the student's checklist and:

1500 OHT *The history of the universe*

1600 OHT *Hubble's law and the age of the universe*

2000 OHT *The "age" of the universe*

2100 OHT *The cosmic microwave background radiation*

Summary check-up

Distance scale ✓

- Astronomical distances in the solar system can be measured by radar
- Distances to nearby stars can be found by parallax
- Some larger distances can be estimated by the apparent brightness of standard candles, e.g. Cepheid variables and Type Ia supernovae

Special theory of relativity ✓

- No matter what the uniform motion of an observer, physical laws are unchanged
- No matter what the uniform motion of an observer, the speed of light is the same
- Distance is measured by light travel-time, and all velocities are measured relative to the speed of light
- The Doppler shift of wavelength of a source receding at relative velocity v is

$$k = \frac{\lambda + \Delta\lambda}{\lambda} = \sqrt{\frac{1 + v/c}{1 - v/c}} = \gamma(1 + v/c)$$

- The factor γ is the relativistic time dilation

$$\gamma = \frac{1}{\sqrt{1 - v^2/c^2}}$$

At low speeds, γ is very close to one and the Doppler shift is very close to $(1 + \frac{v}{c})$

Properties of distant objects ✓

- The cosmological distance scale is still subject to uncertainty
- Velocities of astronomical objects can be established by the Doppler shift with $\frac{\Delta\lambda}{\lambda} = \frac{v}{c}$ for v much less than c
- Masses of astronomical objects can be found from the velocities of objects orbiting them at known distances or in a known time
- Spectra of distant objects over a wide range of wavelengths provide knowledge of their chemical composition

Evolution of the universe ✓

- Red shifts of distant galaxies give evidence of the expansion of the universe. A red shift $z = \frac{\Delta\lambda}{\lambda}$ corresponds to an expansion in scale of $\frac{R_{now}}{R_{then}} = 1 + z$.

- Current estimates of the expansion time scale of the universe put it at about 13.7 ± 1 Gyr
- Evidence that the universe has evolved from an initial uniform, hot, dense state comes from the existence of cosmic microwave background
- There are still fundamental problems in explaining the major features of the universe

Questions

1. **Figure 1 shows the intensity trace of the spectrum of light from the quasar Q1215+333. The peak in the spectrum is the red shifted ultraviolet line of atomic hydrogen, wavelength 121.6 nm.**

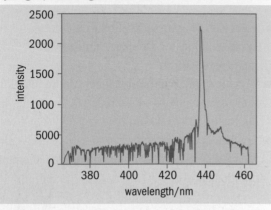

(a) What is the red shifted wavelength of the line?
(b) By how much has the wavelength changed?
(c) What is the value of the red shift $z = \Delta\lambda/\lambda$
(d) By what factor has the universe expanded since the light was emitted?
(e) What are the photon energies $E = hf$ of the light when it was emitted and when it was received?
(f) What is the ratio of these two energies? How is this connected to the answer to (d)?

2. **The Earth travels round the Sun in 365 days at a distance of 150 million km. This information is enough to work out the mass of the Sun.**
(a) What is the speed of the Earth in its orbit?
(b) What is the acceleration of the Earth towards the Sun?
(c) What is the gravitational field of the Sun at the distance of the Earth's orbit? How is this answer related to the answer to (b)?
(d) Write down the relation between the mass of the Sun, the distance from it and the gravitational field at that distance.
(e) What mass must the Sun have to give the gravitational field from (c) at the distance of the Earth's orbit?

3. **The Hubble constant is about 210 km s^{-1} per million light years.**
(a) Show that the expansion speed is of the order $\frac{1}{10}c$ for distances of a few hundred million light-years.
(b) Show that even at this speed, the error in using the non-relativistic Doppler equation $\frac{\Delta\lambda}{\lambda} = \frac{v}{c}$ to find the speed of recession is less than 1%.

4. **Two observers look at the Moon against the background of fixed stars, from opposite sides of the Earth. The radius of the Earth is 6400 km. The radius of the Moon's orbit is 60 times as large as the radius of the Earth.**
(a) What is the difference in the angle in radians at which the two observers see the Moon?
(b) What is this angle in degrees?
(c) Could the observers arrange to look at the Moon at the same moment? If so, how?
(d) Calculate the time of travel of a radar pulse to the Moon and back.
(e) Make some estimate of the accuracy with which a radar method using 1 GHz waves might measure the distance to the Moon.
(f) Why is the radar signal not Doppler-shifted appreciably by the motion of the Moon in its orbit?

5. **Figure 2 shows the history of the expansion of the universe, from a few seconds old to now.**

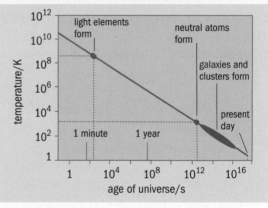

(a) On what kind of scale are the temperature and age plotted? Why was this necessary?
(b) Assume that, as the temperature changes, the product λT of wavelength and temperature is constant. If the photons at the time neutral atoms formed had a wavelength of about 1 μm, what would have been their wavelength when the light elements formed?
(c) By what factor did the universe expand between these two times?
(d) What is the photon energy $E = hf$ of the photons at the time the light elements formed? Find the value in electron volts. How does this compare with the energy achieved by modern accelerators (chapters 16 and 17)?

13 Matter: very simple

Gases are perhaps the simplest possible form of matter. Much of their behaviour depends only on the motion of their molecules, nothing else. We will explain:

- the simple relationships between pressure, volume, temperature and amount of gas
- how these relationships can be explained simply from the motion of molecules
- how to calculate changes of energy when things get hotter or colder

13.1 Up, up and away

At 9.05 a.m. on 1 March 1999 a balloon as tall as the Leaning Tower of Pisa lifted off from a field in the small Swiss village of Chateau-d'Oex. The crew were Swiss psychiatrist Bernard Piccard and British balloon expert Brian Evans. In *Breitling Orbiter 3* for 21 days, they became the first to circumnavigate the globe in a free-flight balloon, relying for buoyancy on both hot air and helium.

The physics of ballooning is simple. It was discovered two and a half millennia ago by Archimedes: any object in a fluid that displaces a volume of the fluid that weighs more than the object does, will float upwards in the gravitational field of the Earth. The problem is making the object stay at the desired height – and making it travel in the desired direction.

The brothers Montgolfier

It was recognised in the 17th century that the atmosphere is like an ocean of air. Things float in an ocean of water, so why not in air? Smoke rises, so why not fill a large, light envelope with smoke and let it carry the envelope upwards with it? Brothers Joseph-Michel and Jacques-Etienne Montgolfier first made successful use of this idea in June 1783. A large sphere of cloth and paper was placed over a very smoky fire and held down by eight strong men until it filled, lifted off into the

The Breitling Orbiter

Breitling Orbiter takes off from Chateau-d'Oex, Switzerland (top). It is as high as the Leaning Tower of Pisa and the envelope would hold the equivalent of seven Olympic swimming pools.

Weather conditions faced by *Breitling Orbiter*

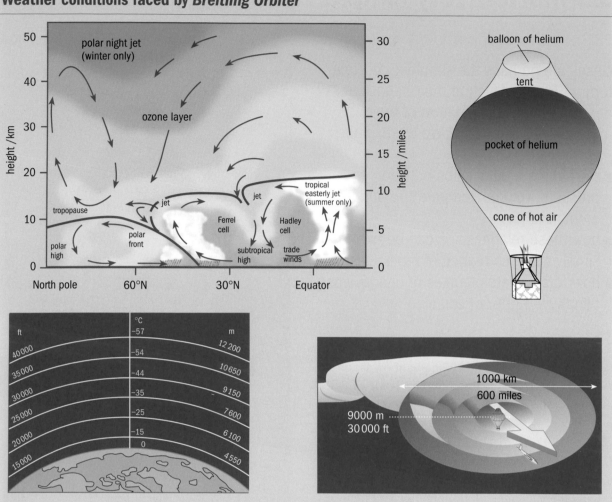

The weather conditions encountered during the *Breitling Orbiter 3* flight, illustrated above, required flying at a high altitude to take advantage of jet streams and to avoid potential storm zones. At these heights the low temperatures and low pressures mean that a pressurised cabin is required. Jet streams form above warm fronts where the large global convection cells meet — another case of hot air rising. The *Breitling Orbiter 3* was carried round the world by these jet streams. They travel at up to 300 km h^{-1} at heights of 7000–12 000 m. The *Breitling Orbiter 3* is a mixed, or Rozier, balloon — a combination of a hot-air and a gas balloon, characterised by a helium cell placed within the hot-air envelope.

summer sky and was blown away from their home town of Annonay in France.

The second major Montgolfier balloon flight took up a trio of animal passengers: a sheep, a duck and a cockerel. On landing they were pronounced fit to eat – not the ideal reward for being the first aeronauts. The first humans to leave the Earth and live to tell the tale were the would-be balloonist François Pilatre de Rozier and François Laurent Marquis d'Arlandes. This intrepid pair lifted off from Paris on the afternoon of 21 November 1783. In spite of their balloon fabric catching fire they descended safely 25 minutes after taking off, having travelled eight kilometres.

Laws for gases

A gas is the simplest form of ordinary matter. The first gas law was discovered by Robert Boyle in 1660. He was interested in what he called "the spring of the air", that is, how hard it pushes back when compressed. You pump up your cycle tyres until the air inside holds the rubber walls taut enough to support your weight. Boyle showed that decreasing the volume V of a fixed quantity of air by half doubles its pressure p. He proved that air obeyed a simple law, which became known as Boyle's law:

$$pV = \text{constant}$$

Key summary: Boyle's law and gas density

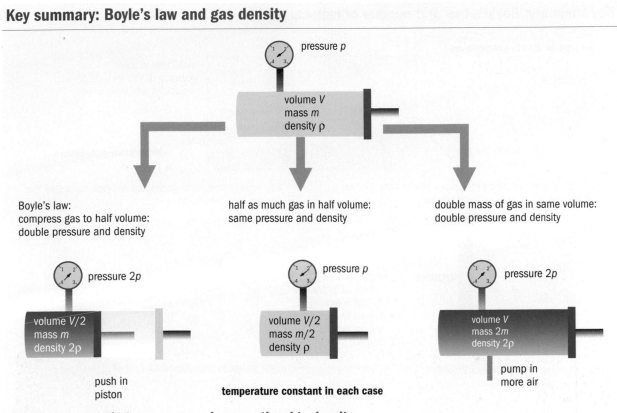

pressure p

volume V
mass m
density ρ

Boyle's law:
compress gas to half volume:
double pressure and density

half as much gas in half volume:
same pressure and density

double mass of gas in same volume:
double pressure and density

pressure $2p$

volume $V/2$
mass m
density 2ρ

push in
piston

pressure p

volume $V/2$
mass $m/2$
density ρ

temperature constant in each case

pressure $2p$

volume V
mass $2m$
density 2ρ

pump in
more air

Boyle's law says that gas pressure is proportional to density

provided the temperature and amount of gas stay the same.

There's another, related law. It is that if you pump more air into a fixed volume, at constant temperature, the pressure is proportional to the mass of air inside. Twice the amount of air: twice the pressure. In effect, then, Boyle's law says that:

pressure ∝ density (at constant temperature)

Bernoulli's idea

Boyle started with air but it soon became clear that lots of other gases – hydrogen, nitrogen and others – behave in the same way. Compare the stretching of different materials, say, steel and polythene: they don't behave in the same way as each other. Different gases, however, do. There is a simple explanation: any gas exerts a pressure because it is made of very small, rapidly moving molecules. For example, the pressure in a car tyre is just the result of the molecules bumping into and bouncing off the walls of the tyre. The more often and the harder they hit the walls, the greater the pressure.

In 1738 Daniel Bernoulli published this picture of his idea that a gas exerts a pressure and can hold up a weight by being made of many tiny moving particles that bombard the walls of any container.

Key summary: Boyle's law and number of molecules

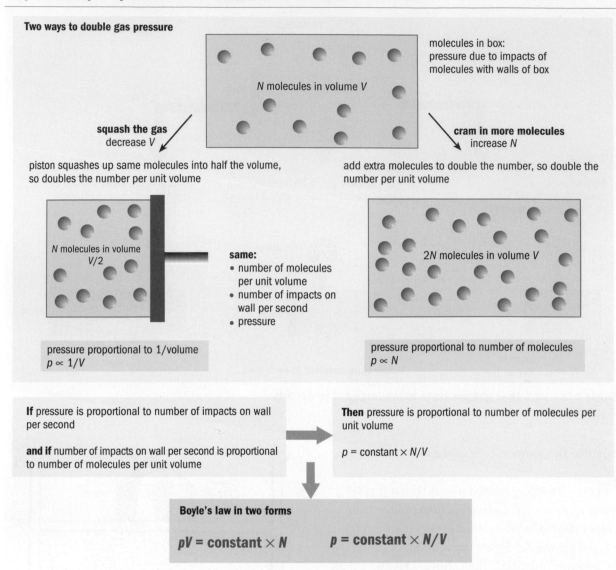

Two ways to double gas pressure

N molecules in volume *V*

molecules in box:
pressure due to impacts of
molecules with walls of box

squash the gas
decrease *V*

cram in more molecules
increase *N*

piston squashes up same molecules into half the volume,
so doubles the number per unit volume

add extra molecules to double the number, so double the
number per unit volume

N molecules in volume
V/2

same:
• number of molecules
 per unit volume
• number of impacts on
 wall per second
• pressure

2*N* molecules in volume *V*

pressure proportional to 1/volume
$p \propto 1/V$

pressure proportional to number of molecules
$p \propto N$

If pressure is proportional to number of impacts on wall
per second

and if number of impacts on wall per second is proportional
to number of molecules per unit volume

Then pressure is proportional to number of molecules per
unit volume

$p = \text{constant} \times N/V$

Boyle's law in two forms

$$pV = \text{constant} \times N \qquad p = \text{constant} \times N/V$$

Boyle's law says that pressure is proportional to crowding of molecules

This was an explanation that had already occurred to Daniel Bernoulli, a Dutch-born mathematician, in 1738, long before the Montgolfier brothers took to the skies in 1783.

Bernoulli's idea suggests why Boyle's law is true. The more crowded the molecules, the greater the number of impacts every second with any nearby wall. So halving the volume of a gas doubles the crowding of the molecules and should double the impacts, doubling the pressure. Another way to double the crowding of the molecules is to double their number in the same volume. This should double the pressure too.

From this, Boyle's law becomes a law about the crowding of molecules. You can now rewrite it in terms of numbers of molecules:

$$pV = \text{constant} \times \text{number of molecules}$$

$$p = \text{constant} \times \text{number of molecules per unit volume}$$

Notice that this explanation has nothing to do with what the molecules are. All they have to do is move about and bump into the walls. This is how completely different gases can have the same behaviour. And the collisions must be small and

Key summary: pressure and volume of gases increasing with temperature

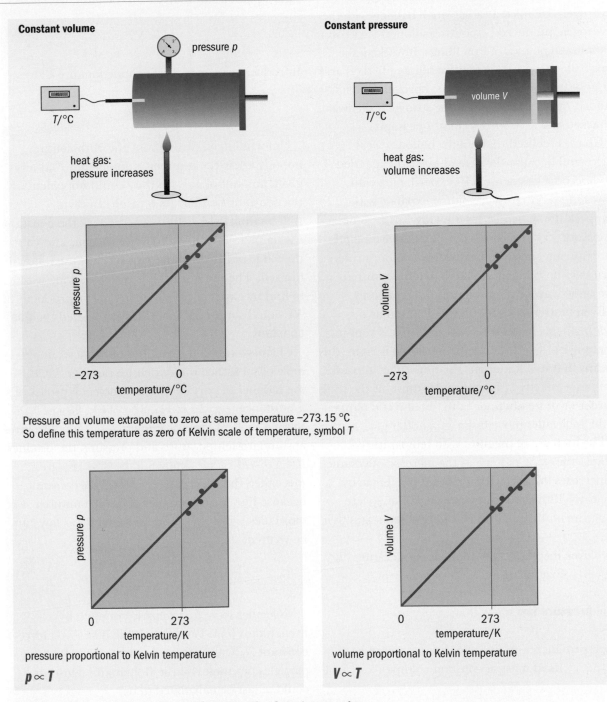

Constant volume

pressure *p*

T/°C

heat gas:
pressure increases

Constant pressure

volume *V*

T/°C

heat gas:
volume increases

pressure *p*

−273   0
temperature/°C

volume *V*

−273   0
temperature/°C

Pressure and volume extrapolate to zero at same temperature −273.15 °C
So define this temperature as zero of Kelvin scale of temperature, symbol *T*

pressure *p*

0   273
temperature/K

pressure proportional to Kelvin temperature

$p \propto T$

volume *V*

0   273
temperature/K

volume proportional to Kelvin temperature

$V \propto T$

Pressure and volume are proportional to absolute temperature

numerous. That's how the pressure of a gas can seem rock-steady even though it is made up of discrete bumps. The air under pressure in the tyres of a car keeps the tyres inflated and the car off the ground because the air molecules continually batter the inside of the tyres.

Gas laws for hot-air balloons

Boyle kept the temperature of the gas constant, but the balloonists used hot air. Why did it work? Within weeks of the Montgolfiers' success, and supported by money from public subscription, the 37-year-old and grandly named Jacques Alexander

Cesar Charles set about conquering the air using the recently discovered very low-density gas hydrogen. He filled a rather small balloon with the hydrogen, produced by pouring sulphuric acid over a thousand pounds of iron filings. It worked: the free balloon, with its expensive filling, lifted off and disappeared over the north-east horizon. When it landed some distance away, the strange object was attacked and destroyed by angry peasants.

Charles realised that hot-air balloons were successful because the heated air had expanded and become less dense. His experiments with gases showed that their volume expands with temperature. Chemist Joseph Gay-Lussac repeated this work 15 years later, claiming better methods for ensuring constant pressure. He demonstrated that volume is linearly related to temperature in the same way for a number of different gases.

Even more interesting, it was found that heating a gas increases its pressure if it is kept in a container of fixed volume. If Bernoulli is right, this means that in a hotter gas the molecular impacts are more effective: the molecules must hit the walls harder or more often, or both. That could happen if the molecules move faster in a hotter gas.

The increase of pressure with temperature follows the same pattern as the increase of volume: it increases linearly with temperature. Thus, by the early 19th century, two more gas laws were experimentally established. **Charles' law** states that:

volume increases linearly with temperature (for fixed mass at constant pressure).

The **pressure law** states that:

pressure increases linearly with temperature (for fixed mass at constant volume).

What was especially important was that different gases expanded in the same way. Graphs of volume and pressure against temperature all gave the same result: the pressure and volume both tended to zero at the same low temperature of $-273.15\,°C$. This is the absolute zero of temperature and is used as the zero point of the Kelvin scale of temperature. Its unit, the kelvin (K), is defined as the same size as the Celsius degree (°C).

One law for all gases: the ideal gas law

One equation summarises all the gas laws:

$$pV \propto NT$$

Introducing a constant of proportionality k gives:

$$pV = NkT$$

Unfortunately, throughout the 19th century, nobody knew the number of molecules N in any given amount of gas. So they could not calculate the constant k either. As a result, Ludwig Boltzmann (1844–1906), after whom the constant k is named, lived his whole life without knowing the value of his constant, or even writing it down himself. The one thing that could be done was to find by experiment the value of the product Nk, equal simply to $\dfrac{pV}{T}$. This was called the **gas constant**.

Chemists called the number of particles in one mole of substance the Avogadro number N_A. It is the number of particles in unit molecular mass of a substance (e.g. 2 g of H_2, 32 g of O_2, 28 g of N_2). Since a mole of gas was known to occupy 22.4 litres at standard temperature and pressure, the product $R = N_A k = pV/T$ was known to be $8.31\,\mathrm{J\,K^{-1}\,mol^{-1}}$. R is called the universal, or molar, gas constant because if there are n moles of gas the number N of molecules is just nN_A, and the general gas law can be written:

$$pV = nN_A kT = nRT$$

Molecules were first reliably counted, by Frenchman Jean Perrin, in 1909. The Avogadro constant N_A was measured to be 6.022×10^{23} particles per mole. So the Boltzmann constant k could at last be calculated to be:

$$k = \frac{R}{N_A} = \frac{8.31\,\mathrm{J\,K^{-1}\,mol^{-1}}}{6.022 \times 10^{23}\,\mathrm{particles\,mol^{-1}}}$$

$$= 1.38 \times 10^{-23}\,\mathrm{J\,K^{-1}}\ \text{per particle}$$

Its value gives a hint of how much energy a molecule actually has. You will see later (p112) that the energy of a molecule is typically of the order kT.

Key summary: one law for all gases

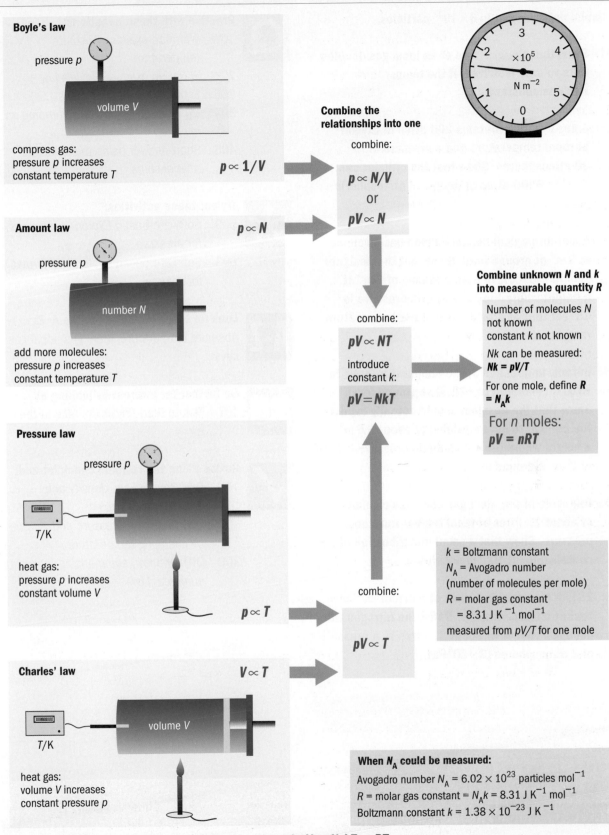

Boyle's law

pressure p

volume V

compress gas:
pressure p increases
constant temperature T

$$p \propto 1/V$$

Amount law

pressure p

number N

add more molecules:
pressure p increases
constant temperature T

$$p \propto N$$

Pressure law

pressure p

T/K

heat gas:
pressure p increases
constant volume V

$$p \propto T$$

Charles' law

volume V

T/K

heat gas:
volume V increases
constant pressure p

$$V \propto T$$

$\times 10^5$

N m^{-2}

Combine the relationships into one

combine:

$$p \propto N/V$$
or
$$pV \propto N$$

combine:

$$pV \propto NT$$

introduce constant k:

$$pV = NkT$$

combine:

$$pV \propto T$$

Combine unknown N and k into measurable quantity R

Number of molecules N not known
constant k not known

Nk can be measured:
$$Nk = pV/T$$

For one mole, define $R = N_{A}k$

For n moles:
$$pV = nRT$$

k = Boltzmann constant
N_{A} = Avogadro number
(number of molecules per mole)
R = molar gas constant
= 8.31 J K^{-1} mol^{-1}
measured from pV/T for one mole

When N_{A} could be measured:
Avogadro number N_{A} = 6.02×10^{23} particles mol^{-1}
R = molar gas constant = $N_{A}k$ = 8.31 J K^{-1} mol^{-1}
Boltzmann constant k = 1.38×10^{-23} J K^{-1}

For N molecules $pV = NkT$. For n moles, $N = nN_{A}$ and $pV = nN_{A}kT = nRT$.

Quick check

Useful data: **A mole is 6×10^{23} particles.**

1. Show that the pressure of an ideal gas doubles if its volume is halved, if the temperature remains constant.

2. A gas cylinder contains 200 litres of oxygen at room temperature and a pressure of 30 atmospheres. Show that the cylinder can provide 6000 litres of oxygen at atmospheric pressure and room temperature.

3. A meteorological helium balloon has a volume of $2\,\text{m}^3$ at ground level. Show that the designer should expect it to have a volume of $8\,\text{m}^3$ at a height where the atmospheric pressure is 25% of that at ground level, if the temperature remains the same.

4. In fact, in question 3, the temperature falls to $-30\,^\circ\text{C}$ compared to $+20\,^\circ\text{C}$ at ground level. Show that the designer should actually expect the balloon to have a volume of about $6.6\,\text{m}^3$ at a height where the atmospheric pressure is 25% of that at ground level.

5. One mole of any ideal gas occupies a volume of about 22 litres at room temperature and pressure. Show that a small matchbox full of air contains about 10^{20} molecules.

6. $28\,\text{g}$ of N_2 contains 1 mole of molecules. Show that at a temperature of $546\,\text{K}$ the nitrogen will occupy a volume of about 22 litres at a pressure of 2 atmospheres ($2 \times 10^5\,\text{Pa}$).

Links to the *Advancing Physics* CD-ROM

Practise with these questions:
10W Warm-up exercise *Absolute temperature*
20W Warm-up exercise *Boyle's law*
20E Estimate *Gases and mass*
30X Explanation–exposition *Pumping up my tyres*
40S Short answer *Using the ideal gas relationships*

Try out these activities:
50S Software-based *Exploring the rules for pressure*
60S Software-based *Exploring the rules for volume*

Look up these key terms in the A–Z:
Absolute temperature; ideal gas; ideal gas laws

Go further for interest by looking at:
10T Text to read *Physicists take to the air*

Revise using the revision checklist and:
10O OHT *Boyle's Law, density and number of molecules*
20O OHT *Changing pressure and volume by changing temperature*
30O OHT *One law, summarising empirical laws*

13.2 The kinetic model

In the late 1850s and early 1860s Rudolf Clausius, and then James Clerk Maxwell and Ludwig Boltzmann, made a bold new step in the theory of gases. Their model of a gas was extremely simplified, using only the most basic assumptions to explain the behaviour of gases and to make important new and counter intuitive predictions.

They reduced a gas to being simply a collection of fast-moving, colliding particles, obeying the laws of mechanics. In particular, the pressure of a gas on the walls of its container was explained as the result of the continual bombardment of molecules hitting and rebounding from the walls. In this way Maxwell and Boltzmann gave the behaviour of gases a purely mechanical explanation. Their idea is a splendid example of how theory works in physics.

You have to remember that, although the idea that matter was made of particles had already been around for over a hundred years, many scientists were deeply sceptical of the very idea that matter is made of atoms or molecules in motion. They were well aware that nobody could say anything much about these supposed particles – about how big they were supposed to be, how massive, or how fast they might move. The whole idea was widely thought of as just a possible hypothesis. There was considerable resistance to taking the moving-particle idea seriously as a picture of how things actually are. It took another fifty years for the kinetic model to win the day.

A simple start: one particle in a box

To start with, imagine the simplest possible model of a gas: just one particle in a box, bouncing backwards and forwards between two end walls. Each time the molecule hits a wall it gives the wall a brief push: it exerts an impulsive force on it. How big an impulse? That depends on the molecule's mass and change in velocity when it collides – on its momentum change $\Delta(mv)$ (chapter 11). The faster the molecule moves, the bigger the impulse. In fact, making it go faster has a double effect: not only is each wall hit harder when the molecule bumps into it, but the hits will happen more often. This is because the molecule covers the distance between walls in a shorter time. Suppose we double the speed: the molecule will hit twice as often – and twice as hard.

Now add more molecules. The total force increases in proportion to the number of molecules. The impacts on the walls come thick and fast and blur out into a steady force or pressure.

Squashing the box

What happens if the molecules are crammed into a smaller space? Think about the walls that are being hit by the stream of molecules. If you squash the sides of the box inwards to make the end walls smaller, the same molecules now hit a smaller area, but as hard as before. The same force is now acting on a smaller area. This means that the pressure on the wall gets bigger.

James Clerk Maxwell (1831–1879) (left) and Ludwig Boltzmann (1844–1906) (right) are thought of (and indeed look like) great authoritative 19th century figures. It is hard now to comprehend how divided people were about the existence of these particles – atoms and molecules – that no-one had ever seen. Several prestigious physicists and chemists treated the atomic hypothesis as the purest speculation, not to be taken seriously. Others were enthusiastic about it.

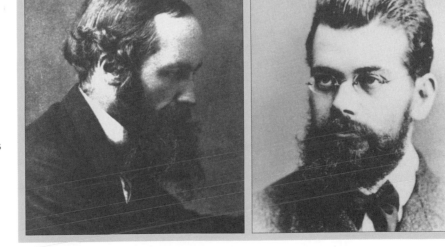

Key summary: kinetic model of a gas

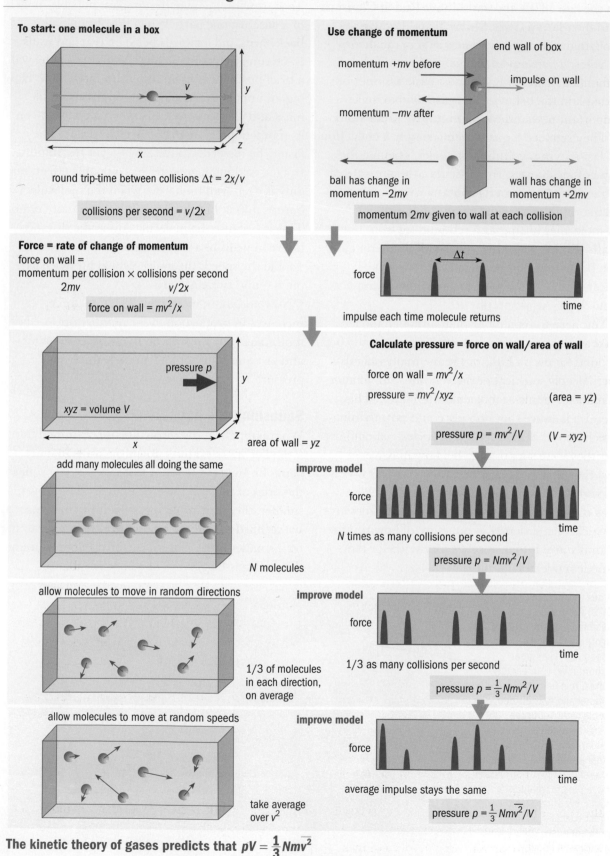

To start: one molecule in a box

round trip-time between collisions $\Delta t = 2x/v$

collisions per second $= v/2x$

Use change of momentum

end wall of box

momentum $+mv$ before

impulse on wall

momentum $-mv$ after

ball has change in momentum $-2mv$

wall has change in momentum $+2mv$

momentum $2mv$ given to wall at each collision

Force = rate of change of momentum
force on wall =
momentum per collision × collisions per second
$2mv$ $v/2x$

force on wall $= mv^2/x$

force

time

impulse each time molecule returns

pressure p

$xyz =$ volume V

area of wall $= yz$

Calculate pressure = force on wall/area of wall

force on wall $= mv^2/x$

pressure $= mv^2/xyz$ (area $= yz$)

pressure $p = mv^2/V$ ($V = xyz$)

add many molecules all doing the same

N molecules

improve model

force

time

N times as many collisions per second

pressure $p = Nmv^2/V$

allow molecules to move in random directions

1/3 of molecules in each direction, on average

improve model

force

time

1/3 as many collisions per second

pressure $p = \frac{1}{3}Nmv^2/V$

allow molecules to move at random speeds

take average over v^2

improve model

force

time

average impulse stays the same

pressure $p = \frac{1}{3}Nm\overline{v^2}/V$

The kinetic theory of gases predicts that $pV = \frac{1}{3}Nm\overline{v^2}$

Now think about the length of the box along which molecules are travelling to and fro. If this is made smaller as well, the molecules will have less distance to go and will cover it more quickly. They will hit the end walls as hard as before but more often, so the force on each wall will increase. In other words, however you make the box smaller you increase the pressure on the walls. This is what Boyle discovered a long time ago.

It's not too hard to turn these ideas into equations. The result is that the pressure p of a volume V of gas containing N molecules each of mass m, travelling at a range of speeds v is:

$$pV = \frac{1}{3}Nm\overline{v^2}$$

where $\overline{v^2}$ is the average of the squares of the velocities.

The ideal gas

The equation $pV = \frac{1}{3}Nm\overline{v^2}$ is the prediction of a model. The model is an idealised version of a gas. Like other models (chapter 10) it concentrates on essentials. The main essentials are that:

- the sample of gas is large enough to contain very many molecules;
- the molecules are moving randomly in all directions, with randomly varying speeds;
- collisions with the walls are elastic, so that molecules gain or lose no energy on collision.

The last assumption won't generally be true molecule by molecule; after all, the walls are made of vibrating molecules too. But it will be true on average if the walls are at the same temperature as the gas. Similarly, collisions between molecules themselves will be elastic.

This model has ignored other things that are less likely to be true in a real gas, and therefore real gases only approximate to the model. They do so better and better when:

- the density is low so that molecules are far apart and the space the molecules themselves occupy is too small to be considered – the model takes this space to be zero; and
- the energy of motion of the molecules is large so that any attractions between them can be ignored in comparison – the model assumes these interactions don't exist.

Key summary: kinetic energy of gas molecules and the Boltzmann constant

compare these

Kinetic model

$pV = \frac{1}{3}Nm\overline{v^2}$

Gas laws

$pV = NkT$

$m\overline{v^2} = 3kT$

One molecule

$\frac{1}{2}m\overline{v^2} = \frac{3}{2}kT$

kinetic energy per molecule $= \frac{3}{2}kT$

Many molecules

$\frac{1}{2}Nm\overline{v^2} = \frac{3}{2}NkT$

total kinetic energy of molecules $= \frac{3}{2}NkT$

for one mole
$N = N_A$
$R = N_A k$
$R = 8.31 \text{ J K}^{-1} \text{ mol}^{-1}$

total kinetic energy of one mole of molecules
$U = \frac{3}{2}RT$

Boltzmann constant k

random thermal energy of one molecule is of order kT

Internal kinetic energy of molecules of one mole at $T = 300$ K

$U = \frac{3}{2}RT$

$U = 3.7 \text{ kJ mol}^{-1}$

Average kinetic energy of a molecule $= \frac{3}{2}kT$

Maxwell and Boltzmann were using the ideas of Newtonian mechanics – of the "clockwork universe" – but in a new way. It was not possible to calculate the motion of a huge number of molecules, and there was no need – averages over randomly varying speeds were enough. Here was the moment when statistical thinking entered into physics. The clockwork universe looked less like pure clockwork.

Temperature and the energy of molecules

The gas laws are summarised in the equation, for N molecules:

$$pV = NkT$$

The kinetic model gives pV in terms of molecular masses and velocities:

$$pV = \frac{1}{3}Nm\overline{v^2}$$

Key summary: speed of a nitrogen molecule

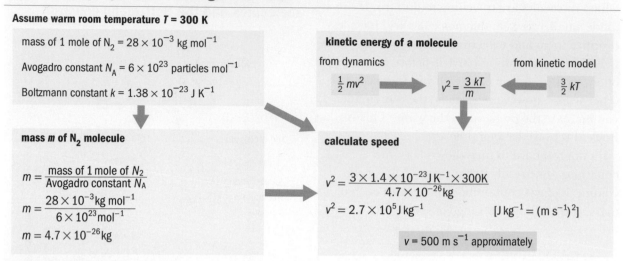

Assume warm room temperature $T = 300$ K

mass of 1 mole of $N_2 = 28 \times 10^{-3}$ kg mol^{-1}

Avogadro constant $N_A = 6 \times 10^{23}$ particles mol^{-1}

Boltzmann constant $k = 1.38 \times 10^{-23}$ J K^{-1}

kinetic energy of a molecule

from dynamics

$\frac{1}{2}mv^2$

$v^2 = \frac{3\,kT}{m}$

from kinetic model

$\frac{3}{2}kT$

mass m of N_2 molecule

$m = \dfrac{\text{mass of 1 mole of } N_2}{\text{Avogadro constant } N_A}$

$m = \dfrac{28 \times 10^{-3} \text{kg mol}^{-1}}{6 \times 10^{23} \text{mol}^{-1}}$

$m = 4.7 \times 10^{-26}$ kg

calculate speed

$v^2 = \dfrac{3 \times 1.4 \times 10^{-23} \text{J K}^{-1} \times 300\text{K}}{4.7 \times 10^{-26} \text{kg}}$

$v^2 = 2.7 \times 10^5$ J kg^{-1} $\qquad$ [J kg^{-1} = (m s^{-1})^2]

$v = 500$ m s^{-1} approximately

Air molecules (mostly nitrogen) at room temperature go as fast as bullets

Key summary: Brownian motion

A simulation of the path of one molecule in a small box of air

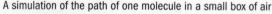

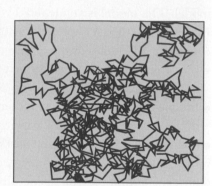

after just 10 steps between collisions $\qquad$ after 100 steps $\qquad$ after 1000 steps

The motion of a molecule repeatedly changes direction at random

The model accounts for all the known gas laws.

Boyle's law predicts that pV will be a constant, proportional to the number N of molecules, as long as the number and speed of the molecules do not change. But the model does more: it gives a new meaning to temperature. The temperature is proportional to the average kinetic energy $\frac{1}{2}m\overline{v^2}$ of a molecule. More exactly:

average kinetic energy of a molecule $\frac{1}{2}m\overline{v^2} = \frac{3}{2}kT$

for this simple model.

Now the model also explains the other gas laws.

Charles' law V is proportional to T when p stays constant – the model makes V proportional to the mean kinetic energy of the molecules. **The pressure law** p is proportional to T when V stays constant – the model makes p proportional to the mean kinetic energy of the molecules.

Energy in a gas

The kinetic model makes it possible to work out just how much kinetic energy the molecules of gas have. The total kinetic energy of all the molecules in a gas is simply N times the average energy of each one. This total energy is usually called the

Key summary: evidence for moving molecules

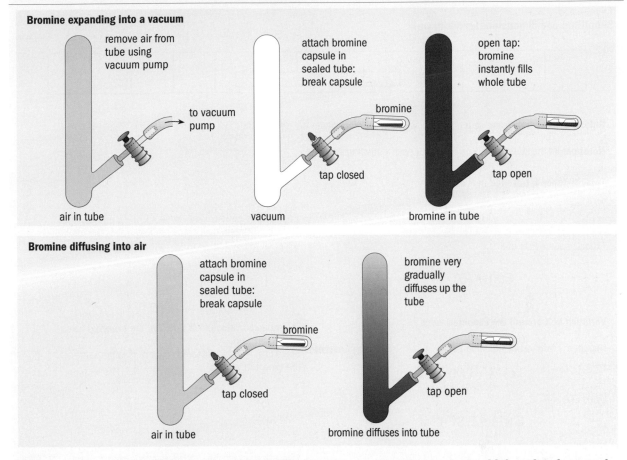

Bromine expanding into a vacuum

remove air from tube using vacuum pump

to vacuum pump

air in tube

attach bromine capsule in sealed tube: break capsule

bromine

tap closed

vacuum

open tap: bromine instantly fills whole tube

tap open

bromine in tube

Bromine diffusing into air

attach bromine capsule in sealed tube: break capsule

bromine

tap closed

air in tube

bromine very gradually diffuses up the tube

tap open

bromine diffuses into tube

Diffusion shows that molecules move. Rapid diffusion into a vacuum demonstrates high molecular speeds.

internal energy U. The kinetic model says that U is given by:

$$U = \tfrac{1}{2}Nm\overline{v^2} = \tfrac{3}{2}NkT = \tfrac{3}{2}pV$$

It works well for monatomic gases such as helium. The molecules of gases like O_2 or N_2 may vibrate or rotate, so they can have more total internal energy, but the part contributed by translational kinetic energy is the same. Remarkably, this part of the internal energy of a gas can be worked out if you know only its pressure and volume at a given temperature – both large-scale properties that are easy to measure. For example, a litre bottle of air at room temperature has pressure p of about 10^5 Pa and volume V of about 10^{-3} m. So the random translational kinetic energy of its molecules is just $\tfrac{3}{2}pV$, about 150 J. If the gas is monatomic this calculation gives the whole internal energy.

Speed of a molecule

If you know the average kinetic energy $\tfrac{3}{2}kT$ of the molecules of a gas, then you can calculate their average speed, knowing that the kinetic energy is $\tfrac{1}{2}mv^2$. For example, air is mostly nitrogen. A simple calculation gives an average speed of about $500\,\mathrm{m\,s^{-1}}$ for air molecules at room temperature. This is as fast as a bullet or a jet aircraft moves. Believable? Well, you can see molecules travelling fast if you let a gas expand into a vacuum with nothing to get in their way.

Better still, the speeds of atoms and molecules – as well as their variation in speed – have been measured directly by timing their flight across a vacuum.

Seeing primordial living motion?

Travel back in time with us once more. In the summer of 1827 the botanist Robert Brown

Key summary: random walk

Simplified one-dimensional random walk

Rule: a particle moves one step at a time, with equal probability to the right or left

Notation: let the steps be x_1, x_2, x_3 etc. with each x equal to +1 or −1

Total distance X travelled in N steps

$$X = (x_1 + x_2 + x_3 + ... + x_i + ...x_N)$$

Expected value E(X) of X

$$E(X) = (0)$$

Each step is equally likely to be +1 or −1.
Thus, on average, over many random walks, the total distance will add up to zero.

Variation of X around the expected value

Departure r from expected value is:

$$r = X − E(X)$$

Since E(X) = 0

$$E(r^2) = E(X^2)$$

The expected value of r is zero. But the expected value of r^2 is not zero.
Since E(X) = 0 the expected value of r^2 is the same as the expected value of x^2

Expected value of r^2

$$E(r^2) = E(X^2)$$

$$= E(x_1 + x_2 + x_3 + ... + x_i + ...x_N)^2$$

The result is simple:
mean square variation

$$E(r^2) = N$$

root mean square variation

$$\sigma = \sqrt{N}$$

When multiplied out this gives two types of term:

squared terms:

$$x_1^2 + x_2^2 + x_3^2 + ... + x_i^2 + ...x_N^2$$

this contains N terms
each equal to +1
expected value = N

$$(+1)^2 = +1$$
$$(−1)^2 = +1$$

mixed terms:

$$x_1x_2 + x_1x_3 + ... + x_ix_j + ...$$

this contains N(N−1) terms
each equally likely to be
+1 or −1

$$(+1) \times (+1) = +1$$
$$(+1) \times (−1) = −1$$
$$(−1) \times (+1) = −1$$
$$(−1) \times (−1) = +1$$

expected value = 0
compare $(a + b)^2 = a^2 + ab + ba + b^2$

Root mean square distance travelled in random walk $= \sqrt{N}$

became fascinated by the continual tiny jiggling motion he could see when looking at fragments of pollen grains in water under his microscope. Nothing stopped it. The pollen fragments just went on moving about, almost as if they were alive, thought Brown. He tried fine particles of other animal and vegetable tissues, and saw the same. He thought he might be seeing the very source of activity that gave life to things. Wisely he tried non-living material such as powdered glass. He even ground up a tiny piece of the Sphinx and tried that. Everything, living or not, showed the same jiggling motion as long as the particles were small enough. Brown began to doubt his first idea.

What Robert Brown was looking at was the first direct visual evidence that molecules in matter actually move. Tiny particles, like smoke particles in air or fine grains in water, get buffeted all the time by the molecules around them. But the buffeting is random and not equal on all sides of the particle, so the particle continually jiggles about. In fact, just like a molecule, such a particle acquires a random kinetic energy of the order of kT. However, the particle is much larger and heavier than a molecule, so its average speed is much slower.

So Robert Brown was right to be astonished that the **Brownian motion** he saw seemed to go on forever, even though he was at first wrong about the reason. He was seeing the perpetual chaotic agitation inside all matter. You ought to see it too – it is one of the most fundamental scientific experiences you can have.

Simulated Brownian motion paths look like tangled wool attacked by a demented kitten. The long path does not get very far, as it randomly twists and turns. Such a path is called a **random walk**. The tangled paths that molecules take explain why gases diffuse quite slowly – why it takes quite some time for the smell of scent to reach you when a bottle is opened. If the scent molecules went direct, they'd be at your nose in a flash. But they don't. Buffeted by air molecules, they reach you very gradually, and then only by chance.

There's a handy rule about this. If a particle zig-zags at random, and makes N steps in all, then on average it will get $\sqrt{N}$ steps-worth from the starting point. Typical figures in air: zig-zag steps are 10^{-7} m long between collisions. Thus 10^{10} steps carry a molecule 1000 m along its tangled path, but the actual distance it gets is only on average $\sqrt{10^{10}} = 10^5$ steps, or just 10 mm. At a few hundred metres per second that would take several seconds. Diffusion is slow, even though molecules move fast. This is because their motion repeatedly changes direction at random.

Real and ideal gases

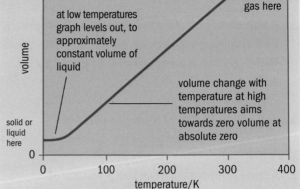

at low temperatures graph levels out, to approximately constant volume of liquid

like ideal gas here

volume change with temperature at high temperatures aims towards zero volume at absolute zero

solid or liquid here

The real stuff

Ideal gases occupy an important place in physics, even though none exist, because the theory is so simple and complete. But real gases do things that ideal gases don't. For example, when cooled they condense to a liquid. Now the space occupied by the molecules becomes important: in the liquid they are nearly touching. The graph (p97) of the volume of an ideal gas shrinking to zero at the absolute zero is not for real. In reality, the graph must level out as the liquid forms because the liquid fills a small but definite volume. But before dismissing ideal gases, remember that it was real gases that pointed in their direction. At higher temperatures many real gases are near enough ideal to provide the essential clues. This is because the volume of the molecules themselves really is only a tiny part of the volume of a real gas, say air or water vapour. A litre of water turns into a thousand litres or so of water vapour. So such a gas really is nearly all empty space between molecules.

You can expect gases like helium, hydrogen, nitrogen and oxygen, which become liquid well below room temperature, to behave very much like ideal gases. But don't expect camping gas (butane), which is liquid under pressure in its cylinder, to behave like an ideal gas.

A way used by Dutch scientist Johannes van der Waals to adapt the ideal gas equation to real gases was to subtract the volume of the molecules from the volume of the gas, and to correct the pressure for the fact that the molecules attract each other.

Quick check

Useful data: A mole is 6×10^{23} particles. The Boltzmann constant $k = 1.4 \times 10^{-23}\,\text{J K}^{-1}$.

1. A ball of mass 0.2 kg travels at $2\,\text{m s}^{-1}$ towards a wall and bounces back at the same speed. Show that it gives a momentum of $0.8\,\text{kg m s}^{-1}$ to the wall.

2. Show that 20 mol of ideal gas contains 12×10^{24} molecules.

3. A mole of hydrogen occupies a volume of $0.024\,\text{m}^3$ at a pressure of $10^5\,\text{Pa}$ and a temperature of 300 K. Show that the mean kinetic energy of a hydrogen molecule is about $10^{-20}\,\text{J}$.

4. Show that the root mean square speed of H_2 molecules (mass about $3 \times 10^{-27}\,\text{kg}$) is about $2.6\,\text{km s}^{-1}$. This is less than the escape velocity from the Earth, which is about $11\,\text{m s}^{-1}$. Explain why hydrogen molecules do nevertheless escape from the atmosphere.

5. Show that a typical molecule travelling at about $500\,\text{m s}^{-1}$ will take only 20 ms to cross a room 10 m wide. Explain why a given molecule will actually take very much longer than this to travel even a few centimetres.

6. Show that in a mixture of oxygen (molecular mass 32) and hydrogen (molecular mass 2) molecules the root mean square speeds of the two kinds of molecule will differ by a factor of four. Which is the faster?

Links to the *Advancing Physics* CD-ROM

Practise with these questions:

60S Short answer *Kinetic theory by numerical example*

70S Short answer *Kinetic theory algebraically*

30W Warm-up exercise *Molecular motion*

40W Warm-up exercise *The ideal gas equation*

90S Short answer *The speeds of gas molecules: Some questions*

Try out these activities:

100S Software-based *Molecules in a box*

120S Software-based *Watching atoms cause pressure*

170S Software-based *Diffusion from random motion*

190S Software-based *Simulated random walk*

Look up these key terms in the A–Z:

Brownian motion; kinetic theory of gases; root mean square speed

Go further for interest by looking at:

30T Text to read *Direct measurement of speeds of atoms and molecules*

Revise using the revision checklist and:

60P Poster *Constructing a model of a gas*

700 OHT *The speed of a nitrogen molecule*

950 OHT *Random walk*

1000 OHT *Boltzmann constant and gas molecules*

13.3 Energy in matter

The development of the steam engine early in the 18th century, obtaining industrial power from steam, made the knowledge of ways to get energy from hot matter of great economic value. The Industrial Revolution, fuelled by newly invented ways of getting hot matter to do work in engines, was well under way before the ideas now called thermodynamics were fully developed. Indeed, this new science came in substantial part from that great social and technological change.

Today, with hindsight from the point of view of the kinetic theory of matter, getting energy into and out of matter is straightforward. The energy inside matter is simply the energy of random thermal motion of particles. To give the particles more energy you just have to hit them hard. There are two ways: hit them yourself or let other particles do the job for you.

You hit molecules yourself and made them go faster the last time you pumped up a bicycle tyre. The moving piston hit air molecules and speeded them up. Result: the air in the pump got hotter. A diesel engine uses the same principle: a piston in the engine compresses the fuel-air mixture until it is hot enough to ignite. To calculate the energy given to the molecules you just multiply the distance moved by the piston by the force on it. That is, you calculate the **work done**. Similarly, to get energy from hot gases inside an engine, the gases push a piston back and energy goes from motion of the molecules to whatever the piston is connected to. Again, the energy passed across is the work done, this time by the gas.

This heating up of a gas when it is compressed is a problem if you need to keep the temperature constant to test Boyle's law. To do this you have to allow time for the energy to leak out again from the warmed gas to the surroundings, until the temperature is back where it was before. Compress the gas slowly enough and the energy leaks out as fast as you put it in.

The second way to get energy into or out of matter is the lazy way – just let the molecules do the job for you. This way has been known ever since humans discovered fire. Glowing coals put energy into cold pans of water, all by themselves. How does it work? The molecules in the hot object

Early industrial steam engines

Top: Matthew Boulton and James Watt's Rotative Beam Engine (1788) could be harnessed directly to rotating machinery, making possible a huge growth in all types of industry that needed power. Bottom left: Boulton and Watt's 1812 pumping engine still operates at Crofton Pumping Station in Wiltshire, alongside an 1846 engine by Harvey of Hayle. Right: the chimney and engine house at Crofton Pumping Station.

are moving about faster or more energetically and, when they collide with a sluggish molecule in some cool stuff, the energy gets shared out more equally. The slow-moving molecules speed up and the fast-moving molecules slow down. The cool stuff gets hotter and the hot stuff gets cooler.

This flow of energy from hot to cold needs no help from outside. The molecules do it all by themselves, just because they are behaving randomly. On average, fast-moving ones pass energy to slow-moving ones until they each have the same average energy. When they have the same average energy – the same temperature – they are said to be in thermal equilibrium. You recall that in the kinetic theory it was necessary to assume that the gas molecules neither gained energy from nor lost energy to the walls of the container. This is simply to assume that they are at the same temperature. If not, the gas molecules gain or lose energy until they are. The

Key summary: transfers of energy to molecules in two ways

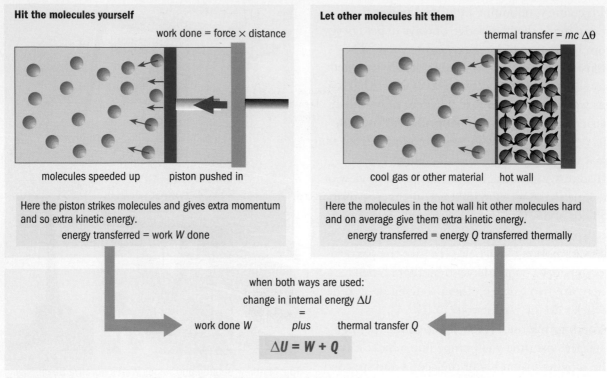

Hit the molecules yourself

work done = force × distance

molecules speeded up   piston pushed in

Here the piston strikes molecules and gives extra momentum and so extra kinetic energy.

energy transferred = work W done

Let other molecules hit them

thermal transfer = $mc\,\Delta\theta$

cool gas or other material   hot wall

Here the molecules in the hot wall hit other molecules hard and on average give them extra kinetic energy.

energy transferred = energy Q transferred thermally

when both ways are used:
change in internal energy ΔU
=
work done W *plus*   thermal transfer Q

$$\Delta U = W + Q$$

Energy can be given to molecules by heating and by doing work

James Joule (1818–1889) kept a steam engine in his Manchester house until the neighbours complained. He devised many different experiments in which a measured amount of work was done on a material and the rise in temperature showing an increase in internal energy was recorded. He showed that in every case a given amount of work corresponded to the same change in internal energy. One of his experiments involved measuring the temperature difference between the top and bottom of a waterfall, carried out while he was in Switzerland on honeymoon.

spontaneous flow of energy from hot to cold is called a thermal transfer of energy.

The internal energy can change in both ways at once. If it does, then change in internal energy of material = work done on material + energy transferred thermally:

$$\Delta U = W + Q$$

This is called the **First Law of Thermodynamics**. It just expresses the law of the conservation of energy. It gave the successors to the early engineers of the Industrial Revolution better ways of calculating the transfers of energy by way of working and heating in their engines. The better the engine, the bigger the profit.

The energy to make things hotter

For most substances there is a simple link between the change in internal energy and the change in temperature. The change in temperature for a given transfer of energy depends on the mass of and the kind of substance being made hotter.

Key summary: how much energy to have a bath?

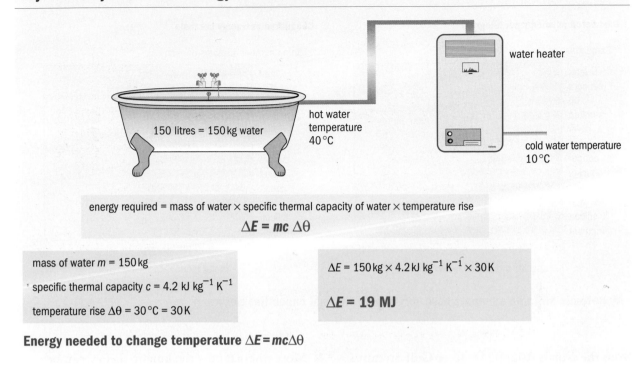

energy required = mass of water × specific thermal capacity of water × temperature rise

$$\Delta E = mc\,\Delta\theta$$

mass of water m = 150 kg

specific thermal capacity c = 4.2 kJ kg^{-1} K^{-1}

temperature rise $\Delta\theta$ = 30 °C = 30 K

ΔE = 150 kg × 4.2 kJ kg^{-1} K^{-1} × 30 K

ΔE = **19 MJ**

Energy needed to change temperature $\Delta E = mc\Delta\theta$

energy transferred = mass of substance × a constant for that substance × temperature rise

At constant volume (so that no work is done by the substance expanding) all the energy used goes to changing the internal energy so that:

$$\Delta U = mc\Delta\theta$$

We have decided to call the constant c the specific thermal capacity. (Its more usual name is specific heat capacity, but that suggests the old meaning of "heat" as the energy inside matter.) Call it what you like, its units are J kg^{-1} K^{-1}. One way to think about specific thermal capacity c is to realise that it is the energy put in or taken out when the temperature of a kilogram of the substance is changed by 1 kelvin (or by 1 °C). An easy way to measure it is to warm up the substance electrically, and measure the temperature rise for a given input of energy.

In a chart of the specific thermal capacities of common materials one substance stands out – water. Water needs a lot of energy put into it to increase its temperature by 1 kelvin. Water covers about three-quarters of the Earth's surface, so the high specific thermal capacity of water has

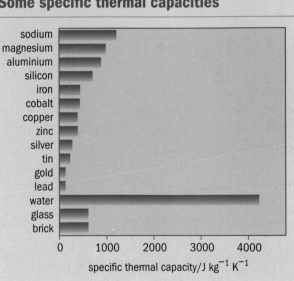

Some specific thermal capacities

specific thermal capacity/J kg^{-1} K^{-1}

large environmental significance. The oceans act as a global temperature regulator. The seas warm up and cool down slowly compared to the land, which has a mean specific thermal capacity about a quarter that of water. In winter the sea is warmer than the land and winds from the sea tend to warm up the land. In summer the effect is reversed, and keeps land near the sea cooler than it would otherwise be. Britain benefits particularly

Key summary: specific thermal capacities of some metals

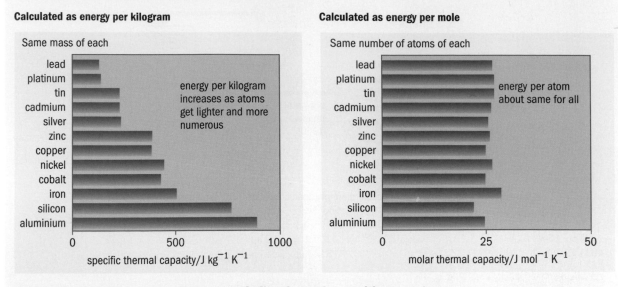

Monatomic metallic elements have very similar thermal capacities per atom

from the North Atlantic Drift (or Gulf Stream), which is a huge ocean current of warm water that moves from the Caribbean to the Arctic and helps to keep winters mild and summers cool (and wet!).

Heating in the home

Quite a bit of the energy used to keep a home warm goes to heating up the air, which has to change a few times an hour. But over the year heating water costs an ordinary family more than all the other heating they do – unless they live in a large and very badly insulated house in a cold climate. It takes about 20 MJ of energy to provide a hot bath. It's one reason to favour a shower, which uses less hot water and therefore costs less. See the chapter "Advances in physics" for more about heating buildings.

Particles have energy roughly equal to *kT*

You heat water for a bath by the litre – in effect by the kilogram – so the value of its specific thermal capacity expressed in joules per kilogram per kelvin makes good practical sense. But the kinetic theory of matter says that things may be simpler if thought about molecule by molecule. For ideal gases whose molecules just whizz about and don't spin or vibrate, the kinetic theory says that the energy per particle is just $\frac{3}{2}kT$.

More importantly, the kinetic theory can be stretched to cover all matter, not just gases, in an interesting if very rough way. The energy of thermal agitation inside any kind of matter is some fairly small multiple of the quantity kT. What that multiple is varies from one kind of stuff to another, and may change a bit with temperature. But if you just calculate kT, you are within striking distance of the energy each particle possesses.

On a nice summer day (27 °C, 300 K) the energy kT amounts to:

$$kT = 1.38 \times 10^{-23}\,\mathrm{J\,K^{-1}} \times 300\,\mathrm{K} = 4.1 \times 10^{-21}\,\mathrm{J}$$

Compare this energy with something else. An electron moving through a potential difference of 1 V changes energy by 1.6×10^{-19} J, an amount called 1 electron volt (1 eV). The random energy kT of thermal agitation is about $\frac{1}{40}$ of an electron volt or $\frac{1}{60}$ of the energy an electron gets from a 1.5 V torch battery.

So here is a general but very approximate rule:

particles in matter at temperature T each have energy of the order of kT

In chapter 14 you can see how this idea can be taken much further.

Quick check

Useful data: specific thermal capacities: water $4170\,\mathrm{J\,kg^{-1}\,K^{-1}}$; milk $4000\,\mathrm{J\,kg^{-1}\,K^{-1}}$; stone $600\,\mathrm{J\,kg^{-1}\,K^{-1}}$.

1. Why is water the best liquid to put in a hot water bottle? Some people use a stone wrapped in a piece of blanket instead. How does this method compare with the genuine hot water bottle?

2. A full bath takes about 200 kg of water. On a winter's day the water has to be heated from 10 °C to 40 °C. Show that about 25 MJ is used to heat the water.

3. A shower delivers 1 kg of hot water every 10 s. The water is heated from 10 °C to 45 °C. Show that only about 9 MJ is needed to provide a 10 minute shower.

4. A jug containing 0.5 kg of milk at a temperature of 20 °C is put into a refrigerator, which cools it down to 5 °C. Show that about 30 kJ of energy is taken from the milk.

5. In the days before metal containers were available, our ancestors heated water by dropping hot stones into a pottery container of water. Show that a rough estimate of the energy given to the water when a stone of mass 1.5 kg at 200 °C is dropped into water at 10 °C is about 170 kJ. Explain why this is an overestimate.

6. During a hot summer the land gets hotter than the nearby sea. Suggest some reasons for this. In calm weather, during the day in summer there is often a cooling breeze coming off the sea. At night this wind changes direction. Explain these effects.

Links to the *Advancing Physics* CD-ROM

Practise with these questions:
110S   Short answer *Specific thermal capacity: Some questions*
130S   Short answer *Thermal transfers in the home*
140D   Data handling *Thermal changes*

Try out this activity:
230P   Presentation *Warming and cooling by pushing and pulling*

Look up these key terms in the A–Z:
Internal energy; kinetic theory of gases; specific thermal capacity

Go further for interest by looking at:
120C   Comprehension *The wonderful oddity of water*

Revise using the revision checklist and:
120O   OHT *Transferring energy to molecules*

Summary check-up

Gas laws ✓

- The gas laws lead to the idea of an absolute zero of temperature $0\,K$ on the Kelvin scale
- For ideal gases $p \propto$ mass; $p \propto \frac{1}{V}$; $V \propto T$; $P \propto T$
- The gas laws can be combined into a single relationship: $pV = NkT$, where N is the number of molecules and k is the Boltzmann constant
- A mole contains 6.02×10^{23} particles, this number being the Avogadro constant N_A
- For n moles of ideal gas, the gas laws can be written $pV = nRT$, where $R = N_A k$ is the molar gas constant

The kinetic model ✓

- A gas can be modelled as a set of moving, elastic particles
- The kinetic model applies to an ideal gas; real gases show departures from ideal behaviour at low temperatures and high densities
- The kinetic model leads to the relationship $pV = \frac{1}{3}Nm\overline{v^2}$, where $\overline{v^2}$ is the mean square speed of the molecules in a gas
- The mean translational kinetic energy of a particle in a gas is equal to $\frac{3}{2}kT$, and so is proportional to the temperature in kelvin
- The internal energy U of an ideal monatomic gas is given by $U = \frac{3}{2}NkT = \frac{3}{2}nRT$

- The root mean square speed of a gas molecule can be estimated using the relationship $\frac{1}{2}m\overline{v^2} = \frac{3}{2}kT$

- Expansion into a vacuum, diffusion and Brownian motion provide evidence that supports the kinetic model for a gas
- A single molecule in a gas moves randomly as a result of the many collisions it makes with other molecules, travelling $\sqrt{N}$ steps after N collisions

Thermal capacity, conservation of energy ✓

- The internal energy U of a material can be changed by doing work W or by thermal transfer of energy Q, with $\Delta U = W + Q$. This expresses the law of conservation of energy.
- Energy ΔE can change the temperature θ of a substance in accordance with the relationship $\Delta E = mc\Delta\theta$, where m is the mass of the substance in kilograms, c is the specific thermal capacity of the substance and $\Delta\theta$ is the change in temperature
- Water has a surprisingly high specific thermal capacity, with important practical and environmental consequences
- Particles in matter at temperature T have energy of the order of kT per particle

Questions

Useful Data: density of air at normal atmospheric pressure and temperature of 293 K is $1.2\,kg\,m^{-3}$. Relative atomic masses: helium (He) 4; xenon (Xe) 131. Specific thermal capacity of water is $4170\,J\,kg^{-1}\,K^{-1}$.

1. List the main assumptions of the kinetic model of gases. Which assumptions are no longer satisfactory when a gas is near the point where it condenses to a liquid?

2. A hot air balloon contains $2500\,m^3$ of air, and is open at the base to allow the air to be heated. The balloon and the gondola beneath it have a total mass of 500 kg. The initial temperature of the air in the balloon is the same as its surroundings at 20 °C.
   (a) The air in the balloon is heated until the balloon just starts to lift off. What is the difference between the mass of air in the balloon at this stage and the mass of air it displaces?
   (b) The rigid balloon casing, open at the bottom, does not expand. What is the average density of the air inside the balloon at lift off?
   (c) Show that to achieve lift-off the air in the balloon must have been heated to about 80 °C.

3. **A gas cylinder contains equal masses of helium and xenon. These are both monatomic gases.**
   (a) Estimate the relative values of the root mean square speeds of the two different atoms.
   (b) A small hole develops in the container, through which the gases can escape. What will happen to the ratio of the numbers of atoms of each element during a short interval after the leak opens? Justify your answer.

4. **A kilogram of water at 20 °C is placed in a refrigerator.**
   (a) How much energy must be removed from the water to make it reach a temperature of 5 °C?
   (b) This cooling takes 20 minutes. What is the average rate at which the energy is removed by the refrigerator?
   (c) Suggest what happens to this energy. Justify your answer.

5. **A mole of any ideal gas occupies a volume of 22.4 litres at a pressure of 1 atmosphere and a temperature of 273 K. 1 litre is $1000\,cm^3$ or $10^{-3}\,m^3$.**
   (a) How many oxygen molecules would there be in a cubic metre of oxygen at the same pressure and temperature?
   (b) What is the average volume available to each molecule?
   (c) An oxygen molecule is roughly spherical with a diameter of $2 \times 10^{-10}\,m$. What volume would 45 moles of oxygen occupy if the gas was condensed into a liquid (with a molecule generally in contact with its neighbours)?

6. The table gives some data for a fixed mass of gas as its pressure is varied at a constant temperature of 27 °C. [Note: you could use a spreadsheet to help you tackle this question.]

Pressure /10^5 Pa	Volume /m^3	Density at 27 °C /$kg\,m^{-3}$	Density at 127 °C /$kg\,m^{-3}$
0.5	0.40		
0.6	0.34		
0.7	0.29		
0.8	0.25		
0.9	0.22		
1.0	0.20	1.24	
1.1	0.18		
1.2	0.17		

(a) Use the data to calculate the mass of gas under test.
(b) Complete the table to show the density of the gas at 27 °C for the different pressures.
(c) Plot a graph of pressure against density at 27 °C. Does this graph show that the sample of gas behaves as an ideal gas over this range of pressures?
(d) An ideal gas conforms to the kinetic model such that $pV = \frac{1}{3}Nmv^2$. N is the number of molecules and m the mass of a molecule. Rewrite this relationship to show how the density is related to pressure.
(e) What does the slope of the pressure-density graph represent?
(f) Use the graph to estimate the root mean square speed of the gas molecules.
(g) The temperature of the gas is raised to 127 °C, and its pressure is varied as before. At a given pressure, will its density be larger or smaller? Justify your answer.
(h) Calculate the density at each pressure for this new temperature and complete the table with these data. Draw a graph of pressure versus density at 127 °C and explain the difference between this graph and the graph from part (c).

14 Matter: very hot and very cold

Matter is made up of an enormous number of particles in continual random thermal agitation. This fact decides the forms in which matter exists at different temperatures and the changes that can occur to them. At temperatures ranging from near absolute zero to the very hot early universe, we will give examples of:

- values of the energy kT per particle
- the importance of the ratio ε/kT
- uses of the Boltzmann factor $\exp(-\varepsilon/kT)$

14.1 The magic ratio ε/kT

Fire destroys; ice preserves. At the highest temperatures matter comes completely apart; at the lowest temperatures many kinds of change slow down and effectively stop. These differences occur because of the very different amounts of random jostling energy shared among the particles at different temperatures.

The coldest temperature created by nature is that of "empty" space: a couple of degrees kelvin. All lower temperatures (and millionths of a kelvin have been reached) are the work of human beings, devising ways to suck yet more energy from the particles of matter – a job that gets harder and

Fire and ice

Some say the world will end in fire,
Some say in ice.
From what I've tasted of desire
I hold with those who favour fire.
But if I had to perish twice,
I think I know enough of hate
To say that for destruction ice
Is also great
And would suffice.

Robert Frost The Poetry of Robert Frost *(1969) Jonathan Cape*

A wall of ice consisting of frozen pillars of water standing upright, seen on the Moreno glacier in Argentina. An ice-face or cliff such as this usually forms at the sides of a glacier by the shearing of the valley sides as the ice moves along.

A view of a large lava river flowing at night. This very fluid lava was produced by the volcano Mount Etna in Sicily.

Left: the sublimation of dry ice. A fog-like vapour spills out of a test tube containing dry ice and pours into a glass beaker.
Right: molten steel being poured into a mould.

Key summary: temperature and energy $\varepsilon = kT$

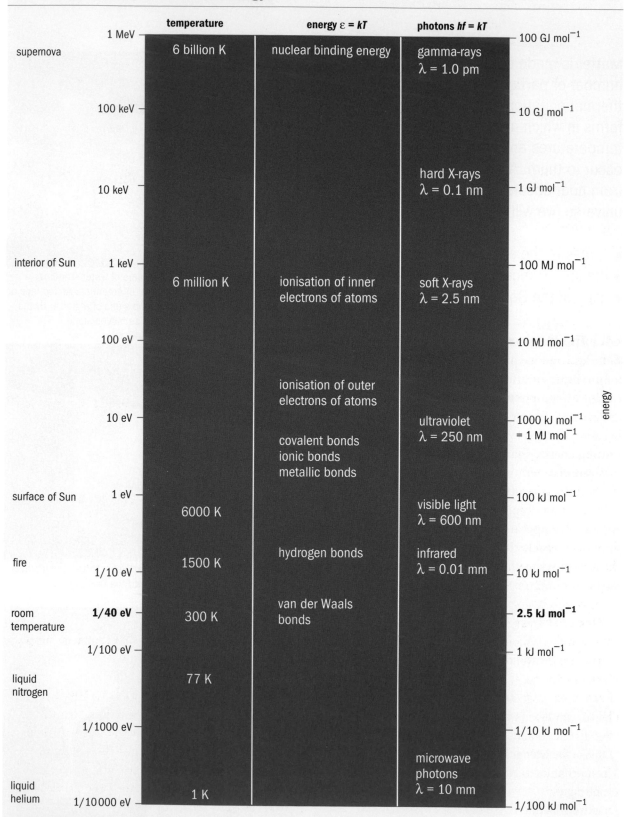

	temperature	energy $\varepsilon = kT$	photons $hf = kT$	
supernova	1 MeV — 6 billion K	nuclear binding energy	gamma-rays $\lambda = 1.0$ pm	— 100 GJ mol^{-1}
	100 keV —			— 10 GJ mol^{-1}
	10 keV —		hard X-rays $\lambda = 0.1$ nm	— 1 GJ mol^{-1}
interior of Sun	1 keV — 6 million K	ionisation of inner electrons of atoms	soft X-rays $\lambda = 2.5$ nm	— 100 MJ mol^{-1}
	100 eV —			— 10 MJ mol^{-1}
	10 eV —	ionisation of outer electrons of atoms	ultraviolet $\lambda = 250$ nm	— 1000 kJ mol^{-1} = 1 MJ mol^{-1}
		covalent bonds ionic bonds metallic bonds		
surface of Sun	1 eV — 6000 K		visible light $\lambda = 600$ nm	— 100 kJ mol^{-1}
fire	1/10 eV — 1500 K	hydrogen bonds	infrared $\lambda = 0.01$ mm	— 10 kJ mol^{-1}
room temperature	**1/40 eV** — 300 K	van der Waals bonds		— **2.5 kJ mol^{-1}**
	1/100 eV —			— 1 kJ mol^{-1}
liquid nitrogen	77 K			
	1/1000 eV —			— 1/10 kJ mol^{-1}
			microwave photons $\lambda = 10$ mm	
liquid helium	1/10 000 eV — 1 K			— 1/100 kJ mol^{-1}

energy

At temperature T the average energy per particle is of the order of kT. Useful approximation: $kT = 1$ eV when $T = 10\,000$ K

harder the colder it gets. Down towards absolute zero, the jostling energy of thermal activity is quietened to a near calm.

The hottest temperatures known occurred in the Big Bang, the origin of the universe, exceeding 10^{15} K. At this temperature, the collisions between particles are as powerful as those in the world's largest particle accelerators, creating all sorts of exotic particles.

Average energy per particle $\approx kT$

A simple general rule links the behaviour of matter at all temperatures. Particles, in all kinds of matter, have an average energy of random thermal agitation of the order of magnitude kT, at an absolute temperature T in kelvin. The constant k is the Boltzmann constant, 1.38×10^{-23} J K^{-1}.

Molecules of an ideal monatomic gas have average energy $\frac{3}{2}kT$ (chapter 13). That's $\frac{1}{2}kT$ for motion in each of three directions in space. If the molecules also have energy from vibrating or spinning, they have more energy per particle. Except at low temperatures, this amounts to a further $\frac{1}{2}kT$ for each mode in which a molecule can have energy, for example both kinetic and potential energy of vibration.

Atoms or ions in solids do the same, vibrating inside the solid with differing amplitudes and in different directions. At high enough temperatures they reach an average energy of $3kT$ each. Even photons of light do it. In the light from a hot object like the Sun, the photons spread over a range of frequencies. The spectrum has a peak of intensity at a given frequency. Photons whose frequency $f_{\max}$ is at the peak have energy $hf_{\max}$ equal to nearly $3kT$.

We want to encourage you to look at the broad picture, not at the details. It doesn't matter for the moment whether the numerical factors in front of kT are $\frac{3}{2}$ or $\frac{5}{2}$ or 3 or some other small number. The big general principle is that particles share roughly energy kT. This broad idea is not less valuable for being only more-or-less true.

You do have to be careful with the idea at low temperatures. Because energy exchanges are quantised (chapter 7, *Advancing Physics AS* and chapter 17), nothing happens if the energy kT on offer from thermal agitation is smaller than the minimum required quantum of energy. Kinds of

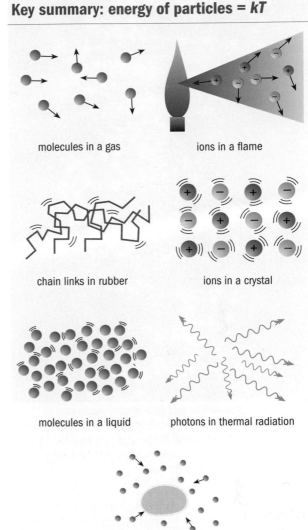

Key summary: energy of particles = kT

molecules in a gas

ions in a flame

chain links in rubber

ions in a crystal

molecules in a liquid

photons in thermal radiation

dust particle in Brownian motion

At temperature T, particles have random thermal energy of the order of kT

motion whose energy quanta are too large simply don't join in the general share-out.

Hotter and hotter: from Earth to the Sun

The ancient four "elements", earth, water, air and fire, are a good model for the states through which matter passes as it is made hotter: solid, liquid, gas and ionised plasma. (The word "gas" is from the Greek meaning "chaos", the primal and undifferentiated). At each change, some structure comes apart. From solid to liquid, the rigid structure holding particles in place; from liquid to gas, the looser structure holding particles together;

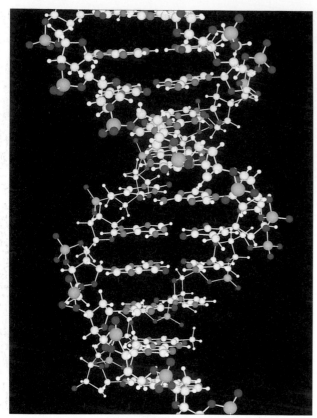

This computer graphic ball-and-stick model represents a segment of DNA. The DNA molecule is made of two strands of atoms twisted into a helical shape and held together by hydrogen bonds. Using enzymes, these come apart just easily enough at 300 K for DNA to replicate.

At 300 K

Energy $kT = 1.38 \times 10^{-23}\,\text{J K}^{-1} \times 300\,\text{K}$

$\approx 400 \times 10^{-23}\,\text{J}$ per particle

$\approx \frac{1}{40}$ electron volt per particle (kT divided by electron charge)

$\approx 2.5\,\text{kJ}$ per mole ($kT \times$ Avogadro number)

Life at $T = 300\,\text{K}$

A water molecule is held to the water surface by an average of two hydrogen bonds, energy ε of the order $20\,\text{kJ mol}^{-1}$ each (0.2 eV). These same bonds hold the chains of DNA together, joining the gene "letters" in pairs. It is essential for the DNA chains to be able to come apart – to be unzipped – so that genetic information can be replicated. But it is also crucial that this doesn't happen too easily, or the DNA molecule wouldn't stay together between replications. With the help of enzymes that unzip the DNA easily, the ratio ε/kT for hydrogen bonds is just about right to manage this balancing act at the temperature of the surface of Earth. But at higher temperatures, ε/kT would be less and DNA would fall apart. The evolution of life as we know it, based on hydrogen bonds, can only occur around 300 K.

Fires and kilns: $T = 1500\,\text{K}$

Early in human history, people found that fire could fuse clay into hard, water-tight pots. Fire could melt metals and extract them from ores. Some refractory materials resisted melting in the hottest fires people could make; these were useful for building and lining furnaces and kilns. Sand mixed with salt would melt and fuse into glass.

Early humans valued fires for light and warmth, just like people do now, but they could not know that fires also conduct electricity. At 1500 K, atoms are beginning to come apart into ions and electrons.

In a fire or kiln things start to glow red hot. Some metals melt. Water, waxes and greases are all gone. Glass softens and melts. Living materials like bacteria are killed and destroyed – making a needle red hot sterilises it. But ceramics stay solid.

Look at some calculations. The average energy kT per particle in a fire is of the order $\frac{1}{10}$ eV. The energy kT in units familiar to chemists is around

from gas to plasma, the particles themselves coming apart into ions and electrons.

From a balmy 300 K on the surface of Earth to 6 million K inside the Sun, the behaviour of matter changes drastically. The average energy of a particle at 300 K is a quantity worth remembering (see box "At 300 K", above). How big is this actually? In a gas the molecules travel at hundreds of metres a second, enough to climb perhaps 10 km against gravity. Does this suggest to you that kT is rather large, or that gravity is rather weak?

Another comparison. Puddles of water evaporate slowly at 300 K. It takes $40\,\text{kJ mol}^{-1}$, 16 times kT, to pull molecules out of water into water vapour (measure it with an electrical heater boiling some water away). So it seems that molecules can, by chance lucky collisions, get a good deal more than the average energy kT. If they didn't, seas wouldn't evaporate, rain wouldn't fall and puddles wouldn't dry up.

In oxyacetylene welding the flame burns hot enough to melt metals without leaving oxides ("rust") in the weld region.

The colourful shiny glaze on this 15th century wine jar is glass melted onto the surface in a hot kiln.

$10\,\text{kJ}\,\text{mol}^{-1}$, similar to the energy of a hydrogen bond. This is why such bonds break and biological materials are destroyed.

Photons with energy $3kT$ are in the deep penetrating infrared (wavelength 0.01 mm), good for aches and pains. But the fire also glows red, so it must emit some photons of energy hf considerably larger than kT: a calculation says up to around 20 times kT.

Sodium and potassium ions in flames colour a fire yellow and red. Metals like these ionise and lose electrons fairly easily – needing only a couple of electron volts, again around 20 times kT. Because of this the flames contain some free ions and can conduct electricity.

Ceramics, however, are typically held together by ionic bonds. These have bond energies ε many times larger than kT, getting on for $100\,kT$. Essentially no particles get enough energy in a fire for these materials to start coming apart. Fire bricks stay solid in a kiln.

The message – another more-or-less true idea – is to expect changes that require energy ε to be starting to happen at an appreciable rate when ε is 10, 20 or 30 times the energy kT. If $\varepsilon = kT$ or less, such changes are over and done with very quickly. If ε is nearer $100\,kT$, don't expect much to happen.

White hot at the Sun's surface: $T = 6000\,\text{K}$

The surface of the Sun is a white-hot furnace at $6000\,\text{K}$. What do similar calculations suggest about what to expect there?

The peak of the Sun's spectrum, where photons have energy $\varepsilon = hf$ of the order $3kT$, is still in the infrared (wavelength 1 mm). At this temperature, even ultraviolet photons have energy only 10 times kT. And indeed the Sun does emit quite a lot of ultraviolet light, much of it absorbed in the upper atmosphere when it reaches the Earth. That which gets through can give you sunburn.

Ceramics don't stand much chance, with bond energies now only $20\,kT$. Indeed, the ceramic tiles protecting a spacecraft as it re-enters the atmosphere are severely ablated by the high temperatures caused by re-entry. Hydrogen atoms need nearly 30 times kT to be torn apart, but this is indeed happening on the surface of the Sun, as evidence from spectroscopes can show.

Hotter still: inside the Sun at $6\,000\,000\,\text{K}$

In the deep interior of the Sun, the temperature reaches 6 million K. The energy kT is $500\,\text{eV}$. It is easy for any kind of atom to get the energy to ionise, and the inside of the Sun is a new kind of matter, a plasma, consisting wholly of ions and electrons. Its electrical conductivity helps give rise

Key summary: processes and energy = *kT* per particle

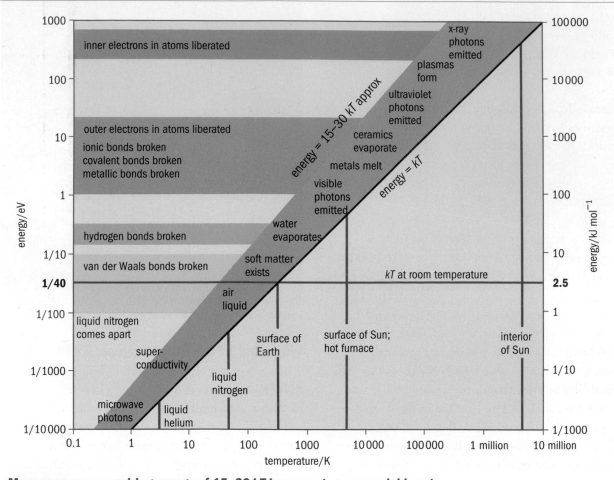

Many processes requiring energy of 15–30 *kT* happen at an appreciable rate

to the Sun's magnetic field.

Light in the deep interior of the Sun is pretty searing stuff. There are lots of soft X-rays, wavelength about the size of a big molecule, and some hard X-rays with photon energies of 10 000 eV, or around 20 times *kT*. But the atomic nuclei mostly stay intact.

Cool, calm and complex

When matter cools down it condenses into liquid or solid, and many reactions tend to go slower or even effectively stop. Complex patterns of molecules, from the ordered arrays of atoms in crystals to the long linked chains of polymers, start to appear.

Really cold stuff is not at all unusual today. Natural gas (methane) becomes liquid at atmospheric pressure at a chilly 112 K, nearly three times colder than room temperature (compare

temperatures by their ratio in degrees kelvin: the ratio of the average energy per particle). It is shipped all around the world in liquid form in pressurised containers.

Making air liquid

In hot dry climates a room can be cooled just by blowing air through a wet screen; when evaporating molecules break their bonds with the water, they take energy from the random thermal agitation and cool the water, thus cooling the air in the room. You get the same effect with a spray used to soothe burns.

Air was first liquefied by German engineer Carl von Linde in 1895. Today, liquid nitrogen costs about the same per litre as beer, and is on tap in hospitals for freezing specimens. Liquid nitrogen boils at 77 K, nearly four times colder than room temperature. kT is small, but so is the energy ε required to pull the molecules apart. The ratio ε/kT comes to about 10.

Liquid nitrogen is very, very cold. Carried in a Thermos flask, it is a transparent, gently boiling liquid. A rubber band dipped into it becomes like glass: hard and brittle. Polythene breaks like a biscuit. Mild steel, tough at room temperature, also becomes brittle. Copper wire conducts electricity quite a lot better than at room temperature. Some of the recently discovered high-temperature superconductors lose their electrical resistance completely at this temperature.

The small value of kT explains why so many things become brittle at low temperatures, a problem for designers of Arctic gear and high-flying planes. Ductile slipping requires atoms to move over one another, and can happen easily when the random energy of thermal agitation can provide the energy for an atom to get "over the hump". When kT is too small, the necessary energy is hard to acquire, and the material loses its ability to flow. Even polythene lunch boxes crack easily after being in the freezer. Chocolate kept in the fridge becomes hard and brittle, more difficult to bite. Syrup becomes very difficult to pour or spoon out of the jar.

As cold as you can get

Helium (^{4}He) liquefies at 4.2 K and large-scale quantum phenomena appear. One of them is superconductivity in metals, first found in mercury by the Dutch physicist Heike Kamerlingh Onnes as early as 1911. It took thirty years to find the quantum explanation for this sudden total vanishing of resistivity. So complete is it that for a demonstration at the Royal Institution in 1932, a lead ring cooled in liquid helium was flown to London from Leiden in Holland, with an electric current started in Holland still flowing when it arrived in the Royal Institution lecture room. Today liquid helium finds use in the superconducting magnets of particle accelerators and hospital scanners.

At these temperatures, typically 100 times lower than room temperature, thermal agitation is very much stilled. The value of the average energy per particle kT has fallen to a tiny $25\,\mathrm{J\,mol^{-1}}$, or

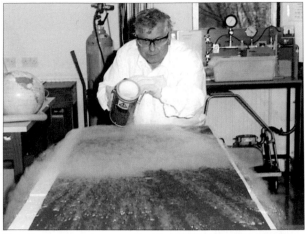

The nitrogen in air becomes liquid at 77 K. It is cold enough to freeze rubber into a glass-like state that will snap if bent. Liquid oxygen fuels some spacecraft.

0.3 milli-eV per particle. Molecular speeds drop to the speed of a ball thrown by hand in a game.

Every year physicists are getting closer and closer to the absolute zero. They have already achieved temperatures lower than $10^{-10}\,\mathrm{K}$.

Summary

We have told the story of how the temperature decides the forms in which matter exists, on a grand scale from the highest to the lowest temperatures. Every physical change involves particles coming apart or coming together – breaking or making bonds between particles. Bonds get broken when the temperature is raised, because the average energy kT of random thermal motion begins to be enough to supply the energy to break a bond. So at high temperatures matter comes apart, breaking up into simpler and simpler constituents. Bonds form when the temperature falls, because forming a bond decreases the potential energy of the particles, and the random thermal energy kT isn't enough to break the bond again. Thus, at low temperatures, complex weakly-bonded arrangements of molecules can exist, including those needed for biological structures.

The first signs of change begin to appear when the bond energy is of the order 10–30 times kT. This is because the thermal energy is randomly distributed and a few particles can always, by chance, have this much more energy than the average. We will explain this more fully in the next section.

Quick check

Useful data: The energy kT at 300 K is approximately 400×10^{-23} J, $\frac{1}{40}$ eV or 2.5 kJ mol^{-1}.

1. Place these energies per particle in increasing order of size:
   (a) the value of kT at 300 K,
   (b) 100×10^{-23} J per particle,
   (c) 10 eV,
   (d) 20 kJ mol^{-1}.

2. Which of the following is the best estimate of the average energy per particle at a temperature of 3000 K?
   (a) 0.25 eV,
   (b) 400×10^{-23} J,
   (c) 0.25 kJ mol^{-1}.

3. Which of (a), (b) or (c) in question 2 is the best estimate of the average energy per particle at 30 K?

4. Show that at a temperature of 12 million K the photon energy kT is of the order of magnitude 1 keV, the energy of soft X-ray photons.

5. Show that at 112 K, the boiling temperature of methane, kT is about 0.9 kJ mol^{-1}. Show that methane boils at a ratio ε/kT of about 9. (Energy of vaporisation of liquid methane = 8.2 kJ mol^{-1}.)

6. Energy of 13.6 eV is needed to strip the electron from a hydrogen atom. Show that this is equal to $30\,kT$ at a temperature of 5400 K. Explain why hydrogen is ionised in a hot flame.

Links to the *Advancing Physics* CD-ROM

Practise with these questions:
10W Warm-up exercise *Molecular energy*
20W Warm-up exercise *Evaporation of water*
30W Warm-up exercise *Thermal radiation*
40W Warm-up exercise *Calculating thermal energies*
20S Short answer *Molecules and change*
45X Explanation–exposition *Matter "comes apart"*

Try out these activities:
10P Presentation *Temperatures everywhere*
100S Software-based *Introducing breakouts*
130S Software-based *Energy per atom in some solids*

Look up these key terms in the A–Z:
Absolute temperature; energy kT; order and disorder

Go further for interest by looking at:
20H Home experiment *Crème brûlée: Carefully controlled changes*
120H Home experiment *Energy for one water molecule to escape*

Revise using the revision checklist and:
10O OHT *Temperature and the average energy per particle*
15S Computer screen *Rough values of the energy* kT
20O OHT *Different processes related to the average energy per particle*
25S Computer screen *The energy it takes*

14.2 The Boltzmann factor exp(−ε/kT)

Getting extra energy by chance

Many different processes – melting, evaporating and ionising, for example – start happening at an appreciable rate when the energy ε per particle needed for them to happen is a moderate multiple of the average energy kT per particle. Many processes therefore depend on some particles having more than the average energy, say 15 to 30 times more.

Only if a particle gets lucky does it acquire energy greater than the average energy of random thermal jostling. It has, just by chance, to be hit by a succession of other particles, or be hit by another unusually energetic particle. So there is a certain probability that any particle will, at a given time, acquire extra energy ε.

A certain small fraction f of molecules will get this extra energy. (This is simply how random events happen – just by chance with a certain probability.) That extra energy has to be taken away from the energy shared by all the other particles around them. What about getting even more energy from the others? As long as there are very many particles altogether, it is the same fraction as before, but this time of those which have already gained extra energy. To gain yet another amount of energy ε the already unusually energetic particle would have to be lucky a third time. It's what gamblers hope for: a run of luck.

Such runs do happen, but rarely. They happen rarely because probabilities multiply. You don't often throw ten heads in a row because the probability of doing it is 1 in 2^{10} (1 in 1024). But keep at it and the probability is that it will happen, one time in a thousand or so. Molecules in matter do keep at it, colliding with one another and exchanging energy a billion or more times a second. And there are a lot of them, so an appreciable number can get very lucky indeed, getting 10, 20 or 30 times the average energy kT.

Flying high

Living on Earth, the air around you seems to be the same at the top of a tower block as at the bottom. But if you climb a high mountain, or go up in a balloon, you soon find it thinning out. Breathing gets difficult and you may need an oxygen cylinder. Seen from space, the atmosphere is just a thin blanket around the Earth. The atmosphere provides an example of molecules gaining the energy to climb up the gravitational hill away from Earth, just by the chance acquisition of extra energy from other molecules around them.

The Earth's gravity pulls air molecules towards the surface. There is a slope of gravitational potential down which they tend to fall. But that would mean all the air falling to the ground, with nothing above it. This can't happen because the air molecules are moving at random, and so some

Key summary: climbing a ladder of energy by chance

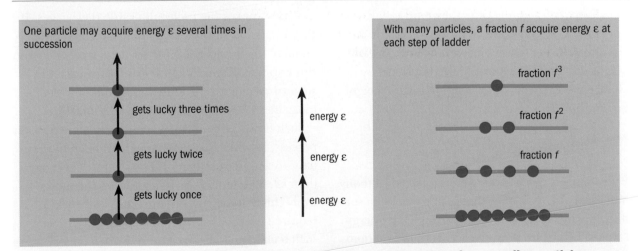

By chance, particles may get extra energy from the random thermal motion of surrounding particles

Key summary: air molecules pulled in two opposite directions

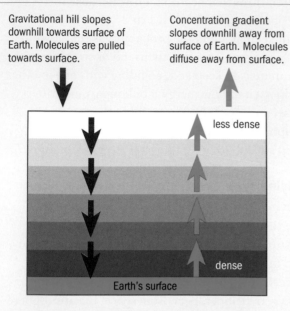

Gravitational hill slopes downhill towards surface of Earth. Molecules are pulled towards surface.

Concentration gradient slopes downhill away from surface of Earth. Molecules diffuse away from surface.

less dense

dense

Earth's surface

Air molecules are pulled towards the Earth's surface by gravity but diffuse away from the surface because of the concentration gradient

Key summary: molecules diffuse from higher to lower concentration

Empty space

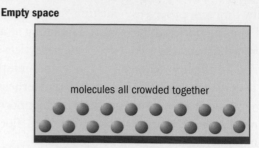

molecules all crowded together

Molecules diffuse into empty space

Random motion takes molecules from where there are many to where there are fewer. Diffusion goes down a concentration gradient.

would inevitably head off back towards the empty space above. Few would be coming down because there would be few of them up there. In other words, the air would diffuse upwards.

The atmosphere settles down with a gradient of concentration. It gets rapidly less dense as you go up. Just by their random motions, molecules continually diffuse away from Earth merely because there are more of them lower down than higher up. At the same time they tend to fall down the gravitational slope towards Earth. In equilibrium, the two effects balance out. At any height the fraction f of molecules that have by chance the extra energy $\varepsilon = mgh$ to be at a height greater by h is

$$f = \exp(-\varepsilon/kT)$$

This idea is much more generally true. In many materials, the ratio of numbers of particles in states differing by energy ε is $\exp(-\varepsilon/kT)$, if the energy ε has to be acquired by chance from the random thermal motions of many particles at a temperature T (assuming quantum effects are not important).

Hotter goes faster

Warm syrup runs easily. Olympic record-breaking swimmers say they prefer warm water, because there is less drag. Oils for car engines have to be designed to flow well enough to prevent wear when the engine is started and is cold, but not to get too runny when it is hot.

How does a liquid flow? Imagine each molecule in a liquid "caged in" by its neighbours. Thermal agitation makes the molecules rattle their cages. If they break out into the neighbouring cages often enough, the liquid flows. Water flows freely because this happens often. Syrup is viscous because the sugar molecules impede the break-out of water molecules, to which they are loosely bonded.

Such flow in liquids is an example of an activation process, in which for something to happen particles have to acquire extra activation energy ε greater than kT. The higher the temperature, the more likely this is to occur, so more particles have the required energy. Raising the temperature makes such a change go faster. The activation energy is like a hill over which the particle has to climb before getting down into the next valley.

Key summary: the Boltzmann factor and the atmosphere

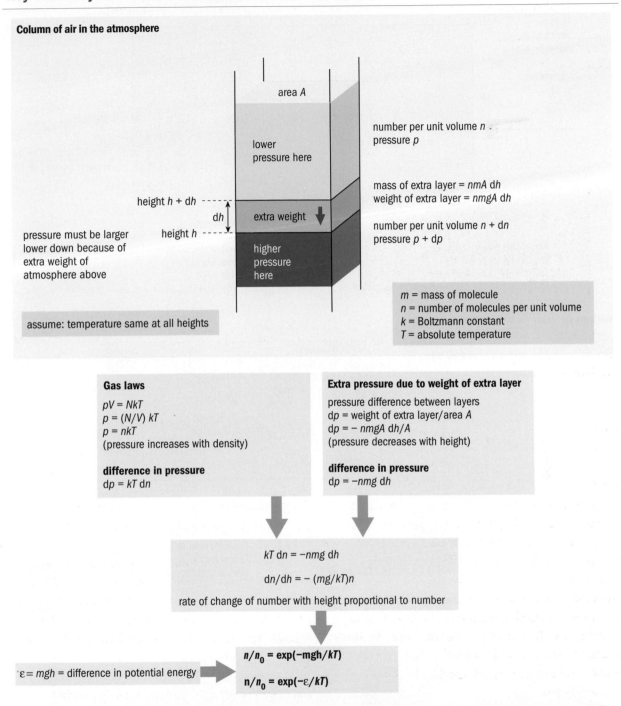

Column of air in the atmosphere

area A

lower pressure here

number per unit volume n
pressure p

mass of extra layer = nmA dh
weight of extra layer = $nmgA$ dh

height h + dh

dh

extra weight

number per unit volume n + dn
pressure p + dp

height h

pressure must be larger lower down because of extra weight of atmosphere above

higher pressure here

assume: temperature same at all heights

m = mass of molecule
n = number of molecules per unit volume
k = Boltzmann constant
T = absolute temperature

Gas laws

$pV = NkT$
$p = (N/V)\,kT$
$p = nkT$
(pressure increases with density)

difference in pressure
d$p = kT$ dn

Extra pressure due to weight of extra layer

pressure difference between layers
dp = weight of extra layer/area A
d$p = -nmgA$ dh/A
(pressure decreases with height)

difference in pressure
d$p = -nmg$ dh

kT d$n = -nmg$ dh

dn/d$h = -(mg/kT)n$

rate of change of number with height proportional to number

$n/n_0 = \exp(-mgh/kT)$

$\varepsilon = mgh$ = difference in potential energy

$n/n_0 = \exp(-\varepsilon/kT)$

Ratio of numbers of particles in states differing by energy ε is equal to the Boltzmann factor $\exp(-\varepsilon/kT)$

As ε/kT increases, the fraction of particles with that energy rapidly gets smaller. In a number of the changes discussed so far (see section 14.1) the ratio ε/kT came out at 15 to 30 or so. The fraction of particles with those energies is in the range one in 10^7 to one in 10^{13}. At first sight it may seem odd that such a tiny minority of particles has such important effects. It seems less odd if you remember that each is attempting a change at billions or more times every second. Molecules in gases collide around 10^9 times a second. Atoms in molecules oscillate around 10^{13} times a second.

Key summary: Boltzmann factor: the minority with energy exp(−ε/kT)

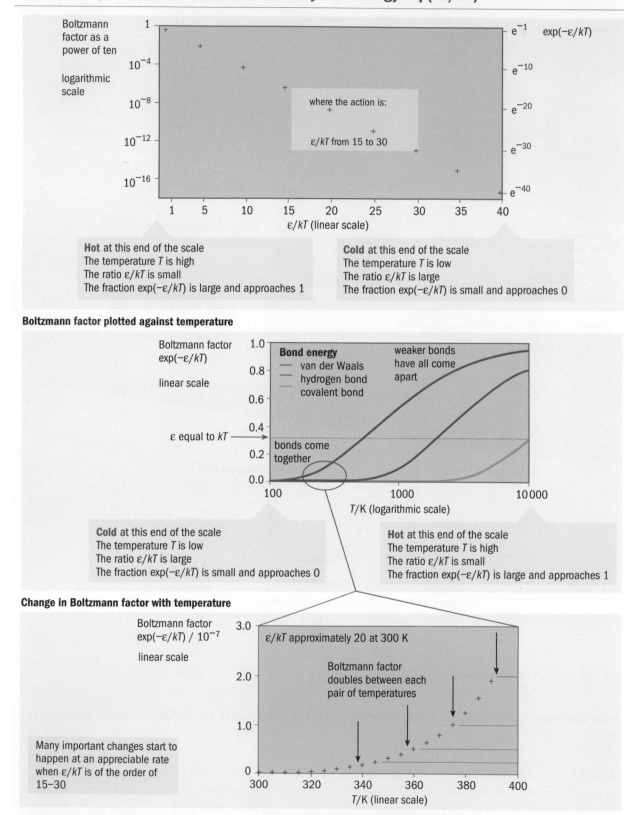

Boltzmann factor as a power of ten

logarithmic scale

1

10^{-4}

10^{-8}

10^{-12}

10^{-16}

e^{-1} exp(−ε/kT)

e^{-10}

e^{-20}

e^{-30}

e^{-40}

where the action is:

ε/kT from 15 to 30

1   5   10   15   20   25   30   35   40

ε/kT (linear scale)

Hot at this end of the scale
The temperature T is high
The ratio ε/kT is small
The fraction exp(−ε/kT) is large and approaches 1

Cold at this end of the scale
The temperature T is low
The ratio ε/kT is large
The fraction exp(−ε/kT) is small and approaches 0

Boltzmann factor plotted against temperature

Boltzmann factor exp(−ε/kT)

linear scale

1.0

0.8

0.6

0.4

0.2

0.0

ε equal to kT

Bond energy
— van der Waals
— hydrogen bond
  covalent bond

weaker bonds have all come apart

bonds come together

100   1000   10 000

T/K (logarithmic scale)

Cold at this end of the scale
The temperature T is low
The ratio ε/kT is large
The fraction exp(−ε/kT) is small and approaches 0

Hot at this end of the scale
The temperature T is high
The ratio ε/kT is small
The fraction exp(−ε/kT) is large and approaches 1

Change in Boltzmann factor with temperature

Boltzmann factor exp(−ε/kT) / 10^{-7}

linear scale

Many important changes start to happen at an appreciable rate when ε/kT is of the order of 15–30

3.0

2.0

1.0

0

ε/kT approximately 20 at 300 K

Boltzmann factor doubles between each pair of temperatures

300   320   340   360   380   400

T/K (linear scale)

The Boltzmann factor exp(−ε/kT) is the ratio of numbers of particles in states differing by energy ε. When ε ≫ kT the Boltzmann factor increases very rapidly with temperature.

Key summary: activation processes

An energy hill has to be climbed before the process can happen.

The energy needed has to be acquired by chance from random thermal agitation of the surroundings.

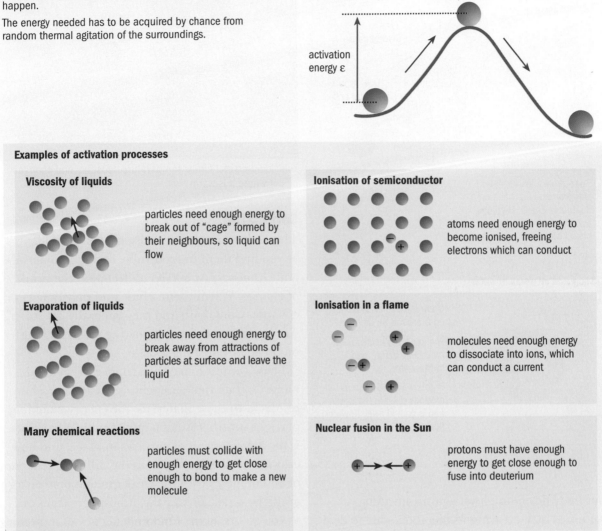

energy needed for process to be possible

activation energy ε

Examples of activation processes

Viscosity of liquids

particles need enough energy to break out of "cage" formed by their neighbours, so liquid can flow

Ionisation of semiconductor

atoms need enough energy to become ionised, freeing electrons which can conduct

Evaporation of liquids

particles need enough energy to break away from attractions of particles at surface and leave the liquid

Ionisation in a flame

molecules need enough energy to dissociate into ions, which can conduct a current

Many chemical reactions

particles must collide with enough energy to get close enough to bond to make a new molecule

Nuclear fusion in the Sun

protons must have enough energy to get close enough to fuse into deuterium

The energy needed for a process to happen has to be acquired by chance from the random thermal agitation of the surroundings

Thus such changes happen on the human time scale of seconds when between one in 10^9 and one in 10^{13} particles have the required energy.

Calculations using the Boltzmann factor work for a wide variety of activation processes. The rate depends on what a particle has to do for the process to happen, and on how often it tries doing it. The frequency of attempts depends to some extent on the temperature, since particles move less at low temperatures. But, to a reasonable first approximation (in other words, more-or-less true),

for many activation processes:

rate of reaction is proportional to $\exp(-\varepsilon/kT)$

The Boltzmann factor helps to explain fevers. Reactions in the body can increase their rate dramatically if the temperature increases, and decrease their rate dramatically if it falls. Cold-blooded creatures like lizards bask in sunlight to stay warm, and grow sluggish if they get chilled. Getting hot and sweaty with a fever is actually

Soap and water

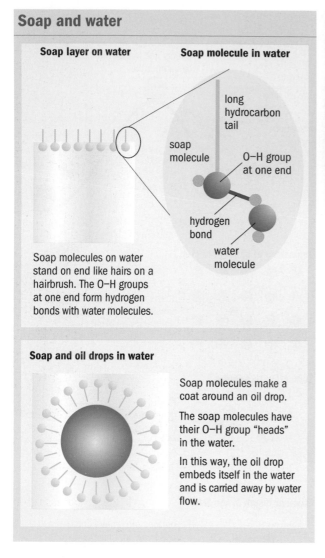

Soap layer on water

Soap molecule in water

long hydrocarbon tail

soap molecule

O–H group at one end

hydrogen bond

water molecule

Soap molecules on water stand on end like hairs on a hairbrush. The O–H groups at one end form hydrogen bonds with water molecules.

Soap and oil drops in water

Soap molecules make a coat around an oil drop.

The soap molecules have their O–H group "heads" in the water.

In this way, the oil drop embeds itself in the water and is carried away by water flow.

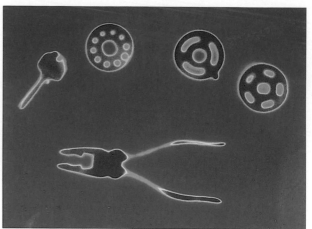

This liquid-crystal film is temperature-sensitive and changes colour where it is warm.

Soft matter

Think about greases, emulsions, jellies and foams as you find them in foodstuffs, cleaning materials and cosmetics. At 300 K, all teeter on the brink of change. Warm them a little and they liquefy or collapse. Cool them and they generally get stiffer. All are easy to deform – whether spreading butter on bread or polishing a table.

These common materials, under the general name "soft matter", are a lively area of current research. In these materials, the energy ε holding particles together is often quite small. What matters is the arrangement of particles in space, and how easy it is to change it. Just as the diffusing tendency of air competes with the weak gravitational energy gradient to decide how the atmosphere thins out as you go up, so the random tendency of molecules to coil up or point in different directions competes with changes in energy. Examples include the behaviour of polymers like polythene or rubber, and of liquid crystals like those used in computer displays or which make up the cell membranes of your body. Soft matter abounds in the kitchen: in emulsions like mayonnaise; and in foams like beaten egg white.

Soap, cell membranes and wristwatches

Water just runs off oily hands. Soap helps because the soap molecules form up side by side like soldiers, surrounding the oil drops. Their "heads" make hydrogen bonds with the water, and stick the oily drops to it. By lining up, the soap molecules have done something that they ought not to do

your body defending itself against invading organisms by increasing its metabolic rate to deal with them more rapidly.

Many physical and chemical reactions can double their rate for only a 10 or 20 K rise in temperature. You can get experimental evidence of the value of the activation energy by observing how much the rate increases for a given increase in temperature.

This is the end of the story, so far as the Advancing Physics course is concerned, but it is just the beginning of the story of how matter changes and of how natural processes happen. The whole picture has to include ways in which molecules are randomly arranged in space as well as how they randomly share energy between themselves. The remainder of this chapter offers some insights into how that story develops.

if they are just moving at random – they have spontaneously got into a less random arrangement. But this does happen. When the tip of each soap molecule clamps on to a water molecule, making a hydrogen bond, the energy of the hydrogen bond formed is spread out among the water molecules, and their extra random motion compensates for the unnatural order of the soap molecules.

One of the miracles of evolutionary design is the cell membrane. Flexible and permeable, yet stiff enough to hold its shape and protect its contents. Like a soap film, it is made of rod-like molecules with one end that bonds to water. In a cell membrane a double layer forms, with the water-adhering ends facing outwards on both sides.

Cell membranes are an example of liquid crystals – ordered but not rigid. Liquid crystals made of rod-like molecules are familiar in liquid-crystal displays used in wristwatches and in the screens of laptops and TVs. Changes in colour due to rearrangement of the molecules can be so sensitive to temperature that liquid crystals make excellent visual thermometers.

The molecules don't care

One simple theme underlies this whole chapter. It is that a change occurs because it is the result of what the molecules, behaving randomly, just end up doing. If you freeze water into ice cubes or extract iron from ore, they are being neither helpful nor unhelpful. The molecules don't care; they just act at random. You have to fix the conditions so that what the molecules do at random results in whatever you want to happen actually happening. If you want diffusion you need a concentration gradient. If you want thermal flow of energy you need a temperature gradient. If you want to freeze things you need a low temperature. If you want to melt them you need a high temperature. Given the right conditions, the uncaring molecules will end up on average doing what you want. But what actually happens is what would have inevitably happened purely through the random dance of the molecules.

Boltzmann found a way of expressing all this in one equation. The idea is that random changes inevitably try out all possible arrangements for molecules and for the energy shared among them.

Boltzmann's tombstone in Vienna carries an equation.

No random changes will ever decrease the number of arrangements. Writing W for the number of ways the molecules and the energy they share can be arranged, Boltzmann's idea comes down to the simple principle:

W never decreases, and generally increases.

The same idea is often expressed in terms of a quantity, called the entropy, which never decreases and generally increases. This principle is called the **Second Law of Thermodynamics** and governs the direction of physical and chemical change.

Few tombstones carry a scientific equation. An exception is the equation translating numbers of arrangements of particles W and of energy among them into the entropy, S:

$$S = k \log W$$

This equation is carved on Boltzmann's gravestone in the central cemetery in Vienna.

Quick check

Useful data: The energy kT at 300 K is approximately 400×10^{-23} J, $\frac{1}{40}$ eV or 2.5 kJ mol^{-1}.

1. The scale of ε/kT in the first graph on p128 runs from 1 to 40. Show that for $T = 300$ K this corresponds to energies from $\frac{1}{40}$ eV to 1 eV, while for $T = 3000$ K it corresponds to energies from $\frac{1}{40}$ eV to 10 eV.

2. Use the first graph on p128 to show that
   (a) the Boltzmann factor is about 10^{-7} for water molecules escaping into the vapour at 300 K, energy 40 kJ mol^{-1}; and
   (b) it is about 10^{-12} for the ionisation of hydrogen atoms at 6000 K, ionisation energy 13.6 eV.

3. Use the first graph on p128 to show that the ratio ε/kT is about 22 when the Boltzmann factor is 10^{-10}.

4. Use the last graph on p128 to show that, when $\varepsilon/kT = 20$ at 300 K, the Boltzmann factor increases by a factor of about 3 from $T = 375$ K to $T = 400$ K.

5. Oxygen molecules bind to haemoglobin molecules with an energy ε of the order 0.3 eV. Show that the Boltzmann factor is about 6×10^{-6} for the process of oxygen molecules escaping from haemoglobin at 300 K.

6. The ratio $\varepsilon/kT = 16$ for water molecules escaping into the vapour at 300 K. Show that the Boltzmann factor at 300 K is about 10 times smaller than it is at 350 K.

Links to the *Advancing Physics* CD-ROM

Practise with these questions:

60W Warm-up exercise *Nuclear fusion in the Sun*

90S Short answer *The Boltzmann factor:* $f_B = \exp(-\varepsilon/kT)$

60X Explanation–exposition *Thinking about the Boltzmann factor*

110D Data handling *Resistance and conductance of thermistors*

120D Data handling *Vapour pressure of water*

Try out these activities:

150E Experiment *A race depending only on chance*

160S Software-based *Energy shared among particles*

170S Software-based *Getting lucky: Climbing an energy ladder by chance*

Look up these key terms in the A–Z:
Boltzmann factor; energy kT; thermal activation processes

Go further for interest by looking at:

30T Text to read *Flow in liquids*

40T Text to read *Why you can't get to absolute zero*

Revise using the revision checklist and:

400 OHT *Climbing a ladder by chance*

500 OHT *An exponential atmosphere*

600 OHT *Examples of activation processes*

700 OHT *The Boltzmann factor*

800 OHT *How the Boltzmann factor changes with temperature*

Summary check-up

Thermal activity, bonds and temperature ✓

- Particles have an average energy of thermal activity of the order kT, which may be expressed in joule per particle, electron volt or kJ per mole
- Bonds are characterised by the energy ε needed to break them
- At high temperatures, bonds break and matter comes apart. Atoms come apart into ions and electrons. The ratio ε/kT becomes small even for large ε.
- At low temperatures, thermal activity is feeble, and ε/kT is large except for processes with very small ε. Matter condenses to solid or liquid, and complex structures form.

Boltzmann factor ✓

- The ratio of numbers of particles in two quantum states differing by energy ε is the Boltzmann factor $\exp(-\varepsilon/kT)$
- The origin of the Boltzmann factor is the progressively smaller probability of repeatedly gaining extra energy at random from a large collection of other particles

Rate of reaction ✓

- To a first approximation the rate of a reaction with activation energy ε is proportional to $\exp(-\varepsilon/kT)$, and increases rapidly with temperature if $\varepsilon \gg kT$. Many processes happen at an appreciable rate when ε/kT is in the range 15 to 30.

Changes ✓

- Reactions can involve changes in the number of spatial or orientational arrangements of particles, as well as of their energy

Questions

Useful data: Planck constant $h = 6.6 \times 10^{-34}$ J Hz^{-1}, Boltzmann constant $k = 1.38 \times 10^{-23}$ J K^{-1}, elementary charge $e = 1.6 \times 10^{-19}$ C, speed of light $c = 3.0 \times 10^{8}$ m s^{-1}, mass of a proton $= 1.7 \times 10^{-27}$ kg, Avogadro constant $= 6.02 \times 10^{23}$ mol^{-1}.

1. At what temperature does each of the following photons have photon energy $E = hf$ equal to the energy kT? Comment on the values obtained.
   (a) An X-ray photon, produced by accelerating electrons through 100 kV.
   (b) A photon in a microwave oven, wavelength 100 mm.
   (c) An ultraviolet photon causing sunburn, wavelength 300 nm.
   (d) A gamma ray of energy 0.5 MeV from the annihilation of an electron and a positron.

2. In the early universe the temperature exceeded 10^{15} K.
   (a) Calculate the energy kT at this temperature.
   (b) Show that this energy is similar to the kinetic energy of a 1 mg mosquito flying at 0.2 m s^{-1}.
   (c) Show using the relation $E = mc^2$ that if this energy appeared as the mass of a particle, its mass would be approximately equal to that of 100 protons.

3. Many living substances are held together by hydrogen bonds, with bond energy ε of the order 20 kJ mol^{-1}.
   (a) Calculate this energy in joule per particle.
   (b) Show that at 300 K the ratio ε/kT is about 8.
   (c) Find the temperature at which $\varepsilon = kT$. Comment.

4. The energy to break bonds in a ceramic may be of the order 1 MJ mol^{-1}.
   (a) Explain why ceramics are very stable, even in a furnace at 6000 K.
   (b) At what temperature might a ceramic start to melt or evaporate?

5. If the atmosphere had a constant temperature, the pressure p at height h above the surface of the Earth would be given by $p/p_0 = \exp(-mgh/kT)$, where m is the mass of a molecule, g is the gravitational field and p_0 is the pressure at the surface.
   (a) For nitrogen molecules, relative molecular mass 28, estimate the height at which the pressure is $\frac{1}{10}$ that at sea level, assuming a temperature of 300 K.
   (b) Repeat the calculation in (a) for a temperature of 200 K. Does the height increase or decrease? Explain why.
   (c) Repeat the calculation in (a) for hydrogen molecules, relative molecular mass 2. Comment on the result.
   (d) Repeat the calculation in (a) for nitrogen on the Moon, where the gravitational field is $\frac{1}{6}$ that on Earth. Comment.
   (e) Write a few sentences about the possible atmosphere of the largest planet, Jupiter.

6. When metals are heated, electrons boil off from the surface of the metal. This is called thermionic emission and is the source of electrons in X-ray tubes. It takes an energy ε of the order of 1 eV to remove an electron from a metal. That energy has to come from the random thermal agitation of particles in the hot metal.
   (a) Make up an argument to suggest that the number of electrons emitted per second might be proportional to the Boltzmann factor $\exp(-\varepsilon/kT)$.
   (b) Suggest why the largest current that can be obtained by collecting these electrons may be proportional to $\exp(-\varepsilon/kT)$.
   (c) If the current I is given by $I = A \exp(-\varepsilon/kT)$, where A is a constant, show that for currents I_1 and I_2 at temperatures T_1 and T_2: $\ln(I_2/I_1) = (\varepsilon/k)(1/T_1 - 1/T_2)$.
   (d) For platinum, an experiment shows that the largest current doubles if the temperature is raised from 1480 K to 1520 K. Estimate the energy ε for platinum, in electron volts.

7. Biological enzymes speed up biological reactions by reducing the energy required per particle for the reaction to take place. They are also very specific, altering the energy only for a particular pair of molecules. Suppose that the rate of reaction r is given to a good enough approximation by $r = A \exp(-\varepsilon/kT)$, where A is a constant.
   (a) Show that if the enzyme reduces the energy from ε_0 to ε, the ratio of the rate of reaction r with the enzyme to the rate r_0 without the enzyme is given by:
   $r/r_0 = \exp[(\varepsilon_0 - \varepsilon)/kT]$.
   (b) Now show that the difference in energy required, produced by the enzyme action, is given by $\varepsilon_0 - \varepsilon = kT \ln(r/r_0)$.
   (c) Calculate the difference in energy in terms of kT, if the enzyme produces a 100-fold increase in the rate of reaction.

15 Electromagnetic machines

Generators, transformers and motors –
these electromagnetic machines are the
workhorses of modern industrial society.
They generate, help to distribute and put
into use the electrical power that societies
depend on. We will describe:

● how some types of generator, transformer
  and motor are designed
● how they all depend on intertwined
  electric and magnetic circuits
● how the electromotive force (e.m.f.) in
  an electric circuit depends on the rate of
  change of magnetic flux through it

15.1 An electromagnetic world

The wired world now means those parts of the
world connected to global communications. But
the world was first wired when industrialised
countries built networks to distribute electrical
power to factories and homes, starting in the
19th century. The world today is still divided into
electrical power "haves" and "have-nots".

Think back to the last time you experienced
a power cut. Everything goes dark. The heating
stops working. Your computer is dead. You can't
boil an electric kettle for a cup of tea. You can't
read or work. Then the power comes back on and
your world becomes normal again. But remember:
it's only normal to you. People in places without
electricity live differently, as did everybody in the
world for most of human history. Today you can live
fully illuminated 24 hours a day, air-conditioned
and centrally heated. Leisure time is extended but
people can be required to work night shifts. So lives
have changed, and not always for the better.

Casually switching on a kettle, it's hard to
remember the scale of the demand for electrical

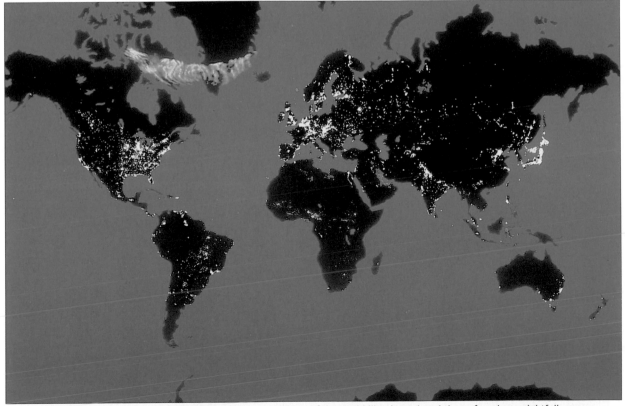

This composite image of the Earth at night shows those regions of the world illuminated and those for whom nightfall means
darkness. Colour-codes: city lights (yellow), oil production flares (red), burning vegetation (purple) and aurora borealis (white).

Electric motors come in a vast array of sizes. Left: the rotor of a very large electric motor. Right: a miniature electric motor.

Facts and figures about power supplies

- A strong, well fed person can keep up a rate of manual labour of around 100 W. So one electric kettle represents the output of 20 workers. A small 100 MW generator is equivalent to a million workers.
- A 1000 MW power station (efficiency 33%) eats up about 300 tonnes of fossil fuel (coal, oil or gas) every hour, which is a truckload every few minutes.
- Power stations are huge buildings. Visit the Tate Modern in London, built in an old power station, to see its vast cathedral-like interior.
- Power station generators are huge machines. To transport them by road requires trucks with many wheels and a police escort to keep other traffic off the road.
- The British national grid distributes power through more than 10 000 km of high-voltage lines, much of it operating at 400 kV or higher.

power, that millions of other people can do the same thing at the same moment.

Electric motors make fast, clean travel possible. Large electric induction motors drive London Underground trains; they also power high-speed locomotives such as the French TGV and the Japanese Bullet train. Your home contains small electric motors of many kinds, each designed specifically for its purpose. Think of some: vacuum cleaner, food mixer, refrigerator, electric drill, washing machine. Think of others: electric shaver, electric toothbrush, electric clock.

The electromagnetic century begins

In the candle-lit world of the year 1800, Italian physicist Alessandro Volta made the first battery – the so-called voltaic pile. For the first time, continuous electric currents of the order of amperes were available to experiment with. Until then, "electricity" had meant only static electricity – large potential differences but little current. The currents from Volta's new source were large enough for their magnetic effect to be noticed. Within thirty years or so, the main laws of electromagnetism had been established by Hans Christian Oersted, Joseph Henry, André Marie Ampère and Michael Faraday.

First came the discovery that electric currents produce magnetic fields. Joseph Henry then showed how to make electromagnets by winding

Magnetic field patterns revealed by iron filings

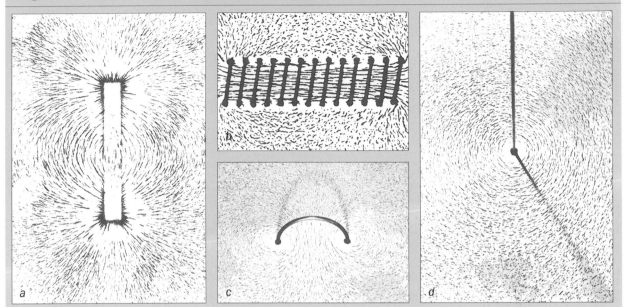

(a) A bar magnet; (b) current in a long coil (solenoid); (c) current in a short coil (wire loop); (d) current in single wire.

A trail of transformers

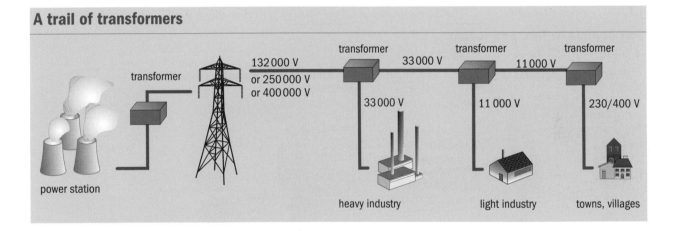

turns of wire round an iron core. Henry found that the magnetic field was stronger for a given current if he used more turns. His electromagnets were also better if their iron cores were as short and thick as possible. Soon, he had an electromagnet able to lift a 1 tonne mass. Furthermore, Henry's electromagnets could be turned on and off by switching the current on and off. He saw how this made electric telegraphy a possibility and helped Samuel Morse to develop the idea.

Most importantly, he had understood a fundamental principle about the production of magnetic flux using electric currents. It was that the amount of magnetic flux increases with the number of current-turns producing it – that is, turns matter as much as current.

Electromagnets, now used in applications as diverse as medical imaging and the scrap yard, began the era of electromagnetic machines. Looking back, the crucial discovery was that electric currents and magnetic fields are intertwined. The current in coils loops round the magnetic field; the magnetic field loops round the electric current. They are like a pair of Olympic rings joined together. So as well as electric circuits wrapped round iron cores, engineers were led to think about magnetic circuits wrapped round electric circuits. They imagined the situation to be as if magnetic flux "flowed" around these magnetic circuits (p138).

Key summary: electric circuits and magnetic flux

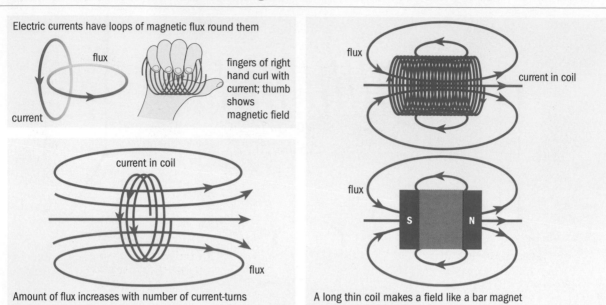

Electric currents have loops of magnetic flux round them

flux

current

fingers of right hand curl with current; thumb shows magnetic field

flux

current in coil

current in coil

flux

flux

S   N

Amount of flux increases with number of current-turns

A long thin coil makes a field like a bar magnet

Flux loops round current. Current loops round flux. Current-turns drive flux.

Key summary: flux and forces

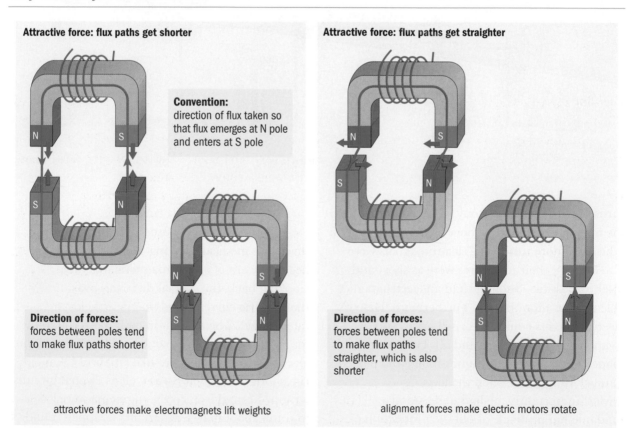

Attractive force: flux paths get shorter

N   S

S   N

Convention:
direction of flux taken so that flux emerges at N pole and enters at S pole

N   S

S   N

Direction of forces:
forces between poles tend to make flux paths shorter

attractive forces make electromagnets lift weights

Attractive force: flux paths get straighter

N   S

S   N

N   S

S   N

Direction of forces:
forces between poles tend to make flux paths straighter, which is also shorter

alignment forces make electric motors rotate

Forces between poles tend to make flux paths shorter and straighter. The flux behaves like an elastic string.

Key summary: Faraday's law of induction

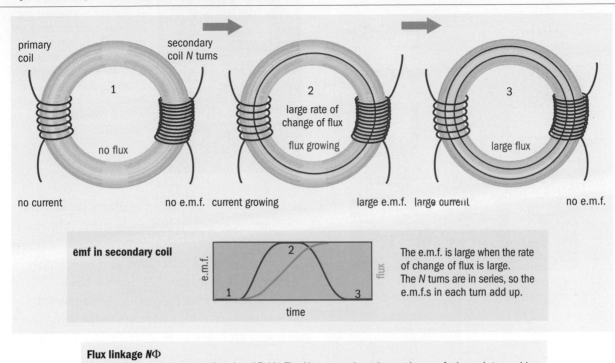

Flux linkage $N\Phi$
The e.m.f. per turn is proportional to $d\Phi/dt$. The N turns are in series so the e.m.f.s in each turn add up.
The e.m.f. across the coil is proportional to $Nd\Phi/dt$. The quantity $N\Phi$ is called the flux linkage.

Faraday's law: e.m.f. is proportional to the rate of change of flux linkage $\varepsilon \propto Nd\Phi/dt$
Lenz's law: the induced e.m.f. opposes the change of flux producing it

Faraday makes electricity from magnetism

Michael Faraday (1791–1867) started as a poor boy apprenticed to a bookbinder, and learned by reading the books he had to bind, especially the *Encyclopaedia Britannica*. After presenting Sir Humphry Davy with bound handwritten notes of Davy's lectures at the Royal Institution, Faraday was appointed to a post there. Starting as a bottle washer, he rose to become its head.

Faraday had a powerful faith that somehow all the forces of nature must be interdependent. He belonged to a tiny fundamentalist religious sect, the Sandemanians, who believed that religious truth was to be found in a simple, honest, consensual reading of the Bible. The Sandemanians saw the working of the whole physical world as an expression of the powers of God. So Faraday saw science as "reading the book of nature", and as a way to understand its unity. Throughout his life he sought unifying connections between light, electricity, magnetism and gravity.

Over almost a decade, Faraday tried experiments in "the hope of obtaining electricity from ordinary magnetism". In 1831 he got measurable electric currents from magnetic fields. Joseph Henry found the same at almost the same time.

Faraday wound coils of many turns on an iron ring. His wire wasn't insulated so he used twine to keep the turns apart and put calico fabric between the layers. One coil was connected to a battery, the other was joined into a complete circuit by a wire passing over a magnetic needle some feet from the ring. He found a remarkable effect. Whenever the battery was connected or disconnected, the magnetic needle deflected momentarily, showing a brief flow of current in the second coil. A steady flow of current had no effect, but a changing current did. Faraday made the modest note that this was "very satisfactory".

This experiment was the key to the connection that Faraday had searched for all that time. Steady magnetic flux in the iron did not produce an

Key summary: measuring changes of flux

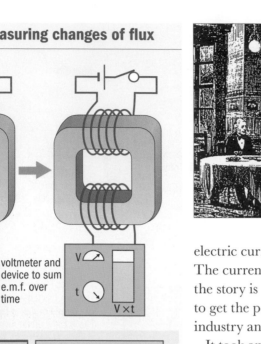

voltmeter and device to sum e.m.f. over time

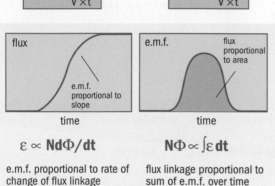

$\varepsilon \propto Nd\Phi/dt$

e.m.f. proportional to rate of change of flux linkage

$N\Phi \propto \int \varepsilon \, dt$

flux linkage proportional to sum of e.m.f. over time

Define weber as the unit of flux:
1 weber is the magnetic flux that induces an e.m.f. of 1 volt in a circuit of one turn when generated or removed in 1 second

Then e.m.f. is numerically equal to rate of change of flux linkage:

$$\varepsilon = -Nd\Phi/dt$$

The minus sign shows that the direction of e.m.f. opposes the change producing it (Lenz's law)

$\varepsilon = -Nd\Phi/dt$

A page from Faraday's notebook (left) and a piece of his apparatus as exhibited at the Royal Institution.

William Armstrong, the Victorian industrialist and entrepreneur, invented high-pressure hydraulic machinery and revolutionised the design and manufacture of guns. He lived at Cragside in Northumberland, where he installed the first electric lighting powered by a hydroelectric generator.

electric current but changing magnetic flux did. The current was tiny, but it was there. The rest of the story is one of how to make the current bigger to get the powerful generators that now supply industry and your home 24 hours a day.

It took another half a century before useful generators and motors were available with power on a scale to begin to rival steam. As a result, the century that had begun with the battery ended with the beginning of domestic and industrial electricity use. In the first half of the 20th century electricity spread even more widely. The gas lamps that lit Victorian streets were replaced by electric light; towns got electric trams; later many railways were electrified. Domestic devices such as vacuum cleaners, cookers, refrigerators and washing machines became popular, and there were fewer domestic servants. Other jobs changed: lamp lighters and candle makers were out of work, but electrician became a new trade. Candles became merely decorations for birthday cakes and romantic dinners, or for infrequent power cuts.

Transformers

Electric power distribution got off to a false start. Thomas Edison set up a public supply in New York based on direct current. Faraday's experiment had shown that the e.m.f. induced in a coil depends on the rate of change of magnetic flux through it. So engineers soon realised that it was much better to provide a supply in which the e.m.f. changed all the time. George Westinghouse, learning from the Croatian inventor Nikola Tesla and buying up his patents, set up a successful alternating current supply. This worked because it went with the grain of electromagnetism: an alternating supply

Key summary: how a transformer works

Use alternating current so that current, flux and e.m.f. are changing all the time

Flux is changing in the primary coil

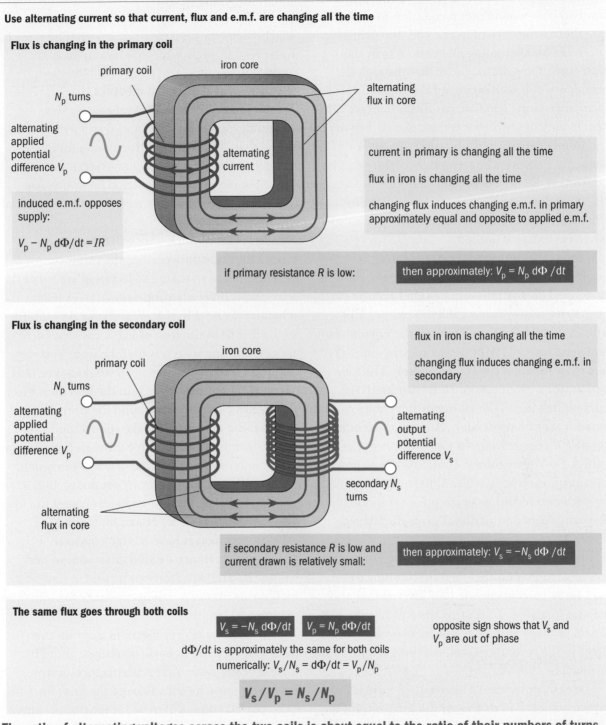

current in primary is changing all the time

flux in iron is changing all the time

changing flux induces changing e.m.f. in primary approximately equal and opposite to applied e.m.f.

induced e.m.f. opposes supply:

$$V_p - N_p \, d\Phi/dt = IR$$

if primary resistance R is low: then approximately: $V_p = N_p \, d\Phi/dt$

Flux is changing in the secondary coil

flux in iron is changing all the time

changing flux induces changing e.m.f. in secondary

if secondary resistance R is low and current drawn is relatively small: then approximately: $V_s = -N_s \, d\Phi/dt$

The same flux goes through both coils

$$V_s = -N_s \, d\Phi/dt \qquad V_p = N_p \, d\Phi/dt$$

opposite sign shows that V_s and V_p are out of phase

$d\Phi/dt$ is approximately the same for both coils
numerically: $V_s/N_s = d\Phi/dt = V_p/N_p$

$$V_s/V_p = N_s/N_p$$

The ratio of alternating voltages across the two coils is about equal to the ratio of their numbers of turns

provides a continually changing e.m.f.

Alternating supplies have a big engineering advantage, which Westinghouse and Tesla exploited. Faraday's iron ring with its two coils could be developed into a device that transforms the e.m.f. of an alternating supply from one value to another. The transformer is simple and efficient. Power can be generated at an alternating e.m.f. that suits the generator design, transformed up to a higher e.m.f. for transmission, and then transformed back down again to a safe value at the supply.

Like Faraday's apparatus, a modern transformer has two coils wound over a common iron core. An alternating current in one coil, the primary, produces an alternating magnetic flux in the iron core. Thus the magnetic flux through the secondary coil is changing all the time and there is an alternating e.m.f. across this other coil. The real beauty of the device is that, merely by varying the numbers of turns in the two coils, the ratio of the e.m.f.s across primary and secondary can be controlled. By design, the alternating e.m.f. across the secondary can be made larger or smaller than the alternating e.m.f. across the primary to whatever extent is needed.

If the transformer steps up the e.m.f., conservation of energy requires that the current is stepped down. This reduces energy losses in electricity transmission cables. Voltages V of 400 kV or more are used so that the current I to transmit say 10 MW of power may be only 25 A, less than that in a car starter motor. The idea is that, for a given power IV transmitted, the current flowing in the transmission cables can be small if the voltage is high. A small current means small I^2R energy losses in the resistance R of the cables. So Westinghouse could distribute power over large distances, while Edison could supply only a small region near his power station. The way was opened to national grids distributing power all over a country.

A transformer can also be used the other way round, stepping up the current to produce the very large currents needed for arc welding and electric furnaces. To do this the e.m.f. is stepped down, perhaps with only one or two very thick turns in the secondary coil, carrying a current of as much as 1000 A.

The early designers of alternating-current power systems had to choose the frequency of the alternating supply. In the USA they chose 60 Hz, in Europe 50 Hz. The rule is the faster the better. This is because of Faraday's law that the induced e.m.f. is proportional to the rate of change of flux. A higher frequency means a faster rate of change of flux for a given maximum flux, but that frequency has to be produced by a spinning generator. If the generator produces one cycle of

alternating e.m.f. per revolution it must turn 50 (or, in the USA, 60) times a second. That is 3000 (or 3600) revolutions a minute – pretty fast for a machine maybe a metre in diameter. The rotor has to be very strongly built so as not to fly apart.

Electric and magnetic circuits

Starting with Joseph Henry, the engineers designing electromagnets and transformers soon found that the way to get the maximum magnetic flux for a given current in a given coil was to make the iron core as short and fat as possible. They noticed an analogy with the conductance of electric circuits.

A short coil of thick wire will have a high electrical conductance, since the conductance G is proportional to the cross-sectional area A of the wire and is inversely proportional to its length L (chapters 2 and 4 in *Advancing Physics AS*). Think of the flux Φ in the iron as like a current, even though nothing travels round the iron. You can define a quantity analogous to conductance that says how much flux there is in the iron for a given number of current-turns round the iron. This is called the permeance. Like conductance, the permeance is larger for large cross sections of iron and smaller for long lengths of iron. A magnetic circuit behaves just as if iron "conducts" flux in closed paths, like circuits conduct current in closed paths. In both cases short and fat is good.

Good conductors have a large conductivity σ – the conductance for unit cross section and length. By analogy, iron can be said to have a high permeability μ – the permeance for unit cross section and length.

This analogy is very useful in guiding the design of electromagnetic machines, though it is far from perfect. The permeance of iron is not the same for all values of the flux, for example. Another difference is that where there is no electrical conductor there is essentially no current. But where there is no iron there is still some flux, in the air (or vacuum) around a coil. Empty space has some permeability; it "conducts" some magnetic flux. The analogy still works for engineering purposes because the permeability of iron is hundreds of times greater than that of air or a vacuum.

Key summary: electric and magnetic circuits

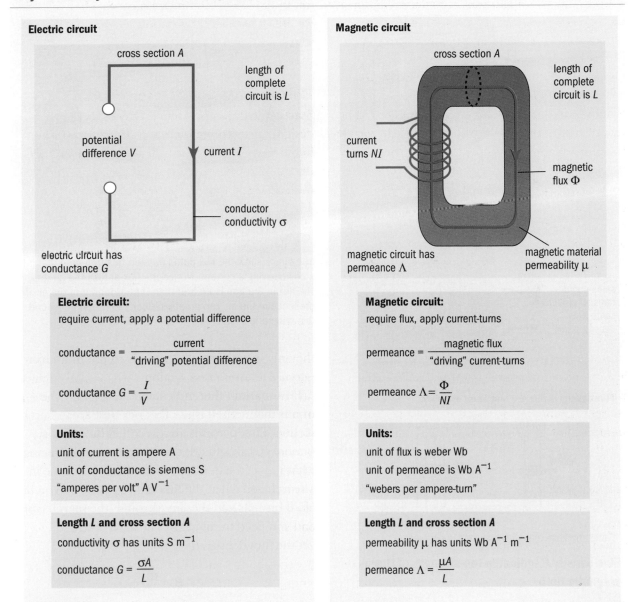

Shorter and thicker is better for both electric and magnetic circuits

Transformer design

How is a good transformer designed? The coils must have a high conductance so that as little energy as possible is wasted in I^2R losses in the coils, which makes them hot. The iron magnetic circuit must have a high permeance to get as much flux in it as possible, so the core should be thick but short. This conflicts with keeping the coils short because they have to go all the way round the thick core. Also, the current in the coils is there to produce the flux, which is greater the

larger the number of turns, again making the conducting coil longer. Compromise is needed. Because copper is a very good conductor of electric current, and iron is only a moderately good conductor of magnetic flux, the advantage lies with a short thick core and longer coils, usually with many turns to help increase the flux. Bigger is also better – the increase in cross section outweighs the increase in length.

It's especially important to avoid breaks (air gaps) in the magnetic circuit. Such a gap, with a

Key summary: flux and flux density

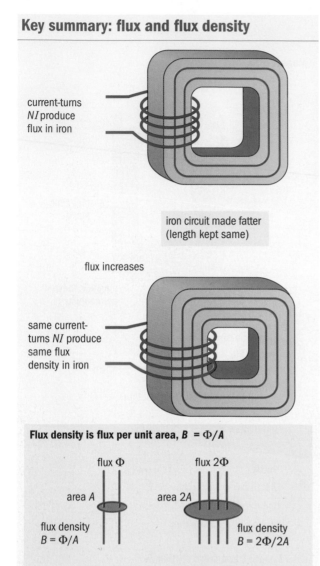

current-turns *NI* produce flux in iron

iron circuit made fatter (length kept same)

flux increases

same current-turns *NI* produce same flux density in iron

Flux density is flux per unit area, $B = \Phi/A$

flux Φ

area A

flux density $B = \Phi/A$

flux 2Φ

area $2A$

flux density $B = 2\Phi/2A$

Fatter iron gives more flux through a larger area

Flux density B indicates the strength of the magnetic field

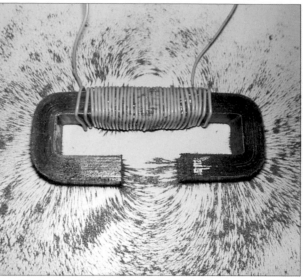

Iron filings show the flux paths outside this iron core that has a coil carrying a current. Notice the flux across the air gap. The flux in the iron is greater than that around it in the air. The presence of the air gap greatly reduces the total flux round this magnetic circuit.

the only way to get more flux is to use more iron – to use a fatter cross section.

The quantity that indicates how strongly the iron is magnetised is the flux per unit area of cross section. Flux per unit area is called flux density, symbol B. Like flux, it can be measured by seeing what induced e.m.f. is produced for what time in a search coil when the flux density is turned on or off. This tells you the flux through the search coil and you need to divide by the area of the coil to get the flux density B:

$$B = \frac{\Phi}{A}$$

The unit of flux density is the tesla: $1\,\text{T} = 1\,\text{Wb}\,\text{m}^{-2}$.

The transformer designer starts by knowing the maximum flux density B that the iron being used can provide. From that, multiplying by the cross-sectional area of the iron, the maximum flux Φ can be found. The number of current-turns NI needed to produce this flux can then be calculated from the permeance Λ of the magnetic circuit, using $\Phi = \Lambda NI$.

In an ideal transformer the power out is equal to the power in. Real transformers do waste some power but come quite close to the ideal case – the power out is often more than 95% of the power in. Then to a good approximation the ratio of the

low permeance in series with the rest of the iron, reduces the permeance of the whole magnetic circuit drastically. In all, magnetic circuits resemble non-insulated electric circuits in salt water. The flux leaks out of the iron circuit, especially at sharp corners. And it does cross air gaps.

Real transformer design has to take into account the extent to which iron can be magnetised. A given piece of iron can't have more than a certain amount of flux through it. Beyond a certain point, more current-turns produce no more flux and the iron is as strongly magnetised as it can be – the magnetisation saturates. Then

Key summary: eddy currents

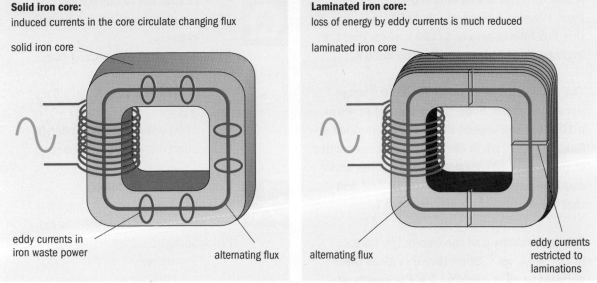

Solid iron core:
induced currents in the core circulate changing flux

solid iron core

eddy currents in iron waste power

alternating flux

Laminated iron core:
loss of energy by eddy currents is much reduced

laminated iron core

alternating flux

eddy currents restricted to laminations

Eddy currents are an example of Lenz's law. They reduce the flux in the core.

voltages across primary and secondary coils is equal to the ratio of their numbers of turns. This means that engineers can afford to transform alternating voltages without much penalty in energy losses.

Suppose that the power out $I_S V_S$ is nearly equal to the power in $I_P V_P$. The engineer can expect the ratios of the voltages and currents in the secondary and the primary to be decided to a good approximation by the ratio of their numbers of turns (p141):

$$V_S/V_P = N_S/N_P$$

$$I_S/I_P = N_P/N_S$$

Real transformers are machines designed to do a job as well as possible. Whether this is to produce large voltages for electricity distribution or large currents for welding, it is important that the power in the first coil is transferred via the linking magnetic circuit with as little loss as possible. A big engineering problem with using iron in the magnetic circuit is that iron conducts electric currents quite well. This means that there are conducting paths in the iron itself round the changing flux in the iron. As a result, wasteful

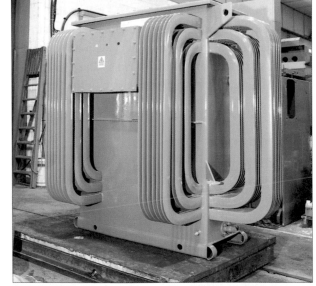

An industrial power transformer with oil-filled cooling tubes.

currents flow in the iron, doing nothing more than making it hot. These are called eddy currents, by analogy with swirls of water in a river. One engineering solution is to cut those conducting paths without cutting the magnetic circuit, by making the core with thin sheets or laminations of iron bolted together but insulated from one another with varnish. For lower powers, the core is sometimes made of sintered grains of magnetic material or of a magnetisable ceramic such as ferrite.

Quick check

1. A 250-turn coil is wound round a closed iron core, carrying a magnetic flux of 4×10^{-4} Wb. If this flux falls to zero in $1/100$ s, show that the average e.m.f. induced in the coil is 10 V.

2. The permeance of an iron core is 10^{-6} weber per ampere-turn. Show that a current of 4 A in a 100-turn coil around the core can produce a flux of 4×10^{-4} Wb in the core. The cross section of the core is 20×20 mm. Show that the cross-sectional area of the core is 4×10^{-4} m^2 and that the flux density in the core is 1 T.

3. The permeability µ of the iron in the core is 10^{-3} Wb A^{-1} m^{-1}. Show that this gives the permeance as in question 2 if the length all round the core is 400 mm, with cross section 20×20 mm.

4. The flux, maximum value 4×10^{-4} Wb, in the core in question 1 now varies sinusoidally at a frequency f of 50 Hz. Show that the maximum rate of change of flux is $2\pi f$ times the maximum flux, and that the maximum e.m.f. in the 250-turn coil is about 30 V.

5. Show that:
   (a) a step-down transformer to reduce a 7.7 kV power supply to a 220 V local supply needs a turns ratio of 35 to 1.
   (b) a step-up transformer to increase the 2.3 kV output of a generator to 138 kV for transmission needs a turns ratio of 1 to 60.

6. Show that a 10-turn secondary used with a transformer having 200 primary turns could provide a current of 300 A from a 240 V, 15 A supply.

Links to the *Advancing Physics* CD-ROM

Practise with these questions:

30S   Short answer *Drawing magnetic circuits*
40S   Short answer *Sketching flux patterns*
60S   Short answer *Changes in flux linkage*
80S   Short answer *Rates of change*
100S Short answer *Transformers*

Try out these activities:

90S   Software-based *Building up a model of electromagnetic induction*
120S Software-based *Modelling transfomers*

Look up these key terms in the A–Z:

Alternating current; electromagnetic induction; electromagnets; Faraday's law of electromagnetic induction; Lenz's law; magnetic flux; transformer

Go further for interest by looking at:

70S   Computer screen *A catalogue of flux patterns*
10T   Text to read *People and electromagnetism: The discoverers*
20T   Text to read *Michael Faraday's vision*

Revise using the revision checklist and:

100   OHT *Electric circuits and magnetic flux*
300   OHT *Flux and forces*
800   OHT *Faraday's law of induction*
900   OHT *Measuring changes of flux*
1100 OHT *How a transformer works*
1300 OHT *Flux and flux density*

15.2 Generators and motors

Picture a summer evening. It's warm and still light so no heating or lighting is needed. Work is over and people are at home watching one of the most popular television programmes. Engineers at Dinorwig power station in north Wales have looked at the television schedule and are ready for what is going to happen. The previous night they took energy from the national grid to pump water up to the reservoir at Marchlyn Mawr. The programme ends; the credits roll. "Fancy a cuppa?", "Time to eat". In millions of homes such words are spoken. Millions of electric kettles and electric ovens go on, all within a few minutes. Demand for power suddenly surges by several thousand megawatts. It's payback time at Dinorwig. The sluice gates are opened and water rushes downhill, spinning the six pumping motors that now become 330 MW generators. Within 12 seconds the Dinorwig generators reach two-thirds of their capacity and are feeding more than 1300 MW into the national grid, helping to cover the sudden surge.

Generators and motors are two sides of the same electromagnetic coin and many machines can function as both. Motion in a magnetic field induces electric current; electric currents in magnetic fields produce motion.

Alternating current generators

The designer of a large power-station generator has to find a way of changing the magnetic flux through a coil to generate an e.m.f. One almost obvious way is to spin a magnet in a gap in an iron core around which a coil is wound. As the magnet spins, flux goes first one way and then the other through the coil – the flux alternates. Such a machine is called an alternator. The dynamo on a bicycle works like this, spinning a permanent magnet as the cycle wheels go round.

A large-scale alternator uses a spinning electromagnet fed with direct current through brushes in contact with slip rings on its rotating shaft. The rotating part of the magnetic circuit is called the rotor. It is usually the DC electromagnet part of the machine. The static part of the machine, which carries the coils from which an alternating e.m.f. feeds power supplies, is called the stator.

The pumped storage facility at Dinorwig – the electric mountain. Turbines, buried in caverns to respect a place of outstanding natural beauty, pump water uphill at times of low electricity demand, storing energy. When demand rises, the water flows downhill through the turbines, driving the electric motors in reverse as generators, feeding power back into the national grid.

Power demand during an eclipse

At 10.30 a.m. on 11 August 1999 electricity demand throughout England and Wales was 35 500 MW. By 11 a.m. this had dropped by 500 MW, and at 11.15 a.m. it was down to 33 150 MW. Minutes later demand had increased by 2000 MW – the equivalent of a million electric kettles being switched on. The Sun had emerged from behind the Moon and people were returning to their normal lives in homes, offices and factories after watching the total eclipse.

Notice that there is an essential engineering problem here. The magnetic circuit must be cut so that part of it can rotate. This introduces an air gap, with low permeance, into the magnetic circuit. As a result, the flux in the iron is less than it might be, so the engineering problem is to shape the rotor and stator so that the air gap is as short as possible and the flux is as large as possible.

In one design, used mainly in smaller generators, the rotor spins between pole pieces shaped on the inside of the stator with coils around them. A disadvantage of this design is that the output e.m.f. is not very sinusoidal

Key summary: transformer into generator

Transformer

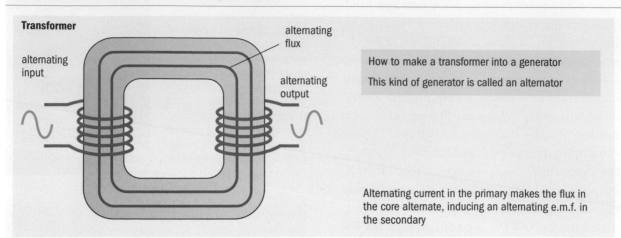

alternating input

alternating flux

alternating output

How to make a transformer into a generator

This kind of generator is called an alternator

Alternating current in the primary makes the flux in the core alternate, inducing an alternating e.m.f. in the secondary

Generator

Cut out a section of the core. Make it an electromagnet with DC current. Spin the electromagnet in the gap in the core.

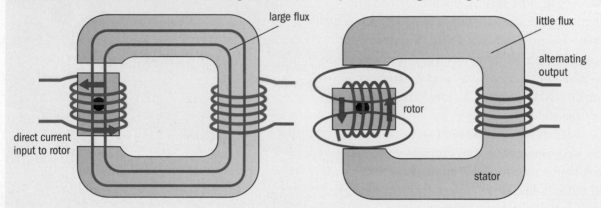

large flux

little flux

alternating output

direct current input to rotor

rotor

stator

As the rotor turns, flux alternates in the stator core, inducing an alternating e.m.f. in the stator winding

Improved magnetic circuit

Wrap the magnetic circuit round the rotor. Shape poles to reduce air gap. Both increase the flux.

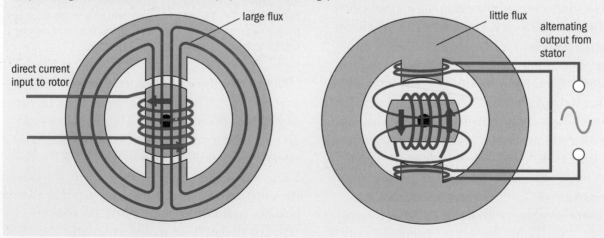

large flux

little flux

alternating output from stator

direct current input to rotor

Alternating flux in the core generates an alternating output. Flux in the core alternates because:
in the transformer, current in the primary coil alternates; and
in the generator, a magnetised section of the core rotates.

Large high-power cylindrical rotor machine

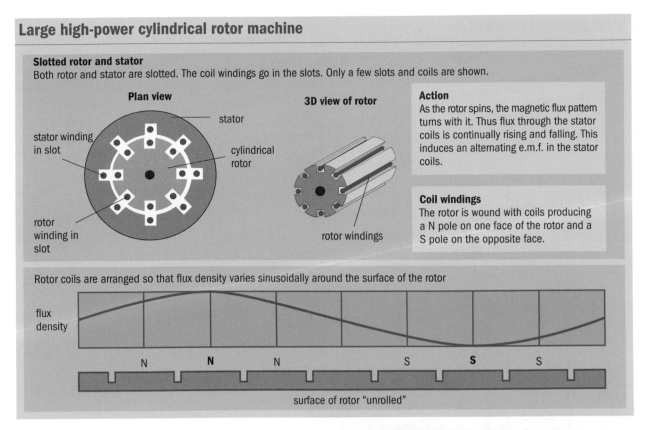

Slotted rotor and stator
Both rotor and stator are slotted. The coil windings go in the slots. Only a few slots and coils are shown.

Plan view

stator

stator winding in slot

cylindrical rotor

rotor winding in slot

3D view of rotor

rotor windings

Action
As the rotor spins, the magnetic flux pattern turns with it. Thus flux through the stator coils is continually rising and falling. This induces an alternating e.m.f. in the stator coils.

Coil windings
The rotor is wound with coils producing a N pole on one face of the rotor and a S pole on the opposite face.

Rotor coils are arranged so that flux density varies sinusoidally around the surface of the rotor

flux density

N   **N**   N   S   **S**   S

surface of rotor "unrolled"

because the flux through the magnetic circuit rises and falls rather sharply as the narrow poles of the rotor pass between the poles of the stator. Another disadvantage is that the magnetic circuit is not very good, but at least it is fairly easy to see how the machine works.

A better design is used in large-scale power-station alternators. The air gap is made small by making the rotor in the shape of a cylinder, spinning inside a cylindrical stator. These are the natural shapes for a rotating machine, but where to put the coils? The clever idea implemented is to cut slots in the surfaces of the rotor and stator and drop the coils into the slots. That way they are protected and the gap between rotor and stator can be as small as engineering tolerances allow.

This design has a further advantage – it makes it easier to get a smooth sinusoidal output, as required for alternating-current power distribution. The rotor coils are arranged around its surface so that the magnetic flux at the surface of the rotor varies sinusoidally. Then, as the rotor turns, the flux passing through the stator coils varies sinusoidally and a good clean alternating output e.m.f. is obtained.

This shows windings being placed in the slots of a stator core.

Three-phase generators

The designers of alternators soon realised that they could get even more out of their machines. In fact, they could build several generators in one. They could add further pairs of coils to the inside surface of the stator so that the rotor turned past first one pair, then the next, and so on. Three pairs of coils is the usual number, arranged at 120° to one another. In principle this gives three alternators in one. But it is better still to connect one end of all the coils together as a common wire and take

Three-phase generator

Stator has three sets of coils arranged at 120° angles around rotor

Each pair is excited in turn as the rotor goes past them

e.m.f. from each pair of coils varies sinusoidally
e.m.f.s differ in phase by 120°

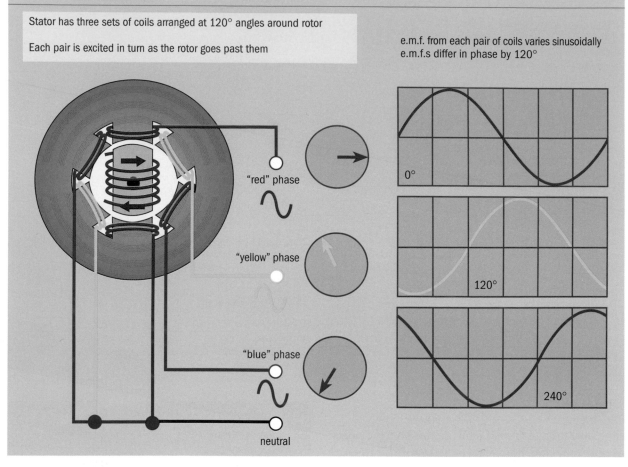

"red" phase

"yellow" phase

"blue" phase

neutral

0°

120°

240°

power from three wires connected to each of the other ends. This method transmits the same power along four wires instead of six.

You might think that the fourth common wire, called the neutral line, would carry three times the current in the others and so have to be extra thick – not so. If the loads on the three power cables can be kept roughly equal the currents with different phases in them add up to approximately zero (see "Three-phase generator", above). The fourth line can be quite thin. There is a saving in wire, and so in cost, for transmission lines. You can spot the thinner fourth cable strung between the pylons of a power line.

Moving field motors

Nikola Tesla, the inventor of three-phase current power systems, was actually given the sack for having the idea. He told his Italian professor that he could make a rotating magnetic field. The professor instructed him to get on with more

serious things but Tesla refused. He went to America and took his ideas to Thomas Edison, who was equally unimpressed. Then transport manufacturer George Westinghouse took them up, and the rest is history.

Tesla's other big idea was to make a rotating magnetic field turn a rotor. Motors that use rotating fields vary in size and power from the smallest to the largest examples. One small low-powered type simply uses a permanent magnet for the rotor. As the poles of the field rotate, the poles of the magnet follow them. This kind of motor is synchronous: the rate of rotation is the same as (or is an integer fraction of) the supply frequency. Similar motors use shaped pieces of soft iron in place of the magnet.

The workhorse motor of industry and transport, the induction motor, is also turned by rotating flux. But now the rotor is an electrical conductor and acts rather like the secondary of a transformer. The rotating field induces currents in the conducting rotor but, instead of supplying electrical power,

Key summary: alternating fields can make rotating fields

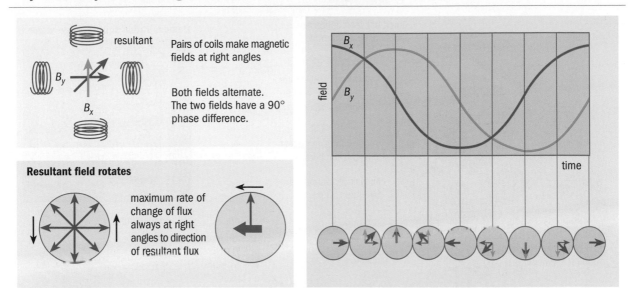

resultant

Pairs of coils make magnetic fields at right angles

B_y
B_x

Both fields alternate. The two fields have a 90° phase difference.

Resultant field rotates

maximum rate of change of flux always at right angles to direction of resultant flux

field

B_x

B_y

time

Two or more alternating fields at different angles with different phases can make a rotating field

Key summary: a rotating field motor

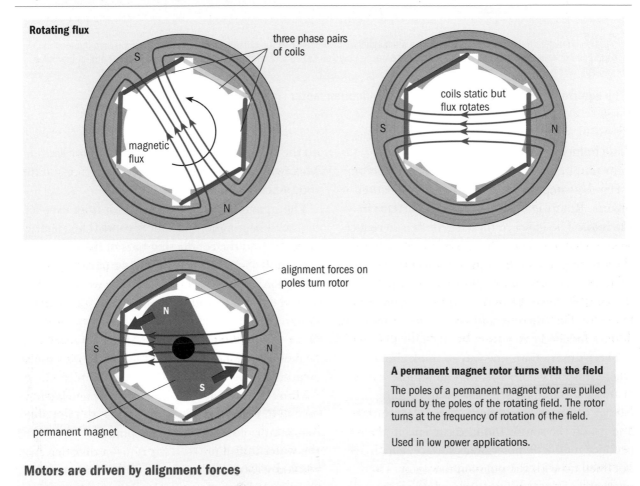

Rotating flux

three phase pairs of coils

S

magnetic flux

N

coils static but flux rotates

S

N

alignment forces on poles turn rotor

N

S

N

S

permanent magnet

A permanent magnet rotor turns with the field

The poles of a permanent magnet rotor are pulled round by the poles of the rotating field. The rotor turns at the frequency of rotation of the field.

Used in low power applications.

Motors are driven by alignment forces

Key summary: the squirrel cage motor

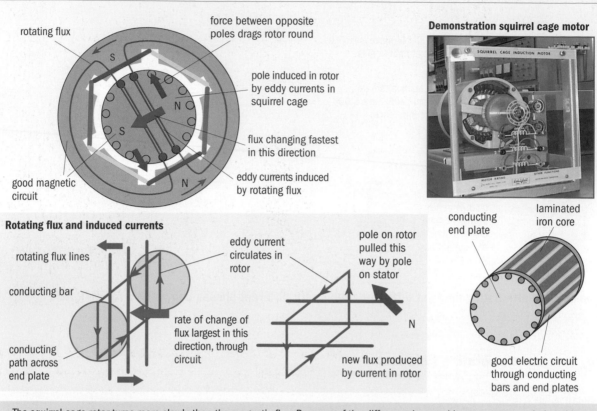

rotating flux

force between opposite poles drags rotor round

pole induced in rotor by eddy currents in squirrel cage

flux changing fastest in this direction

good magnetic circuit

eddy currents induced by rotating flux

Demonstration squirrel cage motor

Rotating flux and induced currents

rotating flux lines

conducting bar

eddy current circulates in rotor

conducting path across end plate

rate of change of flux largest in this direction, through circuit

pole on rotor pulled this way by pole on stator

new flux produced by current in rotor

conducting end plate

laminated iron core

good electric circuit through conducting bars and end plates

The squirrel cage rotor turns more slowly than the magnetic flux. Because of the difference in speed large currents are induced in the good conductors of the rotor. These currents produce poles on the surface of the rotor. The poles on the stator attract the poles on the rotor and drag the rotor round.

The squirrel cage motor is a simple type of induction motor

these induced currents turn the rotor. The induction motor is part transformer, part motor.

Here's a simple way to think about induction motors. Rotating flux induces eddy currents in the rotor. The effect of these currents is to reduce the change producing them (Lenz's law). So the rotor is dragged round, reducing the relative rate of rotation of flux and rotor. The rotor always goes round a bit slower than the flux – if it rotated as fast as the flux there would be no relative rotation, and no induced currents.

There's more than one way to think about where this force comes from. One simple way is to think about magnetic poles. The rotating flux carries a pair of N and S magnetic poles round and round the stator. Induced currents in the rotor produce magnetic poles on its surface, which are not lined up with the poles on the stator. Then magnetic alignment forces try to line up the poles

on the rotor and stator, dragging the rotor round. Meanwhile, the stator poles have moved on, so the rotary action continues.

The squirrel cage induction motor is an excellent design. The magnetic circuit is good. The electric circuit round the conducting bars of the rotor is good. Because the rotor gets its current by induction, there is no need for easily worn brushes to feed current in and out of the rotor. The whole design is very simple and robust. Given a three-phase supply, these motors can deliver a great deal of power. The source of the rotation is simply rotating magnetic flux.

Many similar rotating flux motors have been invented. One example is the simple shaded pole disk motor often used to drive pumps to aerate the water in fish tanks. This is just a disk driven round by rotating flux. It can run silently and reliably 24 hours a day.

Quick check

1. Make an estimate of the rise in demand for electrical power in the UK in the minutes following the end of a programme from the Olympic Games, which many people watch on television.

2. The output coil of an alternating current generator (alternator) is just a coil through which magnetic flux alternates. How can such an alternating flux be produced by a suitable rotor? Compare the alternator with a transformer.

3. A bicycle dynamo has a rotating magnet that induces an alternating e.m.f. in a 300-turn coil on an iron core. Show that, if the dynamo produces a peak e.m.f. of 6 V, the maximum rate of change of flux in the coil is $0.02\,\mathrm{Wb\,s^{-1}}$.

4. The bicycle dynamo in question 3 turns at five revolutions per second and the flux through the coil changes between maximum and near zero every quarter turn. Show that $10^{-3}\,\mathrm{Wb}$ is a reasonable estimate of the maximum magnetic flux in the magnetic circuit.

5. Sketch phasors for two alternating fluxes with a 90° phase difference. If the two fluxes are directed at right angles, show that the resultant flux rotates.

6. A solid cylindrical conducting rotor is placed in a cylindrical stator that produces flux rotating around the axis of the rotor. Explain why currents are induced in the rotor if the rotor is held still. Use Lenz's law to explain why the rotor is dragged around by the rotating flux.

Links to the *Advancing Physics* CD-ROM

Practise with these questions:

140X Explanation–exposition *A bicycle speedometer*

170S Short answer *Graphs of changing flux and e.m.f.*

180S Short answer *Alternating current generators*

200X Explanation–exposition *The induction motor*

220X Explanation–exposition *A variable-speed linkage*

Try out these activities:

190E Experiment *Examining real dynamos and generators*

180S Software-based *Changing flux linkage*

230S Software-based *Making flux rotate*

Look up these key terms in the A–Z:

Electric motor; electromagnetic induction; Faraday's law of electromagnetic induction; generator; Lenz's law

Go further for interest by looking at:

60T Text to read *People and electromagnetism: The inventors and engineers*

180S Computer screen *Large electromagnetic machines*

Revise using the revision checklist and:

1500 OHT *Transformer into generator*

1600 OHT *Large high-power generator*

1700 OHT *Three-phase generator*

2000 OHT *A rotating field motor*

2100 OHT *The squirrel cage motor*

15.3 A question of power

Up to now in this chapter we have described forces in motors in two ways – as due to flux lines shortening or straightening, or to magnetic poles on the surface of iron in the magnetic circuit pulling each other into alignment. These are two ways of saying the same thing. There's another way, which starts from the force on a current-carrying conductor in a magnetic field.

Fields like catapults

Magnetic fields encircle current loops. Add to such a field another uniform field and you get a field shaped like a stretched catapult. If lines of force shorten and straighten, it is easy to see how the force on a current in a conductor arises. But to combine the fields you have to remember that fields are vector quantities and add at each point like vectors.

The result is that the force on a straight length of

Key summary: how a current-carrying wire moves in a magnetic field

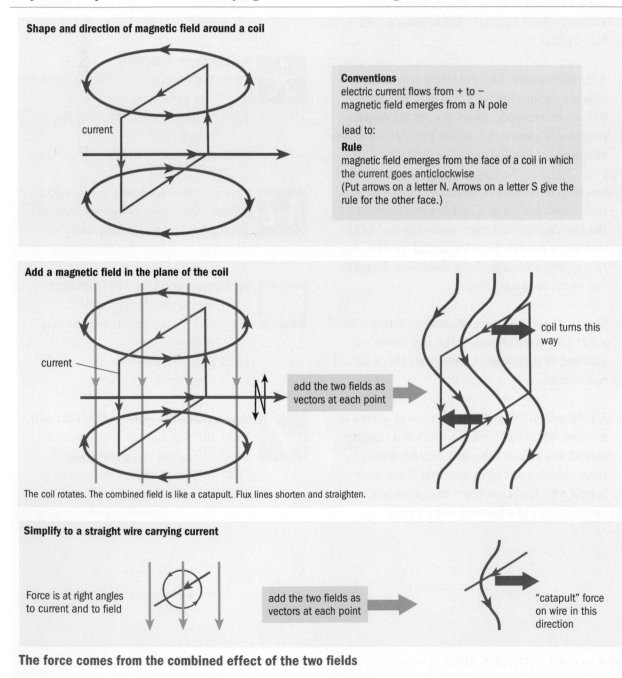

Shape and direction of magnetic field around a coil

current

Conventions
electric current flows from + to −
magnetic field emerges from a N pole

lead to:

Rule
magnetic field emerges from the face of a coil in which the current goes anticlockwise
(Put arrows on a letter N. Arrows on a letter S give the rule for the other face.)

Add a magnetic field in the plane of the coil

current

add the two fields as vectors at each point

coil turns this way

The coil rotates. The combined field is like a catapult. Flux lines shorten and straighten.

Simplify to a straight wire carrying current

Force is at right angles to current and to field

add the two fields as vectors at each point

"catapult" force on wire in this direction

The force comes from the combined effect of the two fields

current-carrying conductor is at right angles both to the conductor and to the field producing the force – a strange kind of force. Like a crab, a wire approached from above looks straight ahead and runs away sideways.

The magnitude of the force F on a wire of length L carrying current I at right angles to a magnetic flux density B is simply written. It is

$$F = ILB$$

We will show you (p157) how this expression can be derived by considering the power input and output of a motor or dynamo. Compare the expression with the simpler force mg on a mass m in a gravitational field g. Now the directions of force and field are more complicated, but the force still has the form "magnitude of thing acted on (IL) multiplied by magnitude of field acting on it (B)".

Flux cutting and flux changing

It's very convenient to imagine magnetic flux running along lines. Then, drawing a flux pattern, you put the lines closest together where the field (the flux density) is largest. In this way the total number of lines going through an area represents the total flux passing through that area. But it's just a picture. There aren't any actual gaps between the lines.

The transformer (p141) shows that when the field changes in strength, the e.m.f. induced in a coil is equal to the rate of change of flux linking the coil. In generators and motors the flux through a coil changes because the flux moves or the coil moves. Picturing lines of flux it's easy to see that the change of flux through a coil is the same as the number of flux lines cut by the coil. As flux lines cross the edge of the coil they go from passing outside the coil to passing through the coil.

Faraday's law can now be put in another way. Instead of

induced e.m.f. is equal in magnitude to the rate of change of flux linked

you can say the same thing another way:

induced e.m.f. is equal in magnitude to the rate of cutting of lines of flux.

Key summary: flux cutting and flux changing

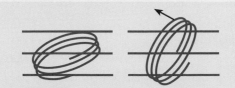

Increasing the flux linking a coil

turn the coil so that more flux goes through it

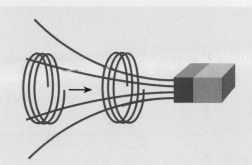

move the coil into a region where the flux density is larger

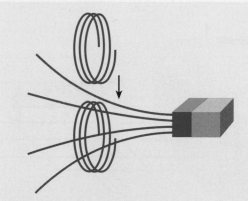

slide the coil so that more flux goes through it

Rotating coil: cutting flux, changing flux linked

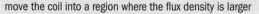

plane of coil parallel to flux
- zero flux linking coil
- maximum rate of cutting flux
- maximum rate of change of flux linking coil

axis

plane of coil perpendicular to flux
- maximum flux linking coil
- zero rate of cutting flux
- zero rate of change of flux linking coil

axis

Cutting flux = change of flux linked
Lines cut = change in lines linking

Key summary: motor and generator

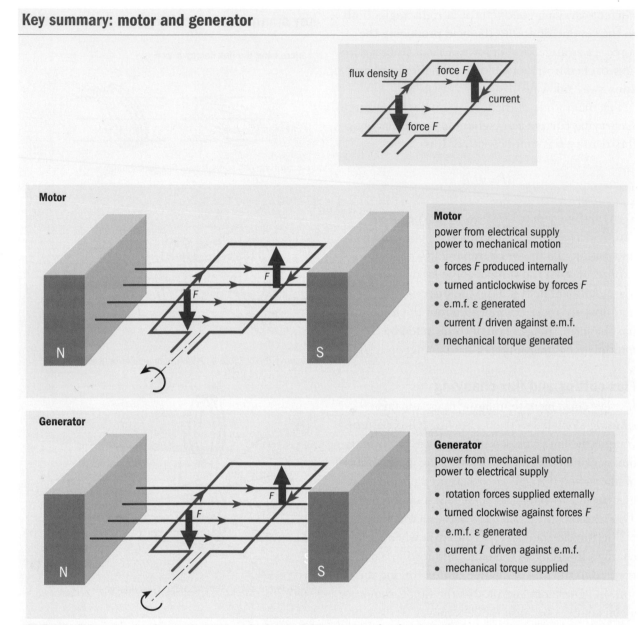

A simple DC motor is also a generator. A simple DC generator is also a motor.

Moving wires cut flux; moving flux cuts wires.

The flux linked is indicated by the product ΦN of flux and turns. Thus Faraday's law can be expressed more generally in the form

$$\varepsilon = -\frac{\mathrm{d}(\Phi N)}{\mathrm{d}t}$$

This means that you can get an e.m.f. by changing the number of turns as well as by changing the magnetic flux. The quantity ΦN is called the **flux linkage**. When number of turns N is constant, as it most often is, the equation can be written as:

$$\varepsilon = -N\frac{\mathrm{d}\Phi}{\mathrm{d}t}$$

as we have done up to now. The essential point is that the e.m.f. per turn is equal to the rate of change of flux through that turn. With turns in series, the e.m.f. per turn adds up.

Where the power comes from

The motors that drive an electric train can also be used as the train's brakes. Instead of having the motors drive the train you let the moving train drive the motors. The motors then act as

Key summary: force on a current-carrying conductor

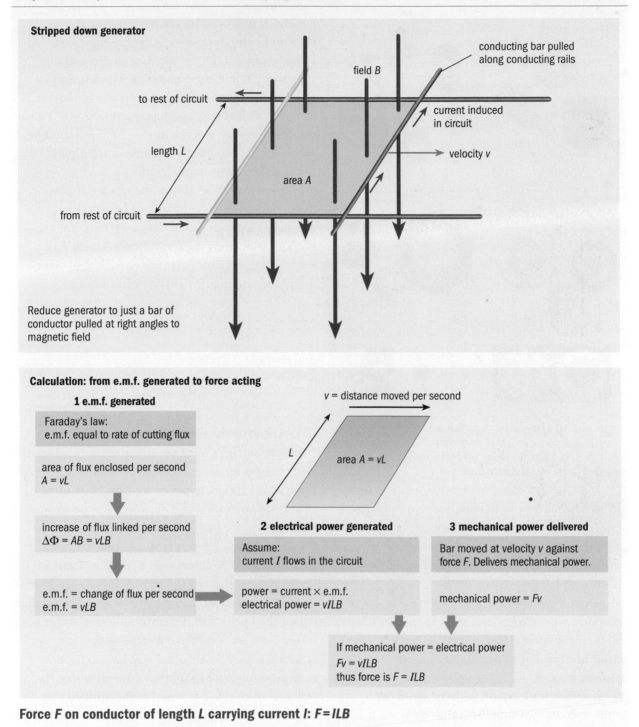

Stripped down generator

conducting bar pulled along conducting rails

field B

to rest of circuit

current induced in circuit

length L

area A

velocity v

from rest of circuit

Reduce generator to just a bar of conductor pulled at right angles to magnetic field

Calculation: from e.m.f. generated to force acting

1 e.m.f. generated

v = distance moved per second

Faraday's law:
e.m.f. equal to rate of cutting flux

area of flux enclosed per second
$A = vL$

L

area $A = vL$

increase of flux linked per second
$\Delta\Phi = AB = vLB$

2 electrical power generated

3 mechanical power delivered

Assume:
current I flows in the circuit

Bar moved at velocity v against force F. Delivers mechanical power.

e.m.f. = change of flux per second
e.m.f. = vLB

power = current × e.m.f.
electrical power = $vILB$

mechanical power = Fv

If mechanical power = electrical power
$Fv = vILB$
thus force is $F = ILB$

Force F on conductor of length L carrying current I: $F = ILB$

generators and power can be taken from them. That power comes from the motion of the train, which slows down. This is called regenerative braking. In a train drawing power from overhead wires, the generated power can be fed back into the grid supply, giving the power stations a helping hand. A less efficient way is to let the current generated heat up resistors connected across the motors, now behaving as dynamos.

Equally, generators can act like motors. When demand on a power-station generator increases, the current drawn gets larger. At the same time the

Towards smallness and precision

A range of miniature electric motors, many of them designed for computer applications such as controlling DVD drives or driving miniature fans. Such low-power precision applications are an important future for electric motors.

turbines have to work harder, supplying the extra power. The current in the generator must produce a turning force on the rotor against which the turbines must act. The dynamo is like a motor that wants to go backwards.

A generator is a motor and a motor is a generator. Crank a generator up to speed by hand and then connect a lamp to it: you can feel the extra force needed to keep it turning when power is drawn from it.

This is best understood with an artificial, stripped-down example. Imagine reducing a generator to a single rod of conductor pulled on conducting rails so that it cuts magnetic flux (see "Key summary: force on a current-carrying conductor" p157). An e.m.f. will be induced. You can work out the value of the e.m.f. from the rate at which the conducting rod cuts flux. Now take power from this simple generator, that is, let a current flow. The power is just the current

multiplied by the generated e.m.f.

As soon as you draw current and power, the generator will need mechanical power to keep the conducting rod moving. There must be a force on the conductor opposite in direction to the movement of the conductor. Work has to be done against such a force to keep the conductor moving. This is how mechanical power is fed into the generator, to emerge as electrical power. The mechanical power is just that force multiplied by the speed of the moving conductor.

Friction and losses apart, the electrical power taken out of the generator must be equal to the mechanical power fed in. By comparing the two you can show that the force F on a length L of a straight conductor carrying current I, at right angles to a magnetic field B, must be

$$F = ILB$$

and that the e.m.f. generated is

$$\varepsilon = vLB$$

if the speed of the conductor is v.

To repeat, there must be such a force, otherwise electrical power generation would come for free. All generators producing power must be hard to turn.

The argument works just as well the other way round. A motor puts out mechanical power via the force ILB. That power must come from an electrical supply. To provide it, the supply must "push the current uphill" against a voltage. That voltage is just the induced e.m.f. in the motor, often called the back e.m.f. Power is delivered to the motor because it acts like a generator.

A motor that is unloaded and has no friction in its bearings will spin up to a speed where the induced back e.m.f. is equal to the applied e.m.f. Such an ideal motor draws no power from the supply and does no work. Negligible current flows. To deliver power the motor must slow down. This reduces the back e.m.f. and so current flows. The turning force produced is proportional to the current in the rotor. It is also proportional to the flux density in which the rotor turns, which depends on the stator current. If rotor and stator are connected in series, both rotor current and flux density are large at low speeds giving high torque.

The design challenge of a motor is to achieve the correct balance of speed of rotation and torque for a particular application. Electric drills and trains need high starting-turning force at low speeds. Fans need rather constant turning force and speed.

The same design features – high conductance electric circuits and high permeance magnetic circuits linked closely together – apply as they do in transformers and generators. A good design requires small air gaps and much iron, together with clever shaping of the electric and magnetic circuits.

Motors today and tomorrow

The motors in your vacuum cleaner or electric drill are probably the most powerful motors you'll have handled yourself. Today, electrical engineers are turning their efforts in a new direction, away from power and towards smallness and precision. They are thinking about how to make motors, often now called actuators, that perform such delicate precision tasks as:

- positioning a read-out head at exactly the right place over a DVD or a computer hard disk;
- spinning a DVD drive at exactly the required speed, which changes as the read-out position changes;
- moving and positioning a robot arm so that a car component is put exactly in place;
- "stepping" from one position to another, for example, to drive a counter or to turn the hands of a watch.

The big powerful machines will go on doing the jobs for which they were so well designed. But these new kinds of precision use make control more important than power. Often the required control electronics will be built into the motor itself. Instead of a mechanical commutator to switch the currents, electronic switching is now often used. This requires sensors to detect the position of the rotor, telling the electronics what to do. So the future looks like small smart motors. They work on a variety of design principles. In some the rotor moves in a straight line instead of turning. In others the job is to move something to an exact position or angle, not to move it continuously. But all of them use the same underlying principles, notably of linked current and flux acting on one another.

Key summary: flux and current intertwined

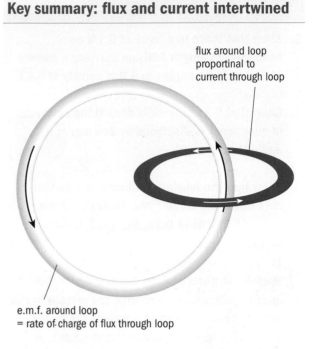

flux around loop proportial to current through loop

e.m.f. around loop = rate of charge of flux through loop

Flux encircles current: current encircles flux

Linked electric and magnetic circuits

Electric and magnetic circuits are necessarily linked. You can't have an electric current in a circuit without having magnetic flux around it, with the flux proportional to the number of current turns NI. And you can't have magnetic flux changing without having an e.m.f. in any circuit surrounding it, with the e.m.f.

$$\varepsilon = -\frac{d\Phi N}{dt}$$

This linkage between electric and magnetic effects goes much deeper than it may seem. It is the reason why, in an electromagnetic wave, a changing magnetic field produces an electric field and the changing electric field produces a magnetic field, so that the two linked fields propagate together as one electromagnetic field. At the very deepest level the fields really do become one, with magnetic fields being just relativistic effects of electric fields of charges moving relative to one another. But that really is another story for another time, showing that electromagnetism is of deep fundamental importance as well as of huge practical significance.

Quick check

1. Show that there is a force of 0.1 N on a conductor of length 100 mm carrying a current of 10 A at right angles to a flux density of 0.1 T.

2. Show that 0.1 J of work is done if the conductor in question 1 is displaced by 100 mm at right angles to its length.

3. Show that the moving conductor in question 2 cuts flux equal to 10^{-3} Wb. Show that if the motion happens in 0.1 s, the e.m.f. induced across the wire will be 10 mV.

4. Sketch the parts of a DC electric motor that uses a commutator to reverse the current in the rotor. Why is the current drawn by the motor much less than V/R, where V is the e.m.f. of the supply and R is the resistance of the motor coils?

5. Sketch the magnetic flux near a current-carrying wire in a uniform magnetic field. How does the shape of the field suggest the way the wire might move?

6. Here are three ways to induce a brief pulse of e.m.f. in a coil, using a uniform magnetic field: (i) remove the coil quickly from the field, (ii) turn the coil through a right angle, (iii) crush the coil to reduce its area. Explain in each case how there is a rate of change of flux through the coil.

7. Give a reason why the turbines driving the rotor of a generator have to deliver more power to the rotor when electrical power is drawn from the stator. How is this related to Lenz's law?

Go to the *Advancing Physics* CD-ROM

Practise with these questions:

Q230S   Short answer *Sketching flux patterns and predicting forces*

Q240S   Short answer *Forces and currents*

Q250S   Short answer *Thinking about the design of a simple DC motor*

Q260S   Short answer *e.m.f. in an airliner*

Try out this activity:

A330P   Presentation *Using an electric drill*

Look up these key terms in the A–Z:

Electric motor; force on a current-carrying conductor; force on a moving charge; Lenz's law; magnetic flux

Go further for interest by looking at:

D290S   Computer screen *A catalogue of motors*

R80T   Text to read *A wide variety of motors*

R140T   Text to read *Relativity drives trains*

Revise using the revision checklist and:

D2300   OHT *How a current-carrying wire moves in a magnetic field*

D2400   OHT *Changing the flux linked to a coil*

D2500   OHT *Flux cutting and flux changing*

D2600   OHT *Force on a current-carrying conductor*

D2700   OHT *Motors and generators*

Summary check-up

Magnetic flux and magnetic circuits ✓

- Faraday's law: the induced e.m.f. is proportional to the rate of change of flux linked:

$$\varepsilon = -\frac{d(\Phi N)}{dt}$$

- Flux density $B = $ flux Φ/area A
- Electromagnetic machines all have at least one electric and one magnetic circuit, linked round one another. The flux in the magnetic circuit is produced by current-turns in NI round it.
- The electric circuit needs a large conductance, increased by increasing its cross section and decreasing its length. Analogously, the magnetic circuit needs a large permeance. This is also increased by increasing the cross section and decreasing the length of the magnetic circuit.

Transformers ✓

- In a transformer, alternating current in the primary coil produces alternating flux in an iron core, which induces an alternating e.m.f. in the secondary coil
- In an ideal transformer the ratio of secondary to primary e.m.f. is equal to the turns ratio $V_S/V_P = N_S/N_P$
- In an ideal transformer the ratio of secondary and primary currents is given by the ratio of numbers of turns $I_S/I_P = N_P/N_S$

Generators ✓

- An induced e.m.f. can be produced in an alternator by moving flux linking a conductor
- An induced e.m.f. can be produced by moving a conductor to cut magnetic flux
- A three-phase system is used for large-scale power generation and distribution

Motors ✓

- Induction motors work by rotating flux-inducing currents in, and thus producing forces on, a conducting rotor
- The force on a current-carrying conductor at right angles to uniform magnetic flux density B is given by $F = ILB$
- A motor is also a generator. A back e.m.f. opposing the supply is generated as the motor turns.

Questions

1. Compare a transformer and an alternator. For each, using diagrams as required:

(a) show the magnetic circuits involved and describe their similarities and differences, especially the presence or absence of air gaps.

(b) state in each case how the induced e.m.f. arises.

(c) for each case state in which coils the current is alternating and in which it is a direct current. Justify your answers.

(d) explain in each case the source of the input power. How do these sources differ?

2. A power transformer has a closed iron core, made of iron of permeability $\mu = 10^{-3}\,\text{Wb}\,\text{A}^{-1}\,\text{m}^{-1}$. The maximum possible magnetic flux density in the iron is 1 T. The core has a square cross section 100×100 mm, and an average length of magnetic circuit equal to 800 mm, as shown.

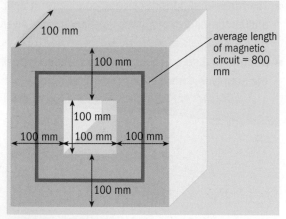

(a) Show that the flux in the iron is 10^{-2} Wb if the flux density is at its maximum value of 1 T.

(b) Show that the permeance of the iron core is $12.5 \times 10^{-6}\,\text{Wb}\,\text{A}^{-1}$.

(c) Show that 10 ampere-turns are needed to create this flux in the core.

(d) If the flux in the core alternates at 50 Hz with the maximum value given in (a), show that the maximum rate of change of flux in the core is approximately $3\,\text{Wb}\,\text{s}^{-1}$.

(e) With the maximum rate of change of flux as in (d), what is the e.m.f. per turn in a coil surrounding the core?

(f) How many turns does the secondary coil need if the output across it is to be 210 V?

(g) Show that the primary coil needs 1000 turns if it is driven by a 3 kV supply.

3. Choose one kind of electric motor and draw a diagram showing the stator and rotor, indicating where coils are wound on one or both of them. For the motor you have chosen:

(a) indicate its magnetic circuit on the diagram.

(b) explain one way that the magnetic circuit could be improved.

(c) state for one position of the rotor how the forces that turn the rotor arise.

(d) explain how the rotor is kept turning continually in this motor.

4. The following questions involve very rough estimates of the rate of change of flux and the e.m.f. involved in a power-station generator. Suppose the generator is a two pole alternator rotating at 50 revolutions per second, generating a 50 Hz supply with a maximum e.m.f. (in one phase) in the stator of 10 kV. Estimate the rotor radius as 1 m and its length as 5 m, as shown. Suppose that flux emerges from an N pole area of about 5 m² of the rotor surface, as suggested in the diagram.

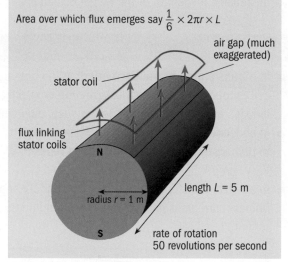

(a) If the iron of the rotor is magnetised to a flux density of 0.1 T, what flux emerges across the N pole area?

(b) Imagine that this flux passes through a winding of the same area on the stator. Estimate the time during which the flux through the stator winding rises from near zero to the maximum value estimated in (a), as the rotor turns.

(c) Estimate the rate of change of flux per turn of the stator winding.

(d) Estimate the number of turns needed in the stator winding to give a varying e.m.f. of the order of 10 kV across the stator.

16 Charge and field

Your dentist has a particle accelerator, so too does your local hospital. Particle physicists use huge accelerators to investigate the ultimate structure of matter. In this chapter we will look at several particle accelerators and explain:

- how charged particles are accelerated by a potential difference (p.d.)
- the relativistic limit to the speed, but not to the energy, of accelerated particles
- how charged particles can be deflected by electric and magnetic fields
- how electric fields can be represented by equipotentials and by field lines
- the ideas of electric field strength and potential

16.1 Accelerating towards the ultimate speed

Did you know that in your dentist's X-ray machine electrons are accelerated to half the speed of light? When the dentist X-rays one of your teeth, the X-rays come from high-speed energetic electrons crashing into a metal target. If you could look inside the X-ray machine, you would see a vacuum tube containing a source of electrons facing a metal target angled to emit X-rays through a window. Most important of all is the electrical circuitry that provides a large p.d., typically 60–70 kV, between source and target. It is this p.d. that gives the accelerated electrons their energy.

A simple linear accelerator

The electrons in an X-ray tube are accelerated by being sent down a potential energy hill. The steeper the hill, the more rapidly the charges accelerate and the higher the hill, the more kinetic energy they gain. The potential energy hill is created by

X-rays for dentistry

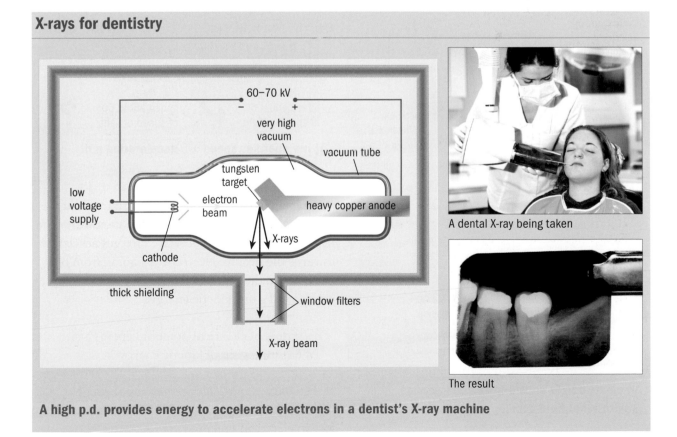

A dental X-ray being taken

The result

A high p.d. provides energy to accelerate electrons in a dentist's X-ray machine

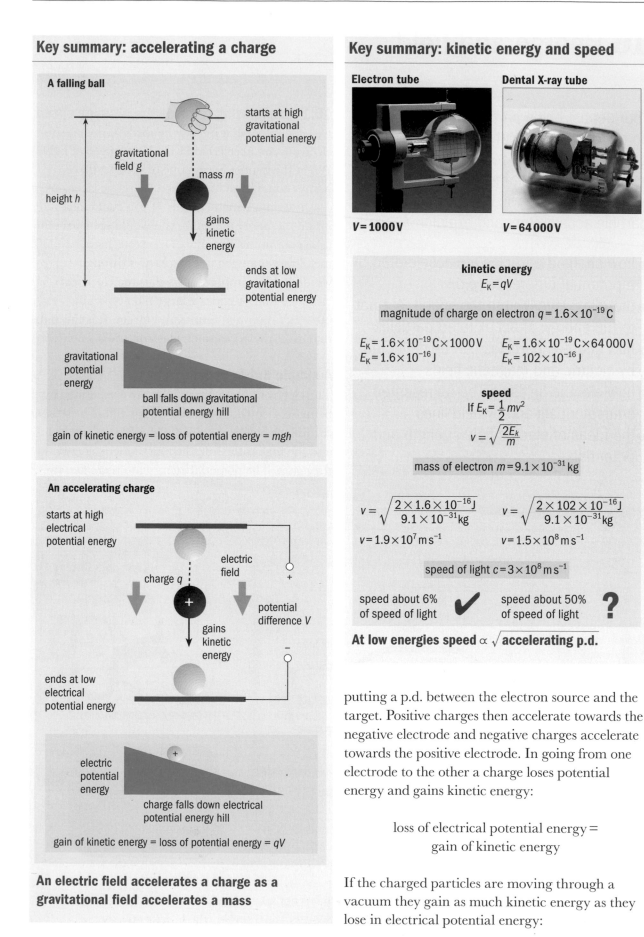

Key summary: accelerating a charge

A falling ball

starts at high gravitational potential energy

gravitational field g

mass m

height h

gains kinetic energy

ends at low gravitational potential energy

gravitational potential energy

ball falls down gravitational potential energy hill

gain of kinetic energy = loss of potential energy = mgh

An accelerating charge

starts at high electrical potential energy

charge q

electric field

potential difference V

gains kinetic energy

ends at low electrical potential energy

electric potential energy

charge falls down electrical potential energy hill

gain of kinetic energy = loss of potential energy = qV

An electric field accelerates a charge as a gravitational field accelerates a mass

Key summary: kinetic energy and speed

Electron tube

$V = 1000\,V$

Dental X-ray tube

$V = 64\,000\,V$

kinetic energy
$$E_K = qV$$

magnitude of charge on electron $q = 1.6 \times 10^{-19}\,C$

$E_K = 1.6 \times 10^{-19}\,C \times 1000\,V$
$E_K = 1.6 \times 10^{-16}\,J$

$E_K = 1.6 \times 10^{-19}\,C \times 64\,000\,V$
$E_K = 102 \times 10^{-16}\,J$

speed
If $E_K = \frac{1}{2}mv^2$
$$v = \sqrt{\frac{2E_k}{m}}$$

mass of electron $m = 9.1 \times 10^{-31}\,kg$

$$v = \sqrt{\frac{2 \times 1.6 \times 10^{-16}\,J}{9.1 \times 10^{-31}\,kg}}$$
$v = 1.9 \times 10^{7}\,m\,s^{-1}$

$$v = \sqrt{\frac{2 \times 102 \times 10^{-16}\,J}{9.1 \times 10^{-31}\,kg}}$$
$v = 1.5 \times 10^{8}\,m\,s^{-1}$

speed of light $c = 3 \times 10^8\,m\,s^{-1}$

speed about 6% of speed of light ✔

speed about 50% of speed of light ?

At low energies speed $\propto \sqrt{\text{accelerating p.d.}}$

putting a p.d. between the electron source and the target. Positive charges then accelerate towards the negative electrode and negative charges accelerate towards the positive electrode. In going from one electrode to the other a charge loses potential energy and gains kinetic energy:

loss of electrical potential energy = gain of kinetic energy

If the charged particles are moving through a vacuum they gain as much kinetic energy as they lose in electrical potential energy:

Key summary: Einstein redefines momentum

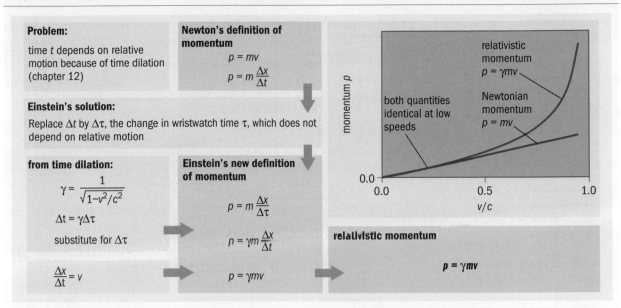

Problem:

time t depends on relative motion because of time dilation (chapter 12)

Einstein's solution:

Replace Δt by $\Delta \tau$, the change in wristwatch time τ, which does not depend on relative motion

from time dilation:

$$\gamma = \frac{1}{\sqrt{1 - v^2/c^2}}$$

$\Delta t = \gamma \Delta \tau$

substitute for $\Delta \tau$

$$\frac{\Delta x}{\Delta t} = v$$

Newton's definition of momentum

$$p = mv$$

$$p = m\frac{\Delta x}{\Delta t}$$

Einstein's new definition of momentum

$$p = m\frac{\Delta x}{\Delta \tau}$$

$$p = \gamma m\frac{\Delta x}{\Delta t}$$

$$p = \gamma mv$$

relativistic momentum $p = \gamma mv$

Newtonian momentum $p = mv$

both quantities identical at low speeds

relativistic momentum

$$p = \gamma mv$$

Relativistic momentum $p = \gamma mv$ increases faster than Newtonian momentum mv as v increases towards c

loss of electrical potential energy $= qV$
gain of kinetic energy $= \frac{1}{2}mv^2$
$$\frac{1}{2}mv^2 = qV$$

To calculate the speed of the particle, rearrange this equation to give

$$v = \sqrt{\frac{2qV}{m}}$$

The amount of energy gained by a particle with charge equal in magnitude to the charge on an electron moving through a p.d. of one volt is called one electron volt (eV). The electron volt is a unit of energy often used in physics.

Towards the ultimate speed

Notice that all particles with the same charge (e.g., an electron or a proton) gain the same kinetic energy when they move through the same p.d., but this doesn't mean that they reach the same speed. For example, a 1 keV electron and a 1 keV proton have the same kinetic energy but the electron moves much faster because its mass is almost 2000 times smaller.

A calculation ("Key summary: kinetic energy and speed") says that electrons in a dental X-ray tube operating at 64 kV would end up travelling at half

the speed of light. Something is wrong here. It seems that the p.d. would only have to be increased some more to get the electrons going faster than the speed of light. The p.d. needed would not even have to be especially large. Since the velocity is proportional to the square root of the p.d., to double the speed to reach the speed of light the p.d. must increase by a factor of $2^2 = 16$. So it seems that increasing the p.d. from 64 000 V to about 250 000 V would be enough to reach the speed of light, but it isn't. You may know that this is impossible: no material particle can travel as fast as light.

Einstein's equation $E_{rest} = mc^2$

The actual speed of the accelerated electrons can be measured by timing short bunches of them as they travel a known distance. As the p.d. is increased, their speed is indeed found to approach, but never exceed, the speed of light. Potential differences of only a few million volts, easily achieved with a Van de Graaff machine, are enough to get very close to the speed of light.

You might think this means that there is also a limit to the energy the electron can be given – not so. The kinetic energy is still equal to the change in electrical potential energy qV but the expressions $\frac{1}{2}mv^2$ for the kinetic energy and mv for

Key summary: Einstein rethinks energy

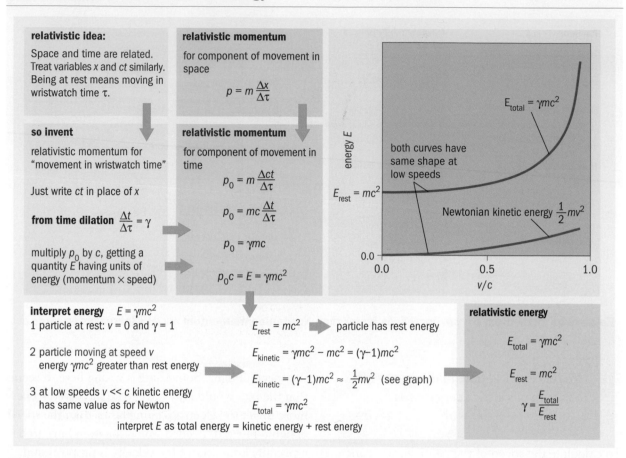

relativistic idea:

Space and time are related. Treat variables x and ct similarly. Being at rest means moving in wristwatch time τ.

so invent

relativistic momentum for "movement in wristwatch time"

Just write ct in place of x

from time dilation $\dfrac{\Delta t}{\Delta \tau} = \gamma$

multiply p_0 by c, getting a quantity E having units of energy (momentum × speed)

relativistic momentum

for component of movement in space

$$p = m\frac{\Delta x}{\Delta \tau}$$

relativistic momentum

for component of movement in time

$$p_0 = m\frac{\Delta ct}{\Delta \tau}$$

$$p_0 = mc\frac{\Delta t}{\Delta \tau}$$

$$p_0 = \gamma mc$$

$$p_0 c = E = \gamma mc^2$$

$E_{\text{total}} = \gamma mc^2$

both curves have same shape at low speeds

$E_{\text{rest}} = mc^2$

Newtonian kinetic energy $\frac{1}{2}mv^2$

energy E

v/c

interpret energy $E = \gamma mc^2$

1 particle at rest: $v = 0$ and $\gamma = 1$

2 particle moving at speed v energy γmc^2 greater than rest energy

3 at low speeds $v \ll c$ kinetic energy has same value as for Newton

$E_{\text{rest}} = mc^2$ → particle has rest energy

$E_{\text{kinetic}} = \gamma mc^2 - mc^2 = (\gamma - 1)mc^2$

$E_{\text{kinetic}} = (\gamma - 1)mc^2 \approx \frac{1}{2}mv^2$ (see graph)

$E_{\text{total}} = \gamma mc^2$

interpret E as total energy = kinetic energy + rest energy

relativistic energy

$$E_{\text{total}} = \gamma mc^2$$

$$E_{\text{rest}} = mc^2$$

$$\gamma = \frac{E_{\text{total}}}{E_{\text{rest}}}$$

Total energy $= \gamma mc^2$. Rest energy $= mc^2$. Total energy = kinetic energy + rest energy.

the momentum are only approximations. They are accurate enough at low speeds but not at high speeds. Different equations, first found by Albert Einstein in his theory of relativity, are needed. Einstein first thought again about the familiar quantity momentum, $p = mv$. He showed that in relativity the momentum has a larger value

$$p = \gamma mv$$

where γ is the relativistic factor

$$\gamma = \frac{1}{\sqrt{1 - v^2/c^2}}$$

that also describes time dilation (chapter 12). The old Newtonian equation $p = mv$ continues to work well at low speeds since in this case $\gamma \approx 1$.

Next, Einstein showed that the energy of a free particle must be thought of as having two

parts – kinetic energy and so-called rest energy. The total energy, kinetic energy plus rest energy, is *the* fundamental quantity and is given by

$$E_{\text{total}} = \gamma mc^2$$

where m is the particle mass. This means that at rest, when $\gamma = 1$, a particle still has energy. This rest energy is

$$E_{\text{rest}} = mc^2$$

The basic idea behind it all is that in relativity there is a similarity between space and time. Not moving in space is a relative idea that just means "travelling with me". Even sitting still you are still travelling forward in time. The meaning of the equation $E_{\text{rest}} = mc^2$ is that the mass of an object is the energy it has by dint of its time travel.

Key summary: energy, momentum and mass

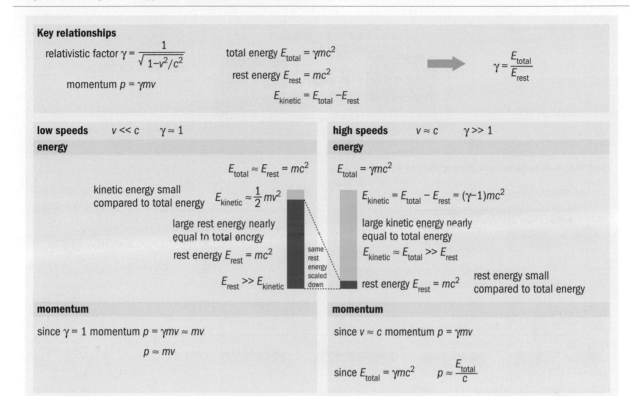

Key relationships

relativistic factor $\gamma = \dfrac{1}{\sqrt{1-v^2/c^2}}$

momentum $p = \gamma mv$

total energy $E_{total} = \gamma mc^2$

rest energy $E_{rest} = mc^2$

$E_{kinetic} = E_{total} - E_{rest}$

$\gamma = \dfrac{E_{total}}{E_{rest}}$

| low speeds | $v \ll c$ | $\gamma \approx 1$ | | high speeds | $v \approx c$ | $\gamma \gg 1$ |

energy

$E_{total} \approx E_{rest} = mc^2$

kinetic energy small compared to total energy $E_{kinetic} \approx \dfrac{1}{2}mv^2$

large rest energy nearly equal to total energy

rest energy $E_{rest} = mc^2$

$E_{rest} \gg E_{kinetic}$

same rest energy scaled down

energy

$E_{total} = \gamma mc^2$

$E_{kinetic} = E_{total} - E_{rest} = (\gamma - 1)mc^2$

large kinetic energy nearly equal to total energy

$E_{kinetic} \approx E_{total} \gg E_{rest}$

rest energy $E_{rest} = mc^2$ rest energy small compared to total energy

momentum

since $\gamma = 1$ momentum $p = \gamma mv \approx mv$

$p \approx mv$

momentum

since $v \approx c$ momentum $p = \gamma mv$

since $E_{total} = \gamma mc^2$ $p \approx \dfrac{E_{total}}{c}$

Kinetic energy small compared to rest energy   **Kinetic energy large compared to rest energy**

Of course, mass and energy are conventionally measured in different units – kilograms (kg) and joules (J). You can think of c^2 as the conversion factor from mass in kg to energy in J.

Particle physicists often express the masses of particles in energy units, frequently electron volts. An electron has a mass of about 0.5 MeV in these units, for example. A proton has a mass of about 1 GeV (2000 times greater). With both sides of the equation in the same units, you can think simply of the mass as being the rest energy, $E_{rest} = m$.

"Key summary: energy, momentum and mass" shows how Einstein's new equations are related. One of the most useful relationships is

$$\gamma = \dfrac{E_{total}}{E_{rest}}$$

because it gives you a different way of working out the relativistic factor γ.

Consider again the electrons in a dental X-ray tube. Their kinetic energy is 64 keV. Their rest energy (mass) is 0.511 MeV, that is, 511 keV. Thus γ,

the ratio of total energy to rest energy, is

$$\gamma = \dfrac{E_{rest} + E_{kinetic}}{E_{rest}} = \dfrac{511\,\text{keV} + 64\,\text{keV}}{511\,\text{keV}} = 1.125$$

This gives a value for the speed $\dfrac{v}{c} = 0.458$, a little bit less than the $\dfrac{v}{c} = 0.5$ that we got previously when ignoring relativistic effects. For the X-ray tube relativity is just starting to be important.

The calculation shows immediately that a p.d. of 511 kV would accelerate electrons to an energy where $\gamma = 2$, and $\dfrac{v}{c} = \sqrt{1 - 1/2^2} = \sqrt{0.75} = 0.866$.

By extreme contrast, in the Large Hadron Collider (LHC) at the European Organisation for Nuclear Research (known as CERN) in Geneva, protons are accelerated to a kinetic energy of 7 TeV = 7000 GeV. Since the rest energy (mass) of a proton is about 1 GeV, the relativistic factor γ has the very large value

$$\gamma = \dfrac{7001\,\text{GeV}}{1\,\text{GeV}} \approx 7000$$

The corresponding speed is extremely close to

Key summary: the principle of a linear accelerator

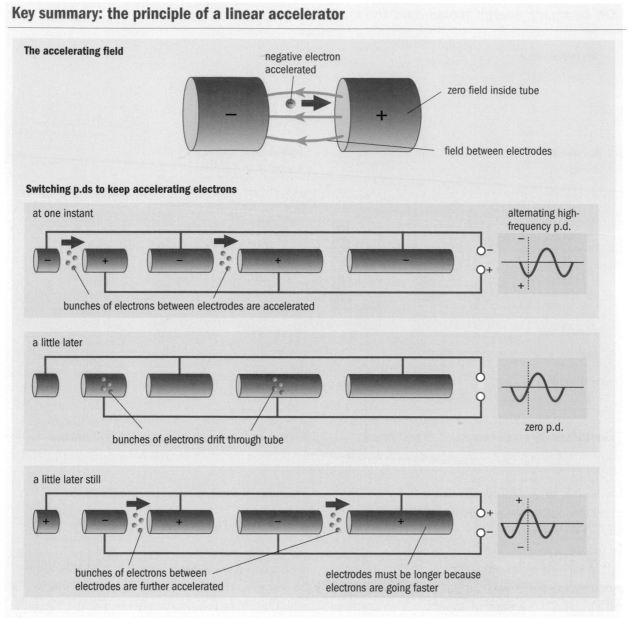

The accelerating field

negative electron accelerated

zero field inside tube

field between electrodes

Switching p.ds to keep accelerating electrons

at one instant

alternating high-frequency p.d.

bunches of electrons between electrodes are accelerated

a little later

bunches of electrons drift through tube

zero p.d.

a little later still

bunches of electrons between electrodes are further accelerated

electrodes must be longer because electrons are going faster

The alternating p.d. switches back and forth, the electrons accelerating as they pass between electrodes

the speed of light, being $0.999999991\,c$. For such highly relativistic particles, knowledge of their energy or momentum is much more important than knowledge of their speed.

An accelerator can't increase the speed of particles above the speed of light, but it can increase their energy and momentum as much as you like. The real point of a particle accelerator is not the final speed of the particles but the energy and momentum that they carry. When such particles collide, some of their energy can become the rest energy (mass) of new, more massive

particles. New forms of matter get created in such collisions.

The electric field

You have seen how useful the ideas of gravitational fields (chapter 9, *Advancing Physics AS* and chapter 11) and magnetic fields (chapter 15) can be. Although invisible and intangible, fields have become as real to physicists as matter itself. Perhaps the clinching point was the discovery that fields could store energy and that, in the form of electromagnetic waves, they could transport energy from place to place. So electrical forces, like gravitational forces, are described in terms of a field: the electric field.

The strength of the electric field E at a point in space is defined to be the force F exerted per unit charge q on a small positive test charge placed at that point:

$$E = \frac{F}{q}$$

Thus the force exerted on a charge by the electric field is just

$$F = qE$$

This is exactly the same as the approach used to define the gravitational field strength, $g = \frac{F}{m}$, except that gravitational fields act on masses so that g is equal to force per unit mass in $\mathrm{N\,kg^{-1}}$ (p43). The unit of electric field E is newtons per coulomb $\mathrm{N\,C^{-1}}$.

The test charge is in danger of changing the field it is supposed to measure. It pushes and pulls on the charges producing the field, so it may alter where they are and thus alter the field they produce. This is why, in the definition, the test charge is said to be small so as to have negligible effect.

It is conventional to say that the electric field is the force that would be exerted on a charge of one coulomb placed in the field. But a charge of one coulomb, even though it is only the charge carried by a current of one ampere flowing for one second, is enormous so far as forces are concerned. A pair of such charges a metre apart would have a force of nearly $10^{10}\,\mathrm{N}$ between them, enough to support a skyscraper-sized block of concrete 30 m square and 300 m high.

It is this very large electrical force of attraction that holds matter together. You are feeling it when you try to break a brick. But ordinary matter shows no sign of the charges on the particles inside it because the positive and negative charges are exactly balanced. A charged body is one that has just a very tiny excess of negative or positive charges. The total charge on all the electrons or protons in a small piece of copper (say 25 g) is more than a million coulombs. The excess charge you can get on it when charging it up in the laboratory

An aerial view of SLAC. Electrons are accelerated down a 3 km-long straight evacuated tube. The tube lies inside the long thin white tunnel which stretches down and to the left across the picture. Its construction is determined by relativistic effects. At the start, where the electrons have low energy, the distances between accelerating gaps have to get larger as the electrons speed up. Towards the end, where the electrons travel at nearly a constant speed c, the gaps must be equally spaced. Relativity theory is thus laid out on the floor of a Californian valley.

will be measured in billionths of a coulomb. The excess is a very tiny fraction indeed, but still exerts quite measurable effects.

The Stanford Linear Accelerator (SLAC)

In an X-ray tube, the electrons are accelerated just once, between the cathode and the anode of the tube. Early in the development of nuclear physics, bigger single stage accelerators like this were built using potential differences of a few million volts. A large linear accelerator, designed to reach energies of say 30 GeV, adopts a simple additional strategy to further increase the energy. The strategy is just: "What you can do once, you can do again." Bunches of electrons are accelerated between a pair of electrodes. Then they travel on to a gap between another pair where they are accelerated again, and so on, again and again. The energy builds up in small doses, its limit set only by the length of the accelerator.

The trick is to time the changing p.d. between each pair of electrodes so that, just at the moment the electrons arrive at the gap, the p.d. is in the right direction to accelerate them. As the electrons

Key summary: field lines and equipotentials

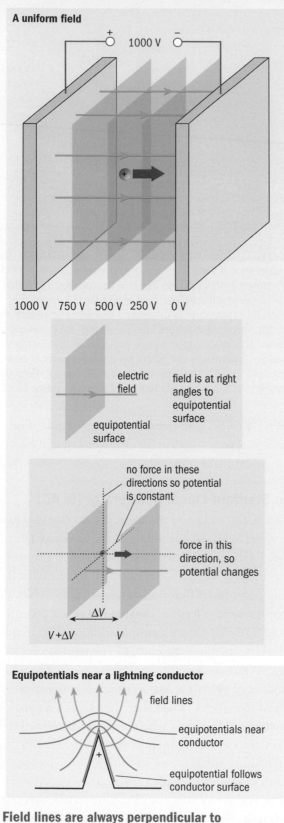

A uniform field

1000 V

1000 V   750 V   500 V   250 V   0 V

electric field

field is at right angles to equipotential surface

equipotential surface

no force in these directions so potential is constant

force in this direction, so potential changes

ΔV

$V + \Delta V$ V

Equipotentials near a lightning conductor

field lines

equipotentials near conductor

equipotential follows conductor surface

Field lines are always perpendicular to equipotential surfaces

go faster, they have to be allowed to travel farther between gaps so that the time from gap to gap is constant. But when the electrons have reached nearly the speed of light, the gap spacings need to be made equal. Relativity thus determines the design of the machine.

Electric potential and equipotential surfaces

When charges accelerate they are "falling down" a potential energy hill. The change of potential energy is given by $q\Delta V$. This obviously depends on the charge q. However, the electrical potential energy per unit charge V depends only on the field. It is called the electric potential:

electric potential at a point in the field
= electrical potential energy per unit charge

$$V = \frac{\text{electrical potential energy}}{q}$$

The units of electric potential are joules per coulomb, that is, volts. Potential difference, or voltage, familiar from electric circuits and the labels on household appliances, is simply a difference in electric potential.

One way to draw a picture of an electric field, to map its strength everywhere, is to draw field lines. A second way is to draw a map of the variation of potential. This is done by drawing lines that connect points of equal potential. It's just like drawing contour lines on a map. And, of course, contour lines connect points of equal height, which are also points of equal gravitational potential (p57).

The field lines necessarily run at right angles to the equipotentials. Moving a charge along an equipotential does not change its electrical potential energy, so there can be no component of force in this direction. If there were, work would be done in moving the charge and its potential energy would change.

Field lines and equipotentials

Think of standing on a hillside. The hardest way up is straight up the steepest slope. If you trip you will start to fall down in the direction of the steepest slope.

The electric field in an accelerator does a similar thing to the charged particles it accelerates. A linear accelerator like SLAC keeps sending electrons down such steep slopes, one after another. The particles are accelerated perpendicular to the equipotential surfaces, straight down the line of steepest descent of the potential. The steeper the potential gradient, the greater the force on the charged particle. Large forces are produced by strong fields so there is a clear link between the potential gradient – the steepness of the slope – and the electric field strength.

Let the change in potential be ΔV. If the electron is accelerating, gaining kinetic energy, the change in potential energy $q\Delta V$ must be negative. The change in kinetic energy is positive, equal in magnitude to the change in potential energy (if the motion is in a vacuum). Thus

$$\text{change in kinetic energy}$$
$$= -\text{change in potential energy}$$

$$= -q\Delta V$$

The change in kinetic energy is produced by a force F acting through a small displacement Δx in the direction of the force. The work done is $F\Delta x$. Thus

$$\text{change in energy} =$$
$$F\Delta x = -q\Delta V$$

This can be rearranged to give an equation for the force on the particle

$$F = -\frac{q\Delta V}{\Delta x}$$

or in words:

$$\text{force} = -\text{potential energy gradient}$$

This also gives a powerful equation for the electric field strength:

$$E = \frac{F}{q} = -\frac{\Delta V}{\Delta x}$$

If the field is changing continuously then it is better to use the limiting form of this equation:

Key summary: potential gradients

Contours and slopes

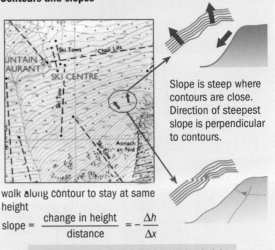

Slope is steep where contours are close. Direction of steepest slope is perpendicular to contours.

walk along contour to stay at same height

$$\text{slope} = \frac{\text{change in height}}{\text{distance}} = -\frac{\Delta h}{\Delta x}$$

negative slope is downhill, decreasing height

Field and potential gradient

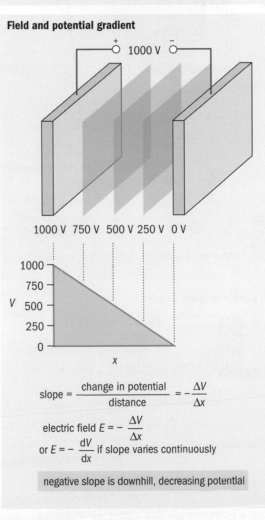

$$\text{slope} = \frac{\text{change in potential}}{\text{distance}} = -\frac{\Delta V}{\Delta x}$$

$$\text{electric field } E = -\frac{\Delta V}{\Delta x}$$

$$\text{or } E = -\frac{dV}{dx} \text{ if slope varies continuously}$$

negative slope is downhill, decreasing potential

Field strength = – potential gradient

Key summary: sparks and ionisation

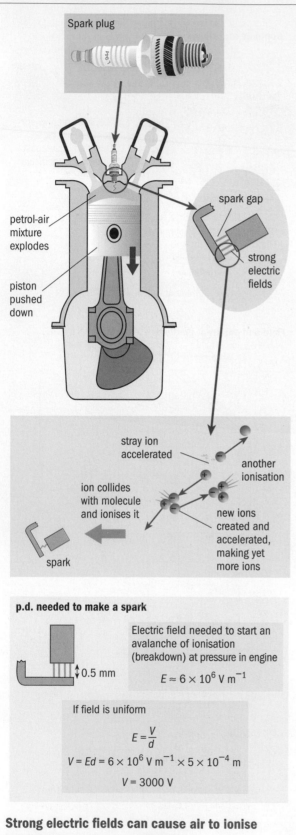

Spark plug

petrol-air mixture explodes

piston pushed down

spark gap

strong electric fields

stray ion accelerated

another ionisation

ion collides with molecule and ionises it

new ions created and accelerated, making yet more ions

spark

p.d. needed to make a spark

0.5 mm

Electric field needed to start an avalanche of ionisation (breakdown) at pressure in engine

$$E \approx 6 \times 10^6 \text{ V m}^{-1}$$

If field is uniform

$$E = \frac{V}{d}$$

$$V = Ed = 6 \times 10^6 \text{ V m}^{-1} \times 5 \times 10^{-4} \text{ m}$$

$$V = 3000 \text{ V}$$

Strong electric fields can cause air to ionise resulting in sparks and lightning

$$E = -\frac{dV}{dx}$$

Again, in words this says

electric field strength = −potential gradient

As we have just explained, the negative sign is there because the force points downhill, and downhill is a negative potential energy gradient. So the positive direction of the field is from high to low potential.

When we defined electric field earlier (p169), we said it was the force per unit charge, with units N C^{-1}, but this is not the easiest way to measure an electric field. To find the uniform electric field between a pair of conducting charged plates, the easy way is to put a voltmeter across the plates to measure the p.d. V and use a ruler to measure the distance d between the plates. The field is uniform, so the equipotentials are all equally spaced and the gradient is the same right across the gap – it's just a straight ramp. This makes the expression for the magnitude of the potential gradient very simple indeed:

$$E = \frac{V}{d}$$

The units of potential gradient are volts per metre, V m^{-1}. The argument above shows that these units are the same as the other unit of electric field, N C^{-1}. They don't look the same, but they are just two ways of saying the same thing. You may recall something similar for the gravitational field (chapter 9 *Advancing Physics AS* and chapter 11). The units of gravitational field can be expressed as either N kg^{-1} or m s^{-2}.

Large electric fields occur naturally in lightning flashes, when potential differences of many millions of volts exist between parts of a thundercloud. Petrol car engines use small lightning flashes to ignite the fuel–air mixture in the cylinder. In the spark plug a typical electric field is a few thousand volts per millimetre, that is, a few million volts per metre. Such electric fields are big enough to ionise the air, creating a spark. Some people like to use ionisers in the home, feeling that they improve air quality. Ionisers use intense electric fields to ionise air.

Summary: drawing field lines and equipotentials

Electric field lines:
- always start and end on charges;
- cannot cross each other;
- are always perpendicular to equipotentials;
- point from higher potential (more positive) to lower potential;
- are close together in strong fields, far apart in weak fields.

You can use these rules to sketch the shapes of various electric fields.

Detecting the charge on just one electron

American physicist Robert Millikan had a clever and simple idea. With basic apparatus that anyone could build he showed that electric charge really did come in discrete lumps – the charge on a single electron or proton. And he could measure the size of that charge too. His idea was based on what other people had tried. They had noticed that expanding moist air in a chamber produces a cloud of tiny water droplets, each forming around one or more charged ions.

Millikan's idea was to watch a single drop, not a cloud. His drops were of oil, chosen because, unlike water drops, they evaporate very slowly. Using a microscope, and with the drops lit from the side, he could see them individually like tiny stars against a dark background. The drops were sprayed into the space between two horizontal plates so that a vertical electric field could be produced in the space. Drops were generally electrically charged when formed in the spray so when the electric field was switched on some drops would be pulled upwards or would fall downwards more slowly.

Millikan measured the force on a drop in the field by noting how fast it was dragged up or down through the air in the cell. Another way of doing it is a bit easier to understand. By varying the p.d., you can try to hold a drop exactly in balance, neither drifting up under the influence of the field nor down under the influence of gravity. Then the upward electrical force qE exactly balances the weight W of the drop acting downwards:

$$qE = W$$

Key summary: discreteness of charge

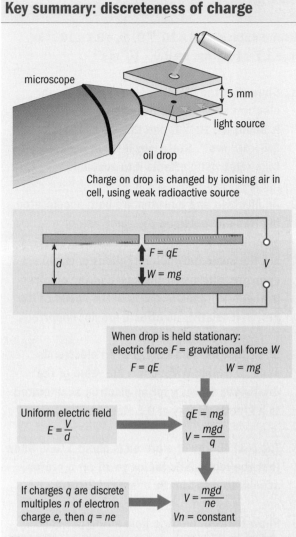

microscope

5 mm

light source

oil drop

Charge on drop is changed by ionising air in cell, using weak radioactive source

$F = qE$

$W = mg$

d

V

When drop is held stationary: electric force F = gravitational force W

$F = qE$ $W = mg$

Uniform electric field
$E = \dfrac{V}{d}$

$qE = mg$

$V = \dfrac{mgd}{q}$

If charges q are discrete multiples n of electron charge e, then $q = ne$

$V = \dfrac{mgd}{ne}$

Vn = constant

Changes can be detected that are due to single electrons leading to a calculated value of e

That's for just one value of the charge. But if you ionise the air in the cell (using a radioactive source, for example), the drop will occasionally change its charge. So it starts moving up or down, no longer in balance. Quickly you adjust the p.d. until the field is the right size to balance the drop again – bigger if the charge got less, smaller if the charge increased.

Millikan found that he could detect the difference made by a change in charge of just one elementary unit, the charge e. More important still, he could show that the charge on the drop was always a whole number multiple of the same elementary unit. Millikan was detecting changes of charge due to one single electron.

Quick check

Useful data: $e = 1.6 \times 10^{-19}$ C, $m_e = 9.1 \times 10^{-31}$ kg, $m_p = 1.7 \times 10^{-27}$ kg, $c = 3.0 \times 10^8$ m s^{-1}

1. Show that the kinetic energy gained by an electron accelerated through a p.d. of 500 V is about 8×10^{-17} J, and that its speed is about 1.3×10^7 m s^{-1}. State why it is not necessary to take relativistic effects into account.

2. An electron and a proton are both accelerated in a vacuum between the same pair of electrodes. Explain why their kinetic energies are the same but their directions of travel are opposite. Show that the speed of the electron is about 40 times larger than the speed of the proton if relativistic effects are not important.

3. Show that the rest energy of an electron is about 0.5 MeV. Write down the value of the relativistic factor γ for an electron accelerated to a kinetic energy of 0.5 MeV.

4. The rest energy of a proton is about 1 GeV. Show that the relativistic factor $\gamma = 30$ for a proton accelerated to kinetic energy of 29 GeV.

5. Show that the electric field strength between parallel plates 5 mm apart with a p.d. of 200 V across them is about 4×10^4 V m^{-1}. Show that a p.d. of 15 kV is needed across the plates to reach the field of 3×10^6 V m^{-1} at which the air ionises and sparks occur.

6. Show that N C^{-1} and V m^{-1} are equivalent units of electric field.

Links to the *Advancing Physics* CD-ROM

Practise with these questions:

10S Short answer *Speed and energy of particles – Newtonian calculation*

20S Short answer *Speed and energy of particles – relativistic calculation*

30S Short answer *Comparing relativistic and Newtonian kinetic energy*

10W Warm-up exercise *The uniform electric field*

60S Short answer *Two uses for uniform electric fields*

70D Data handling *Millikan's oil drop experiment*

Try out these activities:

90S Software-based *Using electric fields to measure electric charge*

Look up these key terms in the A–Z:

Accelerators; electric field; electric potential; electron; invariance of the speed of light; mass and energy; theory of special relativity

Go further for interest by looking at:

30C Comprehension *The Large Hadron Collider (LHC)*

20T Text to read *Atmospheric electricity*

35T Text to read *Why we believe in $E_{rest} = mc^2$*

Revise using the revision checklist and:

850 OHT *Relativistic momentum* $p = \gamma mv$

900 OHT *Relativistic energy* $E_{total} = \gamma mc^2$

950 OHT *Energy, momentum and mass*

300 OHT *Two ways of describing electrical forces*

500 OHT *Field lines and equipotential surfaces*

600 OHT *Field strength and potential gradient*

16.2 Deflecting charged beams

Making charged particles move quickly is not the end of the story. You need to be able to deflect the beams so that they can:

- scan across a screen or a sample, for example in oscilloscopes and electron microscopes;
- hit targets in linear accelerators;
- collide with other beams to make new kinds of matter in circular accelerators.

You also need to be able to focus the beams. All these things are done using electric and magnetic fields.

Oscilloscopes use electrostatic deflection. The electron beam passes between two pairs of parallel plates, the X- and Y-plates. The X-plates create a horizontal electric field that is usually used to make the beam scan repeatedly from left to right across the screen and then jump back. The rate at which this is done is controlled by the timebase, which applies a "sawtooth" potential difference to the X-plates. The Y-plates create a vertical field controlled by the input signal. For example, if the input is an alternating potential difference this will deflect the beam up and down continuously. The effect of the X- and Y-plates together is to make the electron beam draw a graph of the input p.d. against time on the oscilloscope screen. For an alternating signal this gives a sinusoidal curve.

Deflection of an electron beam

Top: electrical deflection of an electron beam. A potential difference between two plates, one above the other, produces a vertical electric field. The beam is deflected along a parabolic path. Bottom: magnetic deflection of an electron beam. A magnetic field at right angles to the beam is produced by a pair of coils. The beam is deflected so as to follow a circular path.

Key summary: deflections of an electron beam by an electric field

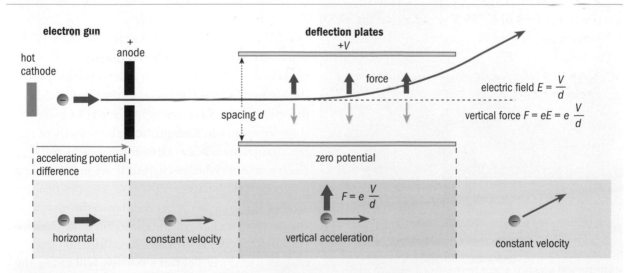

A uniform electric field deflects a charged particle along a parabolic path

Key summary: magnetic deflection

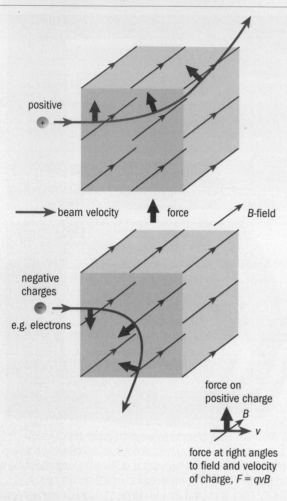

positive

→ beam velocity   ↑ force   ↗ B-field

negative charges

e.g. electrons

force on positive charge

force at right angles to field and velocity of charge, $F = qvB$

Magnetic fields deflect moving charged particles in circular paths

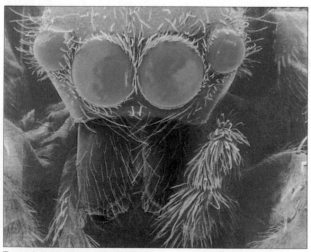

To make this false-colour image, a narrow focused beam of electrons was scanned across the head of a jumping spider. Electrons emitted from the specimen were collected and used to form the image in a scanning electron microscope.

How an oscilloscope works

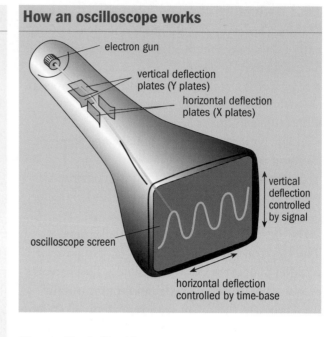

electron gun

vertical deflection plates (Y plates)

horizontal deflection plates (X plates)

vertical deflection controlled by signal

oscilloscope screen

horizontal deflection controlled by time-base

Magnetic deflection

Particle accelerators, and the particle detectors used with them, use magnetic deflection to bend the paths of high-energy charged particles. The curvature of the path inside a detector can then be used to measure the momentum of the particle. The more momentum it has, the harder it is to deflect and the larger the radius of curvature of its path. If the particle slows down, losing energy, its path becomes a spiral (the radius of curvature gets smaller).

Magnetic fields exert forces on moving charges. The magnitude of the force is given by

$$F = qvB$$

where B is the magnetic field strength perpendicular to the motion of the charge, q is the charge and v is the speed of the charge. The direction of the force is perpendicular to both the magnetic field strength and the velocity of the charge. This is the same force that turns electric motors when a current flows in a wire in a magnetic field (chapter 15).

Since the force is perpendicular to the motion of the particle, it changes the direction of its velocity but not its speed. It provides a centripetal force and, in a uniform field, the particle will follow the arc of a circle.

The force is given by:

Key summary: force on a current: force on a moving charge

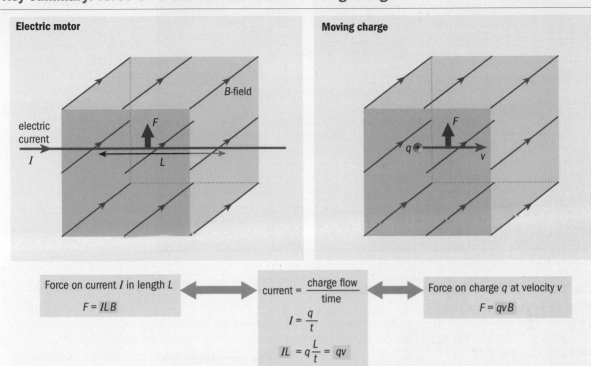

Electric motor

electric current *I*

B-field

F

L

Moving charge

F

q +

v

Force on current *I* in length *L*		Force on charge *q* at velocity *v*
$F = ILB$	current = $\dfrac{\text{charge flow}}{\text{time}}$	$F = qvB$
	$I = \dfrac{q}{t}$	
	$IL = q\dfrac{L}{t} = qv$	

The force that drives electric motors is the same as the force that deflects moving charged particles

This shows particle tracks in a liquid hydrogen bubble chamber. There is a strong magnetic field perpendicular to the plane of the image. Some tracks are sharply curved (lower momentum) whereas others are almost straight (very high momentum).

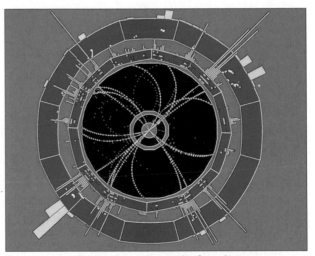

This computer display shows the result of an electron–positron collision inside the ALEPH detector at CERN. The collision occurred at the centre of the frame. The tracks are reconstructed from data collected in the detector.

$$qvB = \frac{mv^2}{r}$$

so the radius of curvature of the path is

$$r = \frac{mv}{qB}$$

Notice that mv is the momentum p. At all velocities, including those close to the speed of light, it remains true that

$$r = \frac{p}{qB}$$

even though the momentum is larger than

Left: the CERN site lies on the borders of France and Switzerland, between the city of Geneva and the Jura mountains. The 27 km-circumference tunnel containing the LHC runs underground. Right: in this interior view of the accelerator tunnel its curvature can just be seen.

Key summary: measuring the momentum of moving charged particles

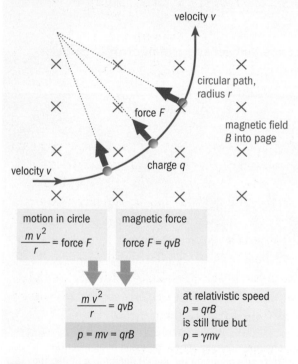

Momentum of the particle is proportional to the radius of curvature of its path

mv and the particle is harder to deflect. Thus the momentum of a particle produced in an accelerator, and so its energy, can always be found from the curvature of its path in a magnetic field, with

$$p = qrB$$

Of course, for particles travelling close to the speed of light, the correct relativistic expression for the momentum p must be used (p165).

Circular accelerators

Linear accelerators have a serious drawback – to reach very high energies the accelerator must be very long. There is an alternative – bend the accelerator into a circle and send the particles round and round, accelerating them again and again with the same pairs of electrodes stationed round the circle. Like the linear accelerator, the principle is still "do it again", but now using the same accelerating electrodes every time. After thousands of trips round the circular accelerator, particles reach very high energies.

How do you make the accelerated beam go in a circle? Magnets stationed round a circular evacuated pipe keep the beam bending round and other magnetic fields keep it in focus. Another advantage of these circular accelerators is that beams of particles can be sent round in opposite directions. If the particles have the same charge, the magnetic fields that steer them round must be in opposite directions. The two counter-rotating beams can then be made to collide inside large detectors at particular points in the ring. This is the principle of a collider.

The Large Hadron Collider (LHC) at CERN is the largest and most powerful accelerator of its kind in the world. It accelerates and collides beams of protons (and also lead ions). It runs in the 27 km-circumference circular tunnel built for its predecessor machine LEP, which collided beams of electrons and positrons. The protons are pre-accelerated to 450 GeV before being injected into the LHC, so are already travelling very close to the speed of light (relativistic factor $\gamma = 450$, see p166). The LHC then increases their energy to a maximum of 7 TeV per particle. Twice this energy – 14 TeV – is available from the collision of oppositely moving particles. Before being brought together to collide, the two beams travel about 200 mm apart in their oppositely directed magnetic fields. Steered by magnetic fields of strength 8.3 T, protons in the beams go all the way round the tunnel about 11 000 times a second.

The acceleration is done at eight places around the ring where the beams pass through pairs of accelerating electrodes tuned to the frequency of a high p.d. alternating supply. The accelerating electric field in the space between the electrodes is 5 MV m^{-1}. The frequency and phase are chosen so that when bunches of charged particles pass through them their electric fields are pointing in the right direction to accelerate the particles. The fact that the particles are travelling very close to the speed of light is a big advantage. It means that they orbit the ring at a constant frequency so the alternating supply to the accelerating cavities can have a fixed frequency (if the particles speed up the frequency would have to increase too).

There are four main experimental stations located around the ring. Each of these houses a large detector and electrostatic deflectors can bring the beams into collision inside the detectors. The resulting explosion sends showers of newly created particles through the detectors, each of which has a layered structure. Each layer records different information about some of the particles that pass through it, for example, the track of the particle or how much energy it deposits in that layer. All of this information is used to work out what happened in the collision.

Charges radiate electromagnetic waves

You may wonder why these accelerators are so large. Couldn't a smaller ring be used? Unfortunately, the answer is no. Bending the beam involves accelerating it towards the centre.

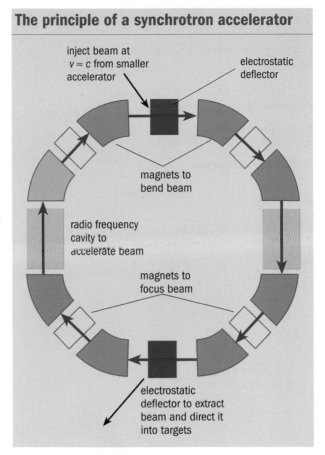

The principle of a synchrotron accelerator

inject beam at $v \approx c$ from smaller accelerator

electrostatic deflector

magnets to bend beam

radio frequency cavity to accelerate beam

magnets to focus beam

electrostatic deflector to extract beam and direct it into targets

Charges radiate when they accelerate. The tighter the ring the larger the centripetal acceleration and the more intense this synchrotron radiation.

Disturbing a charge is rather like wiggling your toes in the bath – ripples spread out in all directions. In a similar way, whenever a charge is accelerated it radiates electromagnetic waves that spread out at the speed of light. Radio and TV transmitters take advantage of this fact, but it is a problem in an accelerator where it takes energy away from the charged particles.

Non-uniform electric fields

The simple Geiger counter, used to detect individual high-energy particles, makes use of the intense electric field close to a fine charged wire. A high-energy charged particle passing by starts off an avalanche of ionisation that you hear as a click from the counter. The intense electric field greatly amplifies the ionisation from a single particle.

What shape must an electric field have close to a fine charged wire? By symmetry the field lines must spread out radially from the wire. This is

Key summary: cylindrical symmetry

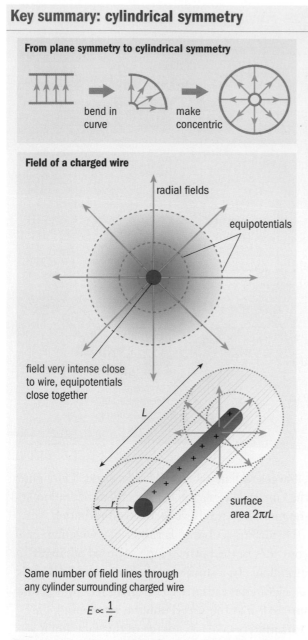

From plane symmetry to cylindrical symmetry

bend in curve

make concentric

Field of a charged wire

radial fields

equipotentials

field very intense close to wire, equipotentials close together

L

surface area $2\pi rL$

r

Same number of field lines through any cylinder surrounding charged wire

$$E \propto \frac{1}{r}$$

By symmetry the electric field at distance r from a long straight wire is proportional to $\frac{1}{r}$

These are the fine wires that make a multiwire detector, seen during construction. It can track particles to within $50\,\mu\text{m}$.

charged particle passes through the tube it leaves ions behind in the gas that fills the detector. Ions and electrons close to the central wire accelerate rapidly, collide with other gas molecules, and create a large number of electrons and positive ions near the wire. The electrons reach the wire and create a pulse of current in an external circuit. This pulse signals that a charged particle has passed through the detector.

The French physicist Georges Charpak took the idea further in his Nobel prize-winning invention of the large multiwire detectors now used in many particle physics experiments. Multiwire detectors contain an array of very fine wires ($20\,\mu\text{m}$ or so in diameter) held at a positive potential relative to a plane of negative wires. When a high-energy charged particle passes close to one of the positive wires it triggers an avalanche of electrons and an electric pulse in that wire. It is possible to work out where along the wire each event took place. There may be thousands of similar wires and the pattern of pulses can be read out from the ends of the array. This information can be used to work out the path that the particle actually followed.

Notice that we used an argument based purely on symmetry to obtain the $\frac{1}{r}$ variation of the field. Symmetry arguments can be very powerful.

enough to work out how the field strength varies with distance from the wire. Imagine a cylindrical surface of radius r centred on the wire. The area of the surface is $2\pi rL$, where L is the length of the cylinder, that is, the length of the central wire. Doubling r doubles this area so the density of field lines falls off as $\frac{1}{r}$.

More to the point, the field close to the wire is very intense, nearly strong enough to ionise the gas around the wire. When a high-energy

Key summary: two ways of saying the same thing

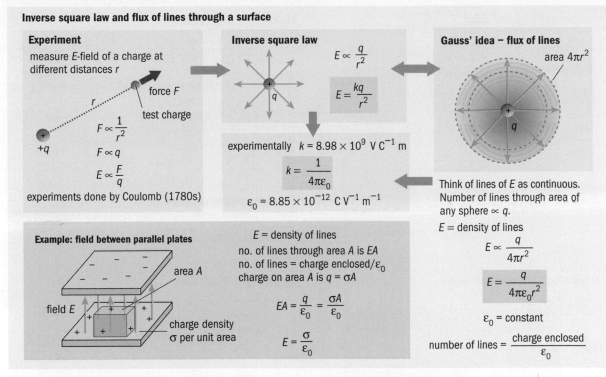

The electric field E can be considered as the density of lines through a surface

Field and potential near a charged sphere

Small charged particles, such as ions, electrons, protons and nuclei, behave pretty much like small charged spheres. Like the gravitational field of a spherical mass

$$g = G\frac{m}{r^2}$$

(chapter 11), the electric field of a spherical charge q also obeys an inverse square law

$$E = k\frac{q}{r^2}$$

This means that the field strength is reduced by a factor of 100 when the distance is increased by a factor of 10.

The German mathematician Carl Friedrich Gauss noticed that there is a different, often simpler, way of thinking about any field that obeys an inverse square law. His idea was to think of the flux of field lines across an area. You have seen how useful this is in magnetic circuits (chapter 15) and we have just used it to think

about the design of particle detectors. The flux of field lines is just the field strength multiplied by the area perpendicular to the field that the lines cross. Gauss's simple idea was that a given charge "sends out" a number of field lines proportional to that charge. The total flux from a charge is proportional to the charge.

It doesn't look like it, but this is just another way of stating the inverse square law. Imagine a busy restaurant spreading butter on toast by firing a spray of butter from a butter gun. Close up, one piece of toast could be covered in a short time. But four pieces of toast could be buttered to a quarter the thickness by holding them just twice as far away. If the butter spray is constant, the rate of buttering toast varies as the inverse square of the distance.

By symmetry, field lines spread out radially from a small spherical charge. Now imagine a spherical surface centred on the charge. All the field lines must pass through it and, again by symmetry, the field strength must be the same at all points on the surface. The surface area of the sphere at radius r is given by $4\pi r^2$ so the density of field lines at

Key summary: gravity and electricity – an analogy

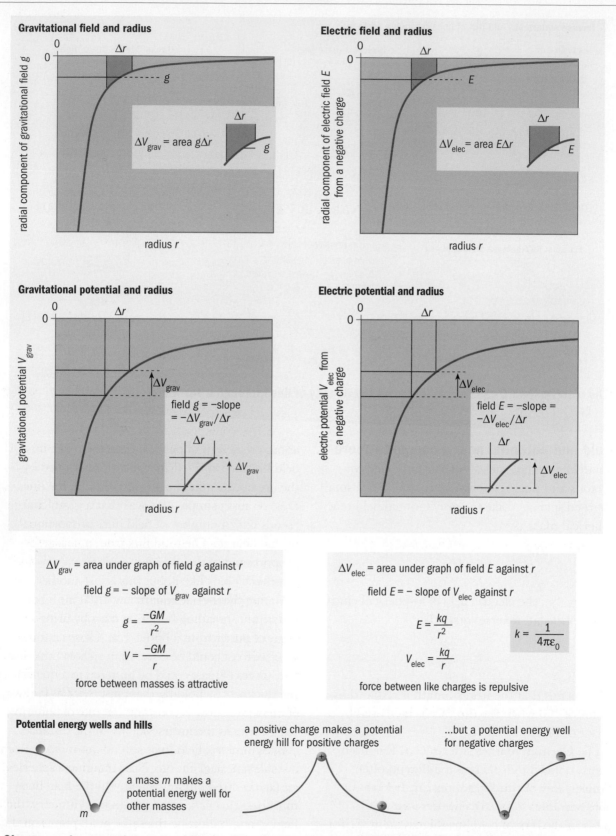

Gravitational field and radius

radial component of gravitational field g

radius r

ΔV_{grav} = area $g\Delta r$

Electric field and radius

radial component of electric field E from a negative charge

radius r

ΔV_{elec} = area $E\Delta r$

Gravitational potential and radius

gravitational potential V_{grav}

radius r

ΔV_{grav}

field g = –slope
= $-\Delta V_{grav}/\Delta r$

Electric potential and radius

electric potential V_{elec} from a negative charge

radius r

ΔV_{elec}

field E = –slope =
$-\Delta V_{elec}/\Delta r$

ΔV_{grav} = area under graph of field g against r

field g = – slope of V_{grav} against r

$$g = \frac{-GM}{r^2}$$

$$V = \frac{-GM}{r}$$

force between masses is attractive

ΔV_{elec} = area under graph of field E against r

field E = – slope of V_{elec} against r

$$E = \frac{kq}{r^2}$$

$$V_{elec} = \frac{kq}{r}$$

$$k = \frac{1}{4\pi\varepsilon_0}$$

force between like charges is repulsive

Potential energy wells and hills

a mass m makes a potential energy well for other masses

a positive charge makes a potential energy hill for positive charges

...but a potential energy well for negative charges

Charges make potential energy hills for like charges and potential energy wells for unlike charges

distance r is proportional to $\frac{1}{r^2}$.

Thus the electric field strength near an isolated charged sphere obeys an inverse square law

$$E \propto \frac{1}{r^2}$$

The field strength must also depend on the amount of charge. Doubling the charge would be like superimposing the same field pattern on itself – it would double the number of field lines and therefore double the electric field strength at all points. Thus the electric field strength is proportional to the size of the central charge

$$E \propto \frac{q}{r^2}$$

In the SI system of units, the flux of field lines leaving a charge q in a vacuum is chosen by convention to have the simple value

$$\text{number of lines} = \frac{q}{\varepsilon_0}$$

The quantity ε_0 is called the permittivity of free space.

Since the area crossed by the lines at radius r is $4\pi r^2$, the density of lines, which is the electric field, is given by

$$E = \frac{q}{4\pi\varepsilon_0 r^2}$$

This links the value of ε_0 to the constant k in

$$E = k\frac{q}{r^2}$$

with

$$k = \frac{1}{4\pi\varepsilon_0}$$

and

$$e_0 = 8.85 \times 10^{-12}\,\mathrm{C^2\,N^{-1}\,m^{-2}}$$

$$k = 8.98 \times 10^{9}\,\mathrm{N\,C^{-2}\,m^{2}}$$

Notice again the use of a symmetry argument.

Why no factor 4π in the corresponding gravitational equations, you might ask? It's just an historical accident of choice. To get $g = G\frac{m}{r^2}$ the flux of gravitational field lines from a spherical

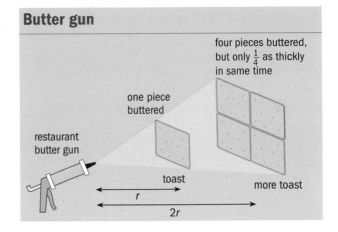

Butter gun

four pieces buttered, but only $\frac{1}{4}$ as thickly in same time

one piece buttered

restaurant butter gun

toast

r

$2r$

more toast

mass has to be written $4\pi Gm$, not Gm. So people who work with gravitational theory have to learn to write $4\pi G$ in place of G in many equations.

Potential near a charged sphere

The analogy between gravitational and electric fields goes even further. The difference in gravitational potential between (say) the orbit of Earth and the orbit of Mars can be used to calculate how much energy is needed per kilogram to get a spacecraft from one to the other. Similarly, the difference in electric potential between (say) a place close to a charged atomic nucleus and a place far away can be used to calculate how much energy is needed, per coulomb of charge, to get an electron from one to the other. That is, it decides how much energy is needed to ionise the atom.

The table on p184 shows just how similar the equations for gravitational and electric fields and potentials turn out to be.

The arguments linking field and potential are the same in each case. The field is the negative slope of the potential $-\frac{\mathrm{d}V}{\mathrm{d}r}$. The potential is the area below the graph of the field against radius. In both cases it is conventional to take the potential as zero a long way from the source of the field (mass or charge).

There is one big difference. Gravity is always attractive so the gravitational potential gets lower the shorter the distance r from the mass. This makes the values of gravitational potential always negative (it is taken to be zero when r is very large). However, like electric charges repel, so a positive test charge is pushed away from a positive charge and runs down the slope of a potential energy hill

Equations for point masses and charges

Gravitational field and potential	Electric field and potential

Radial component of gravitational field:

$$g = -G\frac{m}{r^2}$$

Radial component of electric field:

$$E = k\frac{q}{r^2} = \frac{q}{4\pi\varepsilon_0 r^2}$$

field proportional to $\frac{1}{r^2}$

Gravitational potential:

$$V_{grav} = -G\frac{m}{r}$$

Electric potential:

$$V_{elec} = k\frac{q}{r} = \frac{q}{4\pi\varepsilon_0 r}$$

potential proportional to $\frac{1}{r}$

Force between masses attractive	Force between like charges repulsive, between unlike charges attractive

This image shows a flame probe being used to measure the electric potential close to a charged conducting sphere. Ions in the flame carry charge onto or off the probe, until it reaches the potential of its surroundings. A meter connected to the probe records the potential.

away from the charge. Masses make potential wells for other masses, attracting and trapping them. Charges make potential hills for like charges. But – and this is what holds atoms and matter together – charges make potential wells for unlike charges. That's how the positive nucleus of an atom holds its negative electrons around it.

Being simply a voltage, the potential near a charged sphere can be measured using a suitable probe connected to a voltmeter. In this way the $\frac{1}{r}$ dependence of potential with distance r can be checked experimentally.

Strength of electrical interactions

The electrical force between two charges follows at once from the field due to a charge. The field strength due to charge q_1 at distance r is

$$E = \frac{q_1}{4\pi\varepsilon_0 r^2}$$

The force on a second charge q_2 is

$$F = q_2 E = \frac{q_1 q_2}{4\pi\varepsilon_0 r^2}$$

It is from this expression that on p169 we estimated the enormous force of 10^{10} N between a pair of 1 C charges 1 m apart.

If both charges have the same sign then the radial component of the force is positive, representing repulsion. If they have opposite signs the radial component of the force is negative, an attraction. The forces on each charge are necessarily equal and opposite.

In the 18th century French physicist Charles-Augustin de Coulomb invented a very sensitive torsion balance that he used to make careful measurements of the forces between charged objects. He discovered this inverse square law of force experimentally so the equation for the force between two charged spheres (or point charges) is known as **Coulomb's law**.

Two point charges q_1 and q_2 exert an electrostatic force on one another that is directly proportional to the product of the charges and inversely proportional to the square of their separation:

$$F \propto \frac{q_1 q_2}{r^2}$$

or, in SI notation,

$$F = \frac{q_1 q_2}{4\pi\varepsilon_0 r^2}$$

Coulomb's law is the electrostatic equivalent of Newton's law of gravitation.

Quick check

Useful data: $e = 1.6 \times 10^{-19}$ C, $m_e = 9.1 \times 10^{-31}$ kg, $\varepsilon_0 = 8.9 \times 10^{-12}$ $C^2 N^{-1} m^{-2}$, $c = 3.0 \times 10^8$ m s^{-1}.

1. Describe the paths of moving charges when they are injected perpendicular to
    (a) a uniform electric field and
    (b) a uniform magnetic field.
    Why do steady uniform magnetic fields never make moving charges travel faster?

2. Use the approximation $p = mv$ to show that the momentum p of electrons accelerated by a p.d. of 10 kV is 5.4×10^{-23} kg m s^{-1}. Show that a magnetic field of 3.4×10^{-3} T is needed to deflect them in a curve of radius 100 mm.

3. In the LEP accelerator at CERN (now replaced by the LHC), electrons were accelerated to a total energy of 100 GeV. Use the relativistic approximation $p = E_{total}/c$ to show that their momentum is 5.3×10^{-17} kg m s^{-1}. Show that a magnetic field of about 0.08 T is needed to make them travel in a circle of circumference 27 km.

4. A long straight wire of radius 1 mm is charged positively and the electric field strength 10 mm from the wire is 10^4 V m^{-1}. Show that the field strength at the surface of the wire is 10^5 V m^{-1}.

5. A hollow conducting sphere of radius 100 mm is raised to a potential of 5000 V. Show that the potential 1 m from the centre of the sphere is 500 V.

6. Show that the electric field around a 10 mm conducting ball at a potential of 100 kV is larger than the breakdown field 3 kV mm^{-1} needed to ionise air, at distances up to about 30 mm from the centre of the ball.

Links to the *Advancing Physics* CD-ROM

Practise with these questions:
80W   Warm-up exercise *Getting* F = qvB
90S   Short answer *Deflection with electric and magnetic fields*
100S Short answer *The cyclotron*
180S Short answer *Non-uniform electric fields*
190S Short answer *Charged spheres: Force and potential*
200S Short answer *Using the $1/r^2$ and $1/r$ laws for point charges*

Try out these activities:
240S Software-based *Mapping inverse square vector fields*
250S Software-based *Summing vector fields*
270S Software-based *Radial force, field and potential*

Look up these key terms in the A–Z:
Accelerators; electric field; electric potential; force on a moving charge; inverse square laws; magnetic field

Go further for interest by looking at:
210D Data handling *Testing Coulomb's law*
260E Estimates *Estimating with fields*

Revise using the revision checklist and:
1000 OHT *How an electric field deflects an electron beam*
1200 OHT *How a magnetic field deflects an electron beam*
1300 OHT *Force on current: Force on a moving charge*
1700 OHT *Shapes of electrical fields*
1800 OHT *Electrical fields with circular symmetry*
1900 OHT *Inverse square law and flux*

Summary check-up

Accelerating charges ✓

- When a charge q moves through a p.d. ΔV the work done is $W = q\Delta V$
- When a charge is accelerated through a p.d. V it gains a speed

$$v = \sqrt{\frac{2qV}{m}}$$

as long as v is less than a few per cent of the speed of light

- The rest energy of a particle is $E_{rest} = mc^2$. The total energy is $E_{total} = \gamma mc^2$ where $\gamma = 1/\sqrt{1 - v^2/c^2}$. Thus

$$\frac{E_{total}}{E_{rest}} = \gamma.$$

The kinetic energy is the difference $E_{kinetic} = E_{total} - E_{rest}$.

Electric field and potential ✓

- Electric field strength is force per unit charge: $E = \frac{F}{q}$ measured in $N\,C^{-1}$ or $V\,m^{-1}$

- Electric potential V is electrical potential energy per unit charge. It is measured in $J\,C^{-1}$ or V.
- The electric field E is given by the negative slope of the potential:

$$E = -\frac{dV}{dx}$$

- The difference in potential is equal to the area under the graph of field against distance
- Field lines in a static field start and end on charges. Equipotential surfaces are perpendicular to field lines.

Deflecting charged particles ✓

- The magnetic force on a moving charge is $F = qvB$, at right angles to both the field and velocity
- Charges moving with momentum p perpendicular to a uniform magnetic field B follow a circular path of radius

$$r = \frac{p}{qB}$$

- Charges moving perpendicular to a uniform electric field follow a parabolic path
- Forces on charged drops provide evidence of the discreteness of electric charge

Non-uniform fields ✓

- The electric field strength near a point charge or a charged sphere falls off as an inverse square law:

$$E = \frac{q}{4\pi\varepsilon_0 r^2}$$

- The potential near a charged sphere or point charge falls off as $\frac{1}{r}$:

$$V = \frac{q}{4\pi\varepsilon_0 r}$$

- The force between two point charges separated by a distance r is given by Coulomb's law:

$$F = \frac{q_1 q_2}{4\pi\varepsilon_0 r^2}$$

Questions

Useful data: $e = 1.6 \times 10^{-19}\,C$, $\varepsilon_0 = 8.9 \times 10^{-12}\,C^2\,N^{-1}\,m^{-2}$, $c = 3.0 \times 10^8\,m\,s^{-1}$, $m_e = 9.1 \times 10^{-31}\,kg$, $m_p = 1.7 \times 10^{-27}\,kg$.

1. An electron in an oscilloscope is accelerated through a p.d. of 1 kV.
(a) Write down the kinetic energy in electron volts gained by the electron.
(b) Calculate the rest energy mc^2 of an electron in electron volts. Is the kinetic energy gained by the electron much smaller than its rest energy?
(c) The kinetic energy of a mass m moving at speed v is given by $\frac{1}{2}mv^2$ if the kinetic energy is much less than the rest energy. Show that for a charge q accelerated through potential difference V the speed v is given by $v = \sqrt{2qV/m}$.
(d) The ratio q/m of charge to mass for electrons is $1.8 \times 10^{11}\,C\,kg^{-1}$. Calculate the speed of an electron accelerated through 1 kV.
(e) Protons are 1840 times more massive than electrons. Calculate the speed of a proton accelerated through 1 kV.

2. Sketch equipotentials and electric field lines for the sets of charged conducting electrodes shown.

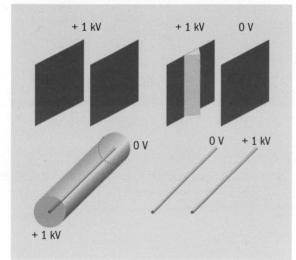

3. Give one example each of the use of an electric field and a magnetic field in a particle accelerator. In each case make sure that you:
(a) describe the type of accelerator briefly;
(b) state the use to which the electric or magnetic field is put;
(c) indicate the shape of the electric or magnetic field;
(d) show the directions of any forces on an accelerated particle;
(e) state how to calculate the magnitude of forces on the accelerated particle.

4. Calculate the electric field, the electric potential and the force on an electron:
(a) at a distance equal to the radius of a hydrogen atom (approximately $0.5 \times 10^{-10}\,m$) from a proton;
(b) at the distance equal to 10 times the radius (approximately $10^{-15}\,m$) of a proton from a proton;
(c) at a point midway between the flat surfaces of the electrodes in a spark plug, spark gap 0.4 mm, p.d. 10 kV.

5. Fill in the missing elements in the relationships in the table below, which compares gravitational and electric field and potential.

Field and potential of spherical masses and charges	
Radial component of gravitational field: $g = [\;]\dfrac{m}{r^2}$	Radial component of electric field: $E = \dfrac{q}{4\pi\varepsilon_0 [\;]}$
Gravitational potential: $V_{grav} = -G\dfrac{m}{[\;]}$	Electric potential: $V_{electric} = \dfrac{q}{[\;]r}$

6. Electric and gravitational fields and potentials can be expressed in various units.
(a) Show from the definition of electric field strength that the units of electric field can be written as $N\,C^{-1}$.
(b) Show from the relationship between field and potential gradient that the electric field can also have the units $V\,m^{-1}$.
(c) Use the fact that the unit V can be written as $J\,C^{-1}$ to show that the unit $V\,m^{-1}$ is the same as the unit $N\,C^{-1}$.
(d) If there were a "gravitational volt" V_{grav}, its unit would be $J\,kg^{-1}$. What would be the unit of gravitational field, in terms of "gravitational volts"?

7. An electron travels at a low speed v at right angles to a uniform magnetic field B. The force on the electron has magnitude $F = evB$.
(a) In what direction is the force $F = evB$? Explain why this force is equal to the centripetal force mv^2/r.
(b) Show that the radius of curvature of the path of the electron is given by $r = \dfrac{p}{eB}$, where p is the momentum of the electron.
(c) Assume that the expression $r = \dfrac{p}{eB}$ in (b) is valid at all speeds. Find the radius of curvature of the path in a magnetic flux density $B = 0.1\,T$ of a relativistic electron of energy $E = 100\,MeV$, for which the relationship $p = \dfrac{E}{c}$ is a good approximation.

17 Probing deep into matter

The world is full of variety and dynamic change, startling beauty and violent destruction. The brute force of nuclear fusion inside stars contrasts with the delicacy of life and the subtleties of thought itself. And yet, beneath the surface there is astonishing simplicity. In this chapter we will describe how:

- matter is made from just a few different types of fundamental particles
- scattering experiments reveal the deep structure of matter
- interactions take place by the exchange of particles
- conservation laws determine what can and cannot happen
- quantum effects result in complex atomic structures

17.1 Creation and annihilation

Antimatter often appears in science fiction. If it comes into contact with ordinary matter there is a violent explosion, in which both matter and antimatter are completely annihilated. It is used to power the Starship *Enterprise* in *Star Trek*. But antimatter is not just the stuff of science fiction – it is also used to try to find out how the brain works.

Positron emission tomography (PET)

Imagine you have agreed to take part in an experiment to look at how the brain works. You are lying down, surrounded by rings of detectors. You are given a task to do, for example memorising some words or looking at some pictures. Detectors surrounding you pick up where your brain is active, indicated by increased blood flow. On a digital map of the brain, the areas that are working hard "light up". A picture of your brain at work is built and the functions of different parts of the brain can be studied.

Scanning the brain

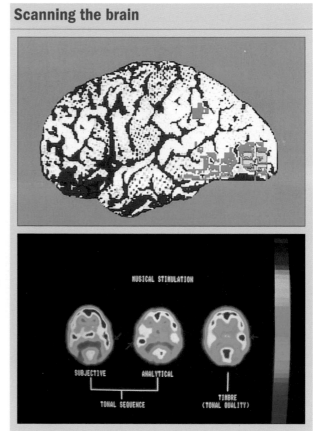

Top: PET scan of a normal brain showing localised activity (green and yellow) in the visual cortex when the subject is reading silently. Bottom: the brains of three people listening to music – activity greatest in the red areas. The reaction of trained musicians is found to be more analytical than that of untrained listeners.

How is it done? The oxygen carried by your blood to the brain has to be labelled with a small amount of the radioactive isotope oxygen-15. This isotope has a short half-life, decaying by emitting a particle of antimatter – a positron or positive electron. The positron does not get far. Within a millimetre or so it encounters an ordinary electron in an atom and the two particles annihilate one another. They simply vanish. Matter has been destroyed but the energy is still there and is carried away by a pair of gamma-ray photons, travelling in opposite directions. The detectors around you pick up these photons. From many such events, a computer-generated map of the activity in your brain can be built up.

Key summary: making PET scans

A pair of gamma-ray photons are emitted in opposite directions as a result of electron–positron annihilation inside the patient

signal processing

Scintillator: captures a gamma-ray photon and emits lower-energy photons into photomultiplier tubes

Photomultiplier: incoming photons create a cascade of electrons, giving an electrical pulse output

γ

γ

One pair of detectors will respond *almost* simultaneously. This near coincidence shows that the two gamma-ray photons came from a common source. The tiny time difference between the two signals is then used to work out *where* they came from along the line between the detectors.

Annihilation of positrons from oxygen-15

Oxygen-15 is an unstable isotope. It is used to "label" a molecule like glucose that is used in body metabolism.

It decays by β^+ decay, emitting a positron (and a neutrino). e^+

The positron (e^+) collides almost immediately with an electron (e^-) in one of the surrounding atoms.

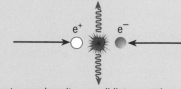
e^+ e^-

The electron and positron annihilate, creating a pair of gamma-ray photons that travel in opposite directions. PET scanners detect these gamma-ray photons.

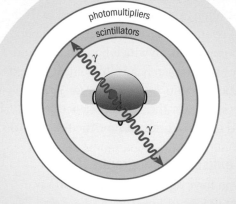

photomultipliers

scintillators

γ

γ

Scintillators are arranged in a grid on the inside surface of the scanner. In any short period of time many detectors will respond to gamma-ray photons from many different annihilations inside the body. A computer produces a slice-by-slice map of activity in the brain.

A positron and electron annihilate each other, emitting a pair of gamma-ray photons travelling in opposite directions. The detection of many of these leads to a map of brain activity being built up.

The positron was the first of a new breed of particle – an antiparticle – to be discovered. Antiparticles are a bit like mirror images of their corresponding particle – everything about a positron is the exact opposite of an electron except for the mass, which is the same. In particular, the two electric charges are equal and opposite.

Not only do positrons and electrons annihilate one another, they are also created together in electron–positron pairs, often from the energy of a gamma-ray photon. Routinely today, when gamma-rays pass through a detector, forked pairs of particle paths are seen, apparently springing out of nothing. The energy of a gamma-ray photon has materialised as a pair of particles.

It's not just the electron that has an antiparticle. All particles have antiparticles, although in some cases (for example photons) a particle is its own antiparticle. The antiproton was first produced experimentally in 1955 and the antineutron in 1956. Positrons and antiprotons are now routinely made and stored for use in collider experiments.

Why antimatter?

The prediction and subsequent detection of antiparticles came as a surprise to physicists, but perhaps the real surprise was that matter itself can be created and destroyed, that energy can be materialised and dematerialised. If this can happen, then the existence of something like antiparticles becomes a logical necessity. If an electron could be created alone, electric charge would be created from nothing and charge would not be conserved. Antiparticles are required that have exactly the opposite values for all conserved properties of their corresponding particles. The existence of conservation laws is why physicists are so sure that every particle has an antiparticle.

Annihilation

Although annihilation involves a violent transformation there are quite a few things that do not change. In any particle annihilation:
- electric charge is conserved;
- linear and angular momentum is conserved;
- total energy (including the rest energy $E_{rest} = mc^2$) is conserved.

Annihilations always happen in pairs of particle

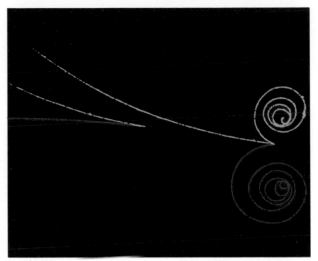

In this false-colour bubble chamber photograph two invisible gamma-ray photons have entered from the right and each has created an electron–positron pair. The electron tracks are coloured green and the positron tracks are coloured red. There is a strong magnetic field in the chamber that deflects electrons upwards and positrons downwards. In the event on the left, with just two tracks, the incoming photon gives all its momentum to the positron–electron pair, which initially move in the same direction as the incoming photon at high speed. Their tracks are only slightly curved. In the event on the right, with three tracks, most of the energy and momentum of the photon is given to an electron knocked out of an atom (the long, slightly curved central track). The positron–electron pair created here has lower energy and their tracks are more strongly curved. Both tracks form the characteristic spiral paths of low-energy electrons and positrons, as each particle gradually loses energy to the atoms in the bubble chamber liquid and slows down.

and antiparticle. Conservation laws restrict what can happen – only changes in which all conserved quantities remain the same can occur. Particle interactions and transformations are not arbitrary events, they are constrained by rules. A major task of particle physics is to work out these rules. One way to do this is to observe as many transformations as possible and note what does and does not happen. Another is to try to imagine underlying principles that would require some quantity to be conserved.

In both cases new ideas have to be tested and these tests are often carried out using high-energy accelerators that can smash particles together to see what happens (chapter 16). Particle physics is a wonderful example of the way experiments have provided new things for theory to explain, and theories have suggested new things to look for experimentally.

Key summary: conserved quantities

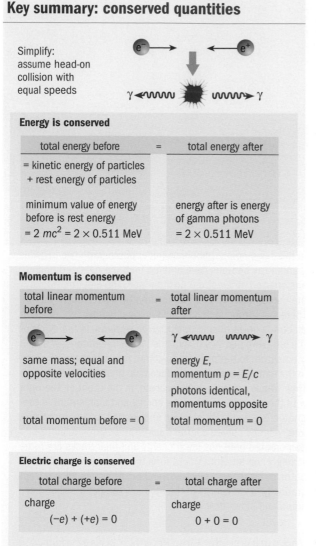

Simplify: assume head-on collision with equal speeds

Energy is conserved

total energy before	=	total energy after
= kinetic energy of particles + rest energy of particles		
minimum value of energy before is rest energy = $2\,mc^2$ = 2×0.511 MeV		energy after is energy of gamma photons = 2×0.511 MeV

Momentum is conserved

total linear momentum before	=	total linear momentum after
same mass; equal and opposite velocities		energy E, momentum $p = E/c$ photons identical, momentums opposite
total momentum before = 0		total momentum = 0

Electric charge is conserved

total charge before	=	total charge after
charge $(-e) + (+e) = 0$		charge $0 + 0 = 0$

Energy, momentum and electric charge are always conserved in electron–positron annihilation

Pair annihilation and creation

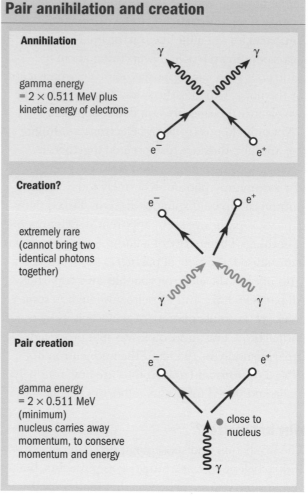

Annihilation

gamma energy = 2×0.511 MeV plus kinetic energy of electrons

Creation?

extremely rare (cannot bring two identical photons together)

Pair creation

gamma energy = 2×0.511 MeV (minimum) nucleus carries away momentum, to conserve momentum and energy

close to nucleus

Creation

It's worth thinking about any interaction that is not ruled out by the conservation laws. For example, imagine reversing an annihilation. Turn the two photons around and let them travel back to a point. What will happen? Is it possible to create an electron–positron pair? The conservation laws won't prevent it because any quantity that stays the same in the annihilation must stay the same in the creation process.

So it could happen but it isn't observed because it is extraordinarily difficult to find two identical photons and aim them back at the same tiny region of space. What about using one photon with double the energy? Could that create an electron–positron pair? Energy and electric charge would be conserved so that's OK but, as the calculations turn out, linear momentum is not.

However, there is a way. If a gamma-ray photon passes close to the nucleus of an atom, the nucleus takes part in the interaction (recoils) and momentum can be conserved. Such a gamma-ray photon must have energy equal to at least $2mc^2$, the rest energy of an electron and a positron, which comes to 1.022 MeV. As with annihilation, creation always involves a particle–antiparticle pair.

Particles and fields

The discovery of processes in which particles are created and destroyed was radical and has led to a completely new way of looking at fields and matter.

Chapters 15 and 16 tell the older, classical story about electric and magnetic fields, similar in many ways to the gravitational field discussed in

Key summary: Feynman diagrams show possibilities to be combined

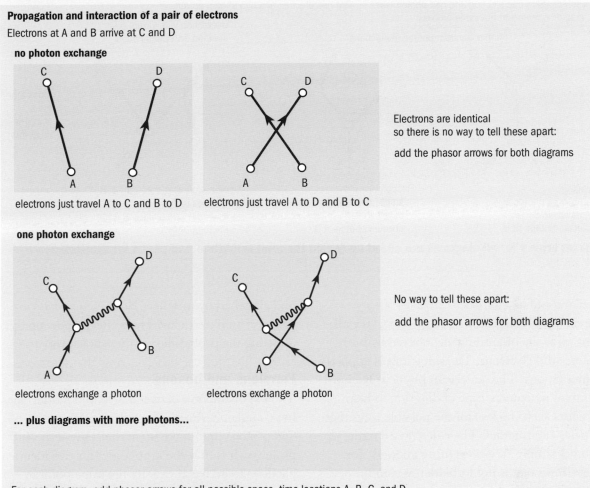

Propagation and interaction of a pair of electrons

Electrons at A and B arrive at C and D

no photon exchange

electrons just travel A to C and B to D

electrons just travel A to D and B to C

Electrons are identical
so there is no way to tell these apart:

add the phasor arrows for both diagrams

one photon exchange

electrons exchange a photon

electrons exchange a photon

No way to tell these apart:

add the phasor arrows for both diagrams

... plus diagrams with more photons...

For each diagram, add phasor arrows for all possible space–time locations A, B, C, and D.
Add total phasor arrows for each type of diagram.

Quantum rule: try all possible ways to interact

chapter 11. In the classical story, charged particles are the source of the electric and magnetic fields (together, the electromagnetic field). In turn, the fields exert forces on other charged particles. In this picture, matter is one kind of thing, built of seemingly indestructible particles located in space. Fields are a quite different kind of thing, an invisible influence spread throughout space.

In the modern quantum field story, the electromagnetic field becomes a creator and destroyer of photons. Photons are created at one point and travel – trying all paths – to another point where they are absorbed. They are created and destroyed by the interaction of the field with charges on particles. In fact, the idea of electric

charge itself gets a new meaning. The length of the quantum arrow (the quantum amplitude, chapter 7 in *Advancing Physics AS*) for the process of creating or destroying a photon is proportional to the charge. So electric charge is now interpreted as a measure of the strength of the interaction between electrons and photons.

How do forces arise in this picture? A pair of electric charges exchange virtual photons, which carry energy and momentum. But electric charge comes in one of two signs – positive or negative. Reversing the sign of one charge reverses the phase of the quantum amplitude to create or destroy a photon. When all the amplitudes are added up, taking account of phase, this leads like charges to

Key summary: ways for an electron to scatter a photon

In each diagram one electron and one photon come in, and one electron and one photon go out.
All diagrams represent the **same** process.

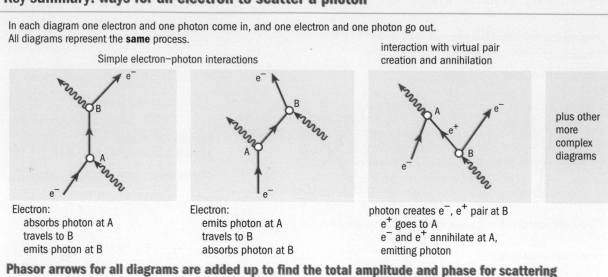

Simple electron–photon interactions

interaction with virtual pair creation and annihilation

Electron:
 absorbs photon at A
 travels to B
 emits photon at B

Electron:
 emits photon at A
 travels to B
 absorbs photon at B

photon creates e⁻, e⁺ pair at B
 e⁺ goes to A
 e⁻ and e⁺ annihilate at A,
 emitting photon

plus other more complex diagrams

Phasor arrows for all diagrams are added up to find the total amplitude and phase for scattering

repel and unlike charges to attract.

In chapter 7 (in *Advancing Physics AS*), "try all paths" was the quantum rule, obeyed by both photons and electrons. The same idea is applied to interactions. The American physicist Richard Feynman invented a type of diagram to help physicists keep track of all the possible ways that particles can interact. The rule "try all paths" becomes simply "try everything allowed" (or "everything that is not forbidden is compulsory").

Each diagram, above, pictures one way for a given process to happen, like a path for that process, and for each diagram the quantum amplitude (phasor arrow) can be written down. Adding these up over all places and times, each diagram gets a total phasor arrow. To find the resultant amplitude for a process, the arrows for all diagrams – all the ways the process can happen – are combined, taking account of their phases. Finally, the probability of an event, such as the creation of an electron–positron pair from a photon, is found by squaring the resultant amplitude for all the diagrams.

It's important to realise that these pictures do not show individual events that happen (or not). They represent different ways in which the same overall process can occur, all of which have to be envisaged. Particles that appear and vanish within the interaction are simply in the mind's eye. You cannot, even in principle, detect them, as they must be exchanged (that is, emitted by one particle

and absorbed by another). So these particles are called virtual particles. Their effects show up by affecting the probability of the whole process.

Fermions and bosons

The old distinction between field and matter does not completely vanish. Particles like electrons and protons that make up matter have a special property. It is that the amplitudes for two identical particles to arrive at exactly the same point in space–time subtract – they add up with opposite phase and the phasor arrows point oppositely. Result: the total amplitude is zero and such particles never come together in exactly the same state. Particles like this are called fermions, after Enrico Fermi who, with Paul Dirac, first identified their behaviour. It's an interesting case of quantum physics predicting a certainty (albeit "never") rather than a mere probability.

This effect makes fermions exclusive – they "don't like" to get together. Atoms become built of electrons, each in their own quantum state, sharing that state with no other electron. Nuclei are built of protons and neutrons doing the same. (This is what tends to keep their numbers roughly equal, since a proton and a neutron can share the same state but two protons or two neutrons can't.)

Electrons, protons and neutrons are all fermions and obey the Pauli exclusion principle that no two particles ever share exactly the same quantum state. In the end this is what makes matter hard – however heavily you sit on a chair, you can't squeeze the particles in it into the same states.

Particles like photons, however, do precisely the opposite. The amplitudes for two identical photons to be at the same point in space–time add up with the same phase. The result is that the total amplitude doubles. So photons are inclusive and positively "like" to be in the same state. The laser is the best known application of this. It gives photons the chance to join others in the same state and they do so enthusiastically, producing a laser beam of a huge number of photons of identical phase and polarisation.

Particles like photons, the exchange of which gives rise to forces, are called bosons after the Indian physicist Satyendra Nath Bose who (with Einstein) first identified this behaviour.

Beta decay and conservation laws

The positrons used in PET scans come from a form of beta decay in which an unstable nucleus emits a positron. This happens if a nucleus has rather too many protons for the number of neutrons. Conversely, a nucleus that has too many neutrons can decay by emitting an ordinary negative electron, leaving it with one more proton. At first these beta decays caused physicists quite a few headaches because they seemed to violate several conservation laws, including the law of conservation of energy.

Strontium-90 (element 38) is a beta source used in school laboratories. It decays to yttrium-90 by emitting a beta particle (electron). A first attempt at an equation for this decay is:

$$^{90}_{38}\text{Sr} \longrightarrow ^{90}_{39}\text{Y} + ^{\ 0}_{-1}\text{e}$$

(which turns out not to be good enough).

Notice how conservation laws work out here. The lower numbers balance, showing that electric charge is conserved. Although a neutron has changed to a proton in the nucleus, the total number of protons and neutrons is the same before and after the decay, and the upper numbers also balance. This suggests inventing the name **nucleon** to include both protons and neutrons, recognising that although they differ in electric charge they are

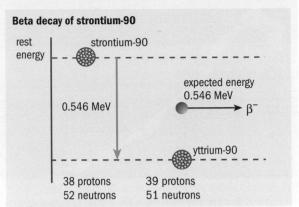

Beta decay of strontium-90

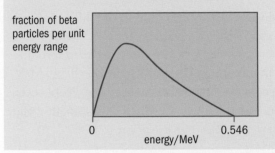

Energy spectrum of beta decay of strontium-90

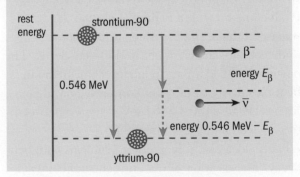

Beta decay of strontium-90, including antineutrino emission

Neutrinos carry away the "missing" energy

otherwise very similar. In these kinds of decays the total number of nucleons – protons plus neutrons – remains the same because the only changes are from proton to neutron or neutron to proton. This idea was later supported by the discovery that neutrons and protons share the same strong nuclear force holding them together, and that both are built from quarks (p204).

However, the above equation for the decay of strontium-90 can't be correct. If it were, the beta particles would have to come out with a single

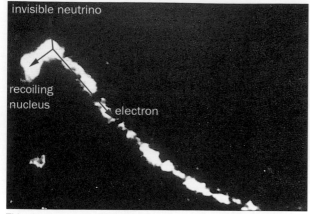

This cloud chamber picture shows the direction of travel of the nucleus and the electron in a beta decay. To conserve momentum an unseen neutrino must have exited vertically.

fixed value of energy, equal to the difference in rest energies of the two nuclei, that is 0.546 MeV. But in fact their energies vary, from near to zero up to a maximum of the expected value 0.546 MeV. So where does the excess energy go? People even wondered aloud whether beta decay might violate the law of energy conservation.

Neutrinos

In 1930 Austrian theoretical physicist Wolfgang Pauli came up with a brilliant solution. He suggested that a third, unseen and almost undetectable particle is emitted in beta decay. The particle is called an antineutrino $\bar{\nu}$ (neutrino in beta-plus decay). Energy is conserved for all beta decays because the 0.546 MeV is shared between the electron and the antineutrino.

Pauli's idea also solved other problems with beta decay. The creation of the antineutrino balances the creation of an electron, in the same way that an anti-electron balances an electron in pair creation. Neutrinos and electrons are assigned to the same particle family, called leptons, and all interactions (as far as is known) conserve lepton number. This works by giving electrons and neutrinos lepton number 1 and their antiparticles (positrons and antineutrinos) lepton number −1.

Perhaps you begin to see more clearly why antimatter is a theoretical necessity. There is more to conserve than just electric charge.

Considerations like these define the properties of Pauli's antineutrino:

- it must be neutral (to conserve electric charge);
- it must be an antilepton with lepton number −1 (to conserve lepton number);
- it must carry away energy;
- it must carry away linear momentum (to conserve momentum);
- it must interact extremely weakly with matter (otherwise it would be detected).

The best initial guess was that its mass – that is, its rest energy – is zero. If so, it travels at the speed of light, as a photon does, with momentum $p = \frac{E}{c}$.

It was 26 years before the neutrino was detected (but by then no one doubted its existence). You may be surprised to learn that billions of neutrinos have passed through you as you read this sentence. But don't worry, they are so weakly interacting that most of them would pass through a light-year of lead without hitting anything.

The full equation for the beta decay of strontium-90 is now:

$$^{90}_{38}\text{Sr} \longrightarrow {}^{90}_{39}\text{Y} + {}^{0}_{-1}\text{e} + {}^{0}_{0}\bar{\nu}$$

with conserved particle properties:

nucleon number: $90 = 90 + 0 + 0$
electric charge: $38 = 39 - 1 + 0$
lepton number: $0 = 0 + 1 - 1$

The weak interaction

Radioactive beta decay is not due to the electromagnetic interaction and the exchange of photons between charged particles. Its cause is a new interaction, called the weak interaction, which changes the nature of particles. Like electromagnetism, the weak interaction works through the exchange of bosons.

There are three new bosons involved in the weak interaction. One, the Z^0, is electrically neutral and can be thought of as a kind of massive photon. The other two, W^+ and W^-, are charged. They too are massive. A neutron can change into a proton by emitting a W^- boson, which then decays into an electron and an antineutrino. Similarly, a proton can change into a neutron by emitting a W^+ boson, which decays into a positron and a neutrino. A neutrino can interact with an electron in an atom, knocking it out of the atom, by exchanging a Z^0 boson. However, by nuclear standards, these weak processes are extremely rare.

Quick check

1. Which of the following particles are leptons: protons, electrons, positrons, antiprotons, neutrinos, neutrons, antineutrinos, antineutrons?

2. Here is an equation for alpha decay. State the conservation laws that apply to this decay and say how they are satisfied. $^{238}_{92}\text{U} \longrightarrow \,^{234}_{90}\text{Th} + \,^{4}_{2}\alpha$

3. Free neutrons are unstable and decay to protons (this is the underlying process in beta decay). Explain why the equation below cannot be fully correct for neutron decay:
$^{1}_{0}\text{n} \longrightarrow \,^{1}_{1}\text{p} + \,^{0}_{-1}\text{e}$

4. Explain why there must be an antineutrino emitted as well as the electron in the neutron decay above.

5. How does pair production of electrons and positrons conserve:
   (a) electric charge,
   (b) energy,
   (c) lepton number?

6. Why are neutrinos so difficult to detect?

Links to the *Advancing Physics* CD-ROM

Practise with these questions:

10S   Short answer *Things that don't change*

20S   Short answer *Beta decay and conservation*

30S   Short answer *Creation and annihilation*

40M   Multiple choice *Particles and interactions*

Try out this activity:

10S   Software-based *Bubble chamber photographs*

Look up these key terms in the A–Z:

Antimatter; electron; mass and energy; neutrino; pair production and annihilation; positron

Go further for interest by looking at:

10T   Text to read *The discovery of beta decay*

Revise using the revision checklist

10S   Computer screen *Annihilation and pair production: Bubble chamber pictures*

30O   OHT *Conserved quantities in electron–positron annihilation*

40O   OHT *Pair creation and annihilation*

100O OHT *Beta decay of strontium-90*

17.2 Scattering and scale

You can't simply look inside an atom, a nucleus or a proton to see what it is made of. Instead, particles are accelerated to high energy and directed at a target. Arrays of detectors surrounding the target track and identify the particles created and scattered in the collision. The target may even be another beam of particles travelling in the opposite direction. Such scattering experiments provide information about the structure and scale of the subatomic objects involved.

The discovery of the nucleus

Hans Geiger and Ernest Marsden carried out the first and best-known scattering experiment under the direction of Ernest Rutherford in 1909. They fired alpha particles at a sheet of thin gold foil and the pattern of scattering produced led to Rutherford's model of the atom with a tiny, dense, positive nucleus and distant orbiting electrons.

Rutherford, Geiger and Marsden knew that alpha particles were helium atoms that had lost their electrons (helium is found if alpha particles are collected). So they knew that alpha particles have a positive electric charge, but they couldn't have any idea how tiny these helium nuclei were, or how small were the gold nuclei at which the alpha particles were being fired. After all, they hadn't yet discovered the existence of the nucleus.

Rutherford and his colleagues did know that alpha particles have a lot of energy, typically 5 MeV – millions of times larger than the few electron volts of energy needed to rip electrons out of atoms. So they expected the alpha particles to go right through the thin gold foil, being barely, if at all, deflected. So sure was Rutherford of this that he and Geiger had picked the experiment as a simple bit of not-very-important practice for the young student Marsden.

They got a big surprise. A few of the alpha particles bounced right back from the foil, being scattered by angles greater than 90°. What could stop an alpha particle and turn it back in its tracks? Answer: it must be something massive. Then why are only very few – something like 1 in 10 000 – turned back like this? Possible answer: both the alpha particle and whatever it hits must be very small so that most alpha particles go straight

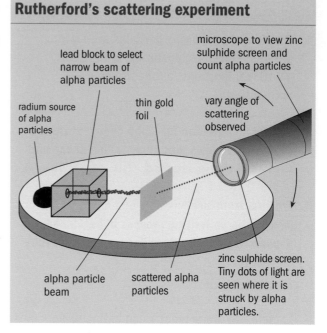

Rutherford's scattering experiment

lead block to select narrow beam of alpha particles

microscope to view zinc sulphide screen and count alpha particles

radium source of alpha particles

thin gold foil

vary angle of scattering observed

alpha particle beam

scattered alpha particles

zinc sulphide screen. Tiny dots of light are seen where it is struck by alpha particles.

through. Final question: what force might be responsible? Likely answer: the electrical repulsion between the positively charged alpha particle and a positively charged core of the gold atoms.

This kind of reasoning led Rutherford to the idea that all atoms consist of a very small, massive, positively charged nucleus, surrounded at much larger distances by electrons. The alpha particle was simply the nucleus of helium.

Several tests of this picture were possible.

- If the alpha particles were slowed down, would more be deflected at greater angles since the nucleus should now more easily turn them back? Yes, they were (Geiger and Marsden used mica absorbers to slow the alpha particles).
- Would nuclei of smaller electric charge scatter alpha particles less strongly, as expected? Yes they did (Geiger and Marsden replaced the gold with lighter metals such as aluminium).
- Would the pattern of numbers of alpha particles scattered at different angles fit the pattern expected from an inverse square law for electrical repulsion? Yes, Rutherford worked it out and the predicted pattern was a good fit.

The really big surprise was how small the nucleus of an atom turns out to be, concentrating very nearly the whole mass of the atom in a tiny core. The large energy of alpha particles means that they can get pretty close to a nucleus before

Key summary: Rutherford's picture of alpha scattering

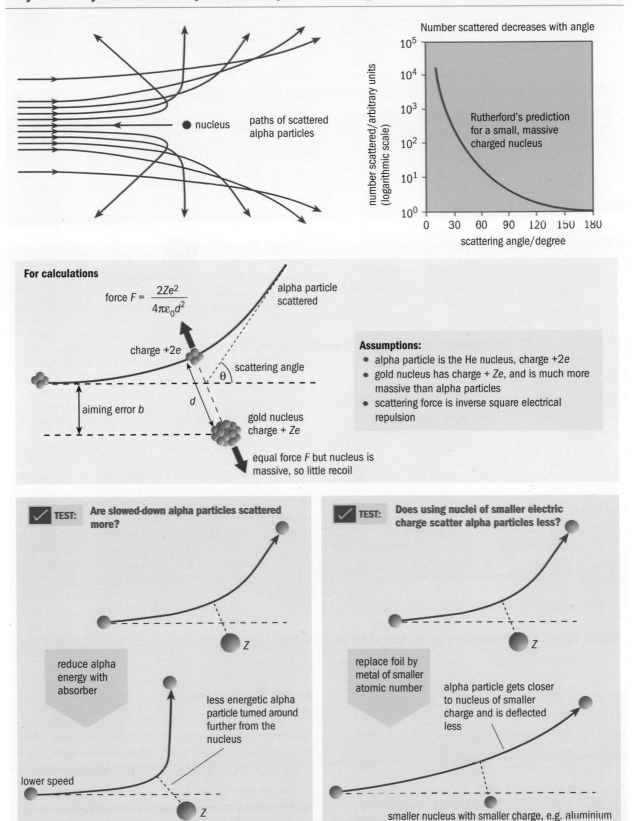

Number scattered decreases with angle

paths of scattered alpha particles

nucleus

Rutherford's prediction for a small, massive charged nucleus

number scattered/arbitrary units (logarithmic scale)

scattering angle/degree

For calculations

$$\text{force } F = \frac{2Ze^2}{4\pi\varepsilon_0 d^2}$$

alpha particle scattered

charge +2e

scattering angle

θ

aiming error b

d

gold nucleus charge + Ze

equal force F but nucleus is massive, so little recoil

Assumptions:
- alpha particle is the He nucleus, charge +2e
- gold nucleus has charge + Ze, and is much more massive than alpha particles
- scattering force is inverse square electrical repulsion

TEST: **Are slowed-down alpha particles scattered more?**

reduce alpha energy with absorber

less energetic alpha particle turned around further from the nucleus

lower speed

Z

Z

TEST: **Does using nuclei of smaller electric charge scatter alpha particles less?**

replace foil by metal of smaller atomic number

alpha particle gets closer to nucleus of smaller charge and is deflected less

Z

smaller nucleus with smaller charge, e.g. aluminium

Careful investigation of alpha scattering supported the nuclear model of the atom

Key summary: distance of closest approach

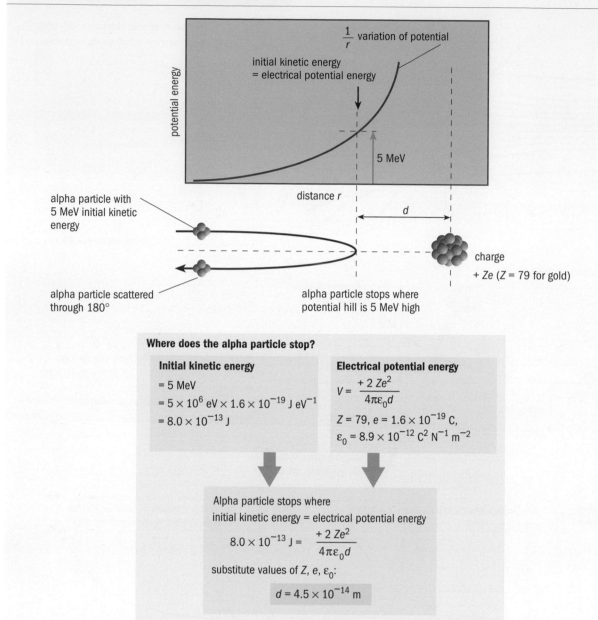

Where does the alpha particle stop?

Initial kinetic energy

= 5 MeV

= 5×10^6 eV $\times 1.6 \times 10^{-19}$ J eV^{-1}

= 8.0×10^{-13} J

Electrical potential energy

$$V = \frac{+ 2\, Ze^2}{4\pi\varepsilon_0 d}$$

$Z = 79$, $e = 1.6 \times 10^{-19}$ C,

$\varepsilon_0 = 8.9 \times 10^{-12}$ C^2 N^{-1} m^{-2}

Alpha particle stops where

initial kinetic energy = electrical potential energy

$$8.0 \times 10^{-13} \text{ J} = \frac{+ 2\, Ze^2}{4\pi\varepsilon_0 d}$$

substitute values of Z, e, ε_0:

$$d = 4.5 \times 10^{-14} \text{ m}$$

The radius of a gold nucleus must be less than 10^{-14} m. Atoms are 10 000 times larger than their nuclei.

being turned back. For 5 MeV alpha particles scattered by gold nuclei, the distance of closest head-on approach turns out to be as little as 4 to 5×10^{-14} m. So the nucleus must be even smaller than that. Remember that atoms are typically 10^{-10} m in size. At around 10^{-14} m, the size of a nucleus is 10 000 times smaller. If the nucleus were the size of a table tennis ball, the atom would be bigger than the dome of St Paul's Cathedral. Atoms really are nearly all empty space.

Inside the nucleus

Scattering and other experiments show that the number of protons, and so the electric charge on the nucleus, is equal to the atomic number Z of the element in the periodic table. By bending beams of ions in electric and magnetic fields (chapter 16) you can measure the masses of atoms, and so of their nuclei (by subtracting the small mass of the electrons in a precision measurement). These nuclear masses are always larger than can be accounted for by the number of protons (except, of course, for hydrogen).

The extra mass can be explained if the nucleus contains two kinds of particle of approximately equal mass – electrically charged protons and uncharged neutral particles called neutrons. The neutrons add mass without adding charge.

This idea also explains why atoms of the same element, with exactly the same chemical properties but slightly different mass, are frequently found. These isotopes have the same number of protons but differ in the number of neutrons. The number of protons decides the positive charge on the nucleus, and so the number of negative electrons to make a neutral atom. The number of electrons determines the chemical properties of the atom. But a small difference in the number of neutrons can make a difference to the mass, and may make an isotope radioactive. Carbon-14 used in carbon dating (p7) is an example of such a radioactive isotope, as is the oxygen-15 used in PET scans (p189).

Electron scattering measures the nucleus

The electrical repulsion between nucleus and alpha particle kept Rutherford's alpha particles at arm's length from the nucleus. From the perspective of the alpha particles he used, the nucleus just looked like a point charge.

Electron scattering has proved to be a very good tool for getting a close-up view of the nucleus. The big advantage of electrons is that, unlike alpha particles, they are not affected by the force that holds nuclei together so they just map the distribution of electric charge in the nucleus.

Exactly like alpha particles, electrons undergo Rutherford scattering (though the force is attractive instead of repulsive), so most pass straight by and a few are scattered through large angles.

But there is something else. As you saw in chapter 7 (in *Advancing Physics AS*), electrons – like photons – can be thought of as having a wavelength. The wavelength is given by de Broglie's relation:

$$\lambda = \frac{h}{p}$$

where p is the momentum and h is the Planck constant.

Key summary: inside the nucleus

		Atomic number Z	Mass number A	Neutron number N
Hydrogen	●	1	1	0
Deuterium	●●	1	2	1
Helium (alpha particle)	●●●●	2	4	2
Lithium	●●●●●●	3	6	3

● proton   ● neutron

Z = atomic number = number of protons. Charge = $+ Ze$.
A = mass number = number of nucleons (protons + neutrons)
$N = A-Z$ = number of neutrons

Isotopes
Nuclei with same charge but different numbers of neutrons. Atoms of isotopes are identical in chemical properties, different in mass.

Examples:

		Z	A	N
$^{12}_{6}C$	carbon–12	6	12	6
$^{14}_{6}C$	carbon–14	6	14	8

		Z	A	N
$^{235}_{92}U$	uranium–235	92	235	143
$^{238}_{92}U$	uranium–238	92	238	146

Protons add mass and electric charge to the nucleus, neutrons simply add mass

Electrons scattered by a small spherical nucleus behave much like photons diffracted by a small disc. The diffraction pattern is a series of concentric rings round a strong central maximum. The angular size of the rings depends on the ratio λ/d, where d is the diameter of the obstacle. So from the size of the rings, the size of the nucleus can be found. The diffraction pattern is superimposed on the Rutherford scattering curve. In practice, it is usually only possible to locate the first minimum of the ring pattern, but this is enough to measure the size of the nucleus. The theory is essentially the same as the diffraction ideas explained in chapter 6, with electrons instead of photons, and with a new kind of diffracting obstacle.

Key summary: the size of nucleus from scattering

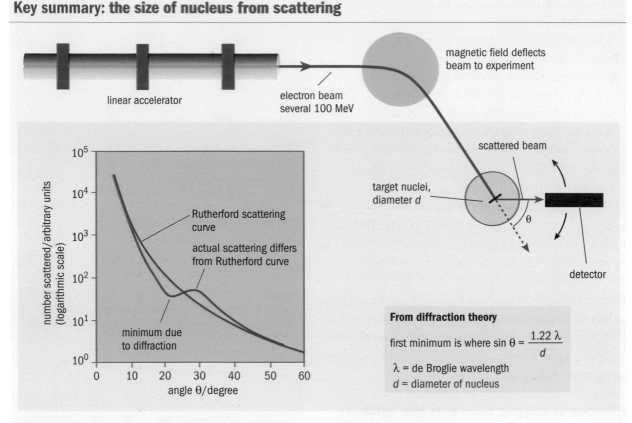

Nuclei diffract electron beams as small particles or apertures diffract light

To get a small enough wavelength, the electrons must be accelerated to several hundred mega-electron volts (MeV), much larger than their rest energy mc^2 of 0.5 MeV. The relativistic calculation of the momentum and so of the wavelength is then very simple. To a good approximation:

$$p \simeq \frac{E}{c} \text{ and so } \lambda = \frac{h}{p} \simeq \frac{hc}{E}$$

For electrons accelerated to 300 MeV, the wavelength comes to about $\lambda = 4 \times 10^{-15}$ m, similar to the dimensions of a nucleus.

Nuclear matter

Measurements by electron scattering of the radii of different nuclei give a striking result. A graph of the volume $\frac{4}{3}\pi r^3$ of a nucleus against the number A of nucleons is a straight line. This means that the volume occupied by each nucleon is much the same in every nucleus. The nucleons must be crammed together like molecules in a drop of liquid. Each extra nucleon adds about the same

extra amount to the volume. And in fact a liquid drop model of the nucleus proves very useful in accounting for nuclear fission (chapter 18).

Since the volume $\frac{4}{3}\pi r^3$ of the nucleus is proportional to the mass number A, the radius of a given nucleus will be proportional to the cube root of A. That is:

$$r = r_0 A^{1/3}$$

The value of the constant r_0, effectively the radius of a single nucleon, is close to 10^{-15} m.

Because of its constant density, you can think of a nucleus as being made of "nuclear matter", as a water drop is made of water. But nuclear matter is no ordinary stuff. For a start, its density is enormous. Divide the mass of a nucleus by its volume and you get a density so great that a matchbox full of nuclear matter has a mass of five billion tonnes.

Nuclear matter does actually exist in nature, and in bulk. When some massive stars (supernovae, chapter 12) run out of nuclear fuel they collapse and explode, leaving behind a core made of neutrons – a neutron star. Spinning rapidly, these cores may be detected using pulses of radio waves that they emit. They are the pulsars found by Jocelyn Bell Burnell and Anthony Hewish (chapter 1, *Advancing Physics AS*). A neutron star is thus rather like a large solid neutral nucleus. It may be only a few kilometres in diameter.

In the nucleus itself, there must be some pretty strong glue to hold it together. Its protons all repel one another, so the electrical forces act to blow every nucleus apart. The nuclear glue is a strong attractive force between all the particles making up the nucleus.

Deeper still: inside the nucleons themselves

In the alpha particle scattering experiments that showed the existence of the nucleus, and the electron scattering experiments that more accurately measured their sizes, no particle creation went on. The nuclei were undisturbed. This is called elastic scattering – no energy is taken from the scattering particle to materialise new particles or to excite nuclei to higher states of energy.

The next big step required energies of the order of a few GeV. There is then enough energy to materialise proton–antiproton pairs and neutron–antineutron pairs. This is because the mass 1.7×10^{-27} kg of such a nucleon gives it a rest energy mc^2 of about 1 GeV.

However, by the 1960s things had become very embarrassing for particle physicists. As accelerator energies rose, they found that they could create large numbers of new kinds of particle, with increasing mass. Names of new "fundamental" particles proliferated – Δ (delta), Σ (sigma), Ξ (xi), and so on. Middleweight particles, less massive than nucleons, such as the π- and K-mesons were already known. All came with various electric charges – positive, negative or zero. Some vanished almost as soon as they were created (in perhaps 10^{-24} s); others lived longer (maybe 10^{-10} s).

Why was this wonderful new zoo of particles an embarrassment? Well, not all of them could be truly fundamental. It seemed obvious that most of them must be combinations of more fundamental components. Patterns were found among the

Key summary: density of nuclear matter

Volume of nucleus increases linearly with number of nucleons

Electron scattering measures radius r of nucleus

calculate volume $= \frac{4}{3}\pi r^3$

[Graph: y-axis "volume of nucleus $= \frac{4}{3}\pi r^3 / 10^{-45}$ m^3" from 0 to 1500; x-axis "number of nucleons" from 0 to 200. Data points labelled ^{1}H, ^{4}He, ^{12}C, ^{16}O, ^{28}Si, ^{59}Co, ^{88}Sr, ^{122}Sb, ^{197}Au along a straight line. Dashed lines mark 100 nucleons at approximately 700.]

Estimate from graph:

100 nucleons in volume 700×10^{-45} m^3

Data:

mass per nucleon u $= 1.7 \times 10^{-27}$ kg

volume per nucleon $= 7 \times 10^{-45}$ m^3

Calculate density:

$$\text{density} = \frac{\text{mass}}{\text{volume}}$$

$$\text{density} = \frac{1.7 \times 10^{-27} \text{ kg}}{7 \times 10^{-45} \text{ m}^3}$$

$$= 2.4 \times 10^{17} \text{ kg m}^{-3}$$

Density of nuclear matter is roughly 2×10^{17} kg m^{-3}

A matchbox full of nuclear matter would have a mass of five billion tonnes

particles and their properties (rather as Mendeleev and others had found patterns among the chemical properties of the elements). But still nobody knew how to think about which particles might be fundamental, not built up out of others.

The answer, when it came, was so surprising that many were very cautious about it. American theoretical physicists Murray Gell-Mann and George Zweig saw that the patterns among these

Key summary: quark models of nucleons

Key summary: quarks explain other particles

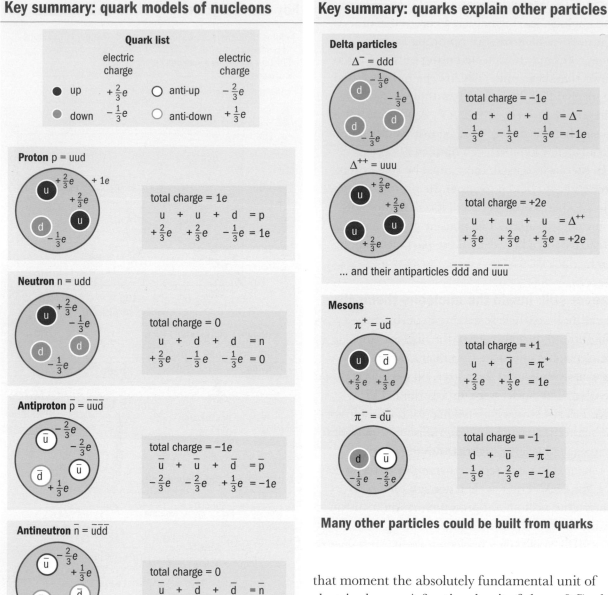

The nucleon family could be built from sets of two kinds of quark

Many other particles could be built from quarks

particles could be explained if *none* of them were fundamental. Instead, they could all be imagined as built out of combinations of new kinds of particles, which Gell-Mann named quarks.

The trouble was that quarks had to be very peculiar indeed, so much so that it was hard to believe in them. The oddest thing about quarks is that they must have electric charges of either $\frac{1}{3}$ or $\frac{2}{3}$ of the charge on an electron – up until

that moment the absolutely fundamental unit of electric charge. A fractional unit of charge? Could there be such a thing, never before seen?

To build protons and neutrons needs two kinds of quark, differing in what is whimsically called flavour. For no good reason, the two flavours are called up (u) and down (d). But protons and neutrons don't use up all the possible combinations of these two quarks. What about the unused combinations "ddd" and "uuu"? These account for two of the remaining zoo – the Δ^- and Δ^{++} particles.

What's more, the middleweight mesons could be described too. They are made in a different way, combining one quark and one antiquark into pairs. So quite a number of different particles can be accounted for, just by shuffling patterns of two kinds of quarks and their antiquarks.

Deep inelastic scattering

Meanwhile, back at the accelerator labs, things were stirring. The question was, are the patterns of quarks just number games, or are there really small quark-like objects inside protons and neutrons? The Stanford Linear Accelerator had enough energy – up to 20 GeV – to find out. Its electrons were good probes of the inside of nucleons because their only interaction with the charged quarks is electromagnetic – through exchange of photons. So they could map the charged particles inside the proton or neutron.

The collision of a high-energy electron with a quark does much more than just deflect the electron through a large angle, as in Rutherford's experiments. Instead, particle creation starts happening. Quark–antiquark pairs materialise out of the energy of the interaction. Following roughly the path of the quark as it is given a huge kick by the electron, they emerge as a jet of new particles. Many of the new particles are mesons, which the theory says are quark–antiquark pairs. This kind of inelastic scattering, with energy going to create sprays of new particles, is obviously more complicated than elastic scattering.

However, Richard Feynman realised that relativistic effects make this complicated situation much simpler. To the accelerated electron, the proton must look like a particle approaching it at nearly the speed of light. Time dilation will slow down the motion of the quarks so that they seem almost at rest. They become sitting targets.

This smart thinking of Feynman's made it much easier to deduce the pattern of electric charges in the nucleon from the pattern of scattering. It was possible to show that the scattering was consistent with the existence of three particles inside a neutron or proton. From the way the amount of scattering depends on the charge on the scattering particle, it was possible to check that quarks do indeed have charges a fraction of the fundamental unit e.

Feynman described the discovery of quarks by deep inelastic scattering as like studying a swarm of bees by radar. Similarly, Rutherford once likened firing neutrons at nuclei to shooting at birds in the dark in a country where there aren't many birds. The discovery of the nucleus by elastic

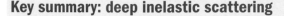

Key summary: deep inelastic scattering

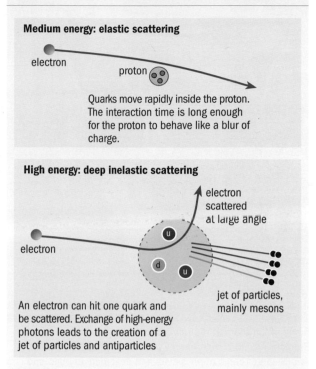

Medium energy: elastic scattering

electron

proton

Quarks move rapidly inside the proton. The interaction time is long enough for the proton to behave like a blur of charge.

High energy: deep inelastic scattering

electron
scattered
at large angle

electron

jet of particles, mainly mesons

An electron can hit one quark and be scattered. Exchange of high-energy photons leads to the creation of a jet of particles and antiparticles

At high energies individual quarks scatter electrons

alpha scattering and the discovery of the quark by inelastic electron scattering show how a similar technique can yield crucial results on radically different scales of size.

Energy and scale

You will have noticed that the smaller the scale you want to resolve, the larger the energy you need to probe it. This is the main reason why particle physicists need to build ever larger and more expensive accelerators.

The detector array in a large accelerator fills an underground hall as big as a block of flats. It consists of layers of detectors to pick up and identify the many energetic particles produced. All this to find out what goes on in a region less than 10^{-15} m across, in which a huge energy was concentrated.

Many scattering experiments involve the electromagnetic interaction between electric charges. If the probe and target carry charges q_1 and q_2 then the kinetic energy needed to approach to within a distance r is given by:

$$\text{kinetic energy required} = \frac{q_1 q_2}{4\pi\varepsilon_0 r}$$

Some of the team at the relativistic heavy ion collider at Brookhaven, USA. High-energy particle physics experiments involve large teams of physicists and engineers to build and maintain the equipment, and to collect and analyse results.

For example, firing protons at protons, an energy of 10 eV gets them to within a distance equal to the atomic scale, around 10^{-10} m. To get them as close as the size of a nucleus, of the order of 10^{-15} m, requires 10^5 times more energy, about 1 MeV.

Of course, the de Broglie wavelengths must also be comparable to or smaller than the structure under investigation, so this is another reason why high energies are needed to probe small-scale structures. A third reason is the need to provide the rest energy $2mc^2$ to create particle–antiparticle pairs of particles of mass m.

Coloured quarks and gluons

Something very strong must hold the quarks together inside a particle. The idea is that, just as electrically charged particles exert forces on one another by exchanging photons, so quarks attract one another by exchanging particles – rather unimaginatively called gluons. They are the glue that keeps protons and neutrons in one piece and they also lead to the further forces that make nucleons attract one another, so keeping the nucleus together. This gluon interaction is known as the strong interaction.

Quark–gluon interaction

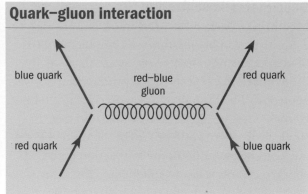

Quarks interact by exchanging gluons, which change the quark colours. Here a red quark and a blue quark exchange a red–blue gluon. The red quark becomes blue and the blue quark becomes red. The quarks exchange energy and momentum.

Quarks carry a new kind of charge called colour. Electric charge comes in two kinds – positive or negative. Colour charge is different and comes in three forms called red, green and blue. The reason for choosing these names is that free particles are always colourless (white) combinations of quarks. Protons and neutrons have one each of red, green and blue quarks. The term colour in no way means that quarks are prettily painted. It's just that the way that the three primary colours combine resembles the way that colour charges work.

There's another way in which gluons are very different from photons. Photons are not electrically charged, so they don't attract one another, but gluons actually carry colour charge themselves – they emit and absorb gluons. The result is that while the electromagnetic interaction via photons gets weaker with distance, the quark colour interaction actually gets stronger with distance. It's like a rubber band – the farther you pull the stronger the force. This is the reason why quarks can't be knocked out of particles to exist on their own. Pulling two quarks apart simply stores more and more energy in the gluon field, until it materialises as a quark–antiquark pair. If you try to kick a quark out of a particle, you get a new combination of quarks.

This explains something that really worried physicists when the quark idea was first suggested – they couldn't find them. At long last, it was understood that free quarks can't exist because they can't be simply pulled apart.

At the deepest level, quarks and gluons do have

an important resemblance to electrons and photons. Quarks and electrons are both matter-like particles. They are fermions. Gluons and photons are both force-like particles, being bosons.

The many kinds of particles, starting with protons and neutrons, that can be built out of combinations of quarks, and of quarks and antiquarks, are collectively called hadrons. The quarks in hadrons are all held together by gluons and all hadrons are affected by the strong interaction.

...and more quarks

We have told you only a small part of the quark story. In 1964, when the idea was invented, the big puzzle was that experimentalists had produced new particles with quite strange properties. Not two but three flavours of quark were needed to account for them. The third was called the strange quark. At the time, just these three were enough to explain all the known members of the zoo of particles, and to successfully predict the existence of further particles.

However, the family of quarks and other fundamental particles has grown since then (p217). Today, physicists believe that there aren't any more of the currently known kinds of fundamental particle left to find. Anything new is expected to be really new, not just more of the same.

...and more creation and annihilation

In 2000 the Large Electron–Positron Collider (LEP) at the European Organisation for Nuclear Research (CERN) in Geneva closed down to make way for an even more powerful accelerator. LEP was designed to collide pairs of electrons and positrons together with as much energy as possible, up to 200 GeV. You already know that this can produce gamma-rays as the particles annihilate (p191) but much more can happen. For example, the pair can produce a virtual photon, which then materialises into a quark–antiquark pair flying in opposite directions. From them come jets of a whole variety of composite particles. Matter is both destroyed and created in one interaction. Quantum fields are indeed creators and destroyers.

The next step is to collide more massive particles than electrons and positrons, so as to achieve higher energies still, and to probe even smaller

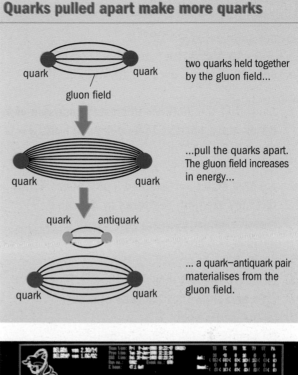

Quarks pulled apart make more quarks

two quarks held together by the gluon field...

...pull the quarks apart. The gluon field increases in energy...

... a quark–antiquark pair materialises from the gluon field.

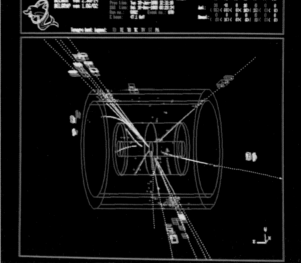

In this event recorded in the LEP's DELPHI detector, a Z^0 particle decays giving two jets. The tracks of two muons can be seen on the right of the image.

scales. This is what the new Large Hadron Collider (LHC) at CERN does.

Collisions of electrons and positrons are rather clean, starting only from their electromagnetic interactions. Collisions between hadrons are much messier. Also, because the interactions are stronger, they are more frequent as well as more complicated. The new detectors have to collect much more data, much more rapidly, than before. The LHC computers have to work extremely fast, and very hard, to process the flow of data.

Quick check

Useful data: $e = -1.6 \times 10^{-19}$ C,
$\varepsilon_0 = 8.9 \times 10^{-12}$ C^2 N^{-1} m^{-2}, $c = 3.0 \times 10^8$ m s^{-1},
$h = 6.6 \times 10^{-34}$ J Hz^{-1}, u $= 1.7 \times 10^{-27}$ kg.

1. Show that the distance of closest approach of a
   6 MeV alpha particle to a nucleus of iron $^{56}_{26}$Fe is
   about 1.2×10^{-14} m.

2. A 5 MeV alpha particle and a proton accelerated
   to 5 MeV are both scattered by the same
   massive nucleus. Which gets closer to the
   nucleus in a head-on collision?

3. Compare the two nuclei thorium $^{234}_{90}$Th and iron
   $^{56}_{26}$Fe and show that the ratio of their:
   (a) masses is about 4.2;
   (b) electric charge is about 3.5;
   (c) neutron number is about 4.8;
   (d) radii is about 1.6.

4. How many quarks are there in an alpha particle?

5. Show that the rest energy of a proton, mass
   about 1 u, is close to 1 GeV.

6. The decay of a neutron to a proton can be
   thought of as changing one down quark in the
   combination udd to an up quark, producing the
   combination uud. Show that the values of the
   quark charges require charge $-e$ to be carried
   away in this process.

Links to the *Advancing Physics* CD-ROM

Practise with these questions:
70S   Short answer *Rutherford
scattering: Energy and closest
approach*
80S   Short answer *Rutherford
scattering: Directions of forces*
90S   Short answer *Electrons "measure
the size of nuclei"*
100S Short answer *Scattering and scale*

Try out these activities:
80S   Software-based *Probes scattered
by a target*
90S   Software-based *Many probes
scattered by a target*

Look up these key terms in the A–Z:
Mass and energy; neutron; nucleon;
proton; quark; strong nuclear force

Go further for interest by looking at:
80T   Text to read *Tracking particles*

Revise using the revision checklist and:
1200 OHT *Rutherford's picture of alpha
particle scattering*
1300 OHT *Distance of closest
approach*
1400 OHT *Density of nuclear matter*
1500 OHT *Deep inelastic scattering*
1600 OHT *Quarks and gluons*

17.3 The music of the atoms

Thanks to the scanning tunnelling electron microscope, it is now possible to see the effects of the de Broglie waves associated with electrons. In particular, standing waves can be seen if electrons are trapped in some kind of box. It's the same with a guitar string (chapter 6, *Advancing Physics AS*). When the string is plucked, standing waves form on the string. The only standing waves that can exist on such a string have wavelengths that let them fit into the length of the string.

The harmonious music of the guitar is down to the wavelengths of the standing waves on strings being in simple numerical relationship. Harmony and simple numbers go together. You will see how the music of the atoms also depends on simple numbers, called quantum numbers, and for the same reason – the existence of standing waves.

A guitar string has standing waves in one dimension. The orange colour of carrots and mangoes is due to one-dimensional electron standing waves on a molecular guitar string. The string is the carotene molecule, which is long and stiff. Some electrons are not tightly bound to the atoms in the molecule (they are delocalised) and spread along the string. Because the molecular string is quite long, the wavelength of a standing wave on it can be relatively large. The carotene

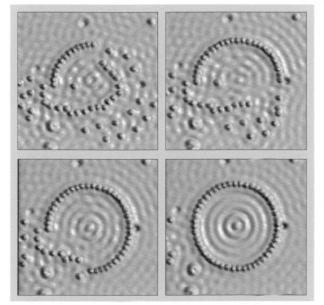

This scanning tunnelling electron microscope image shows individual iron atoms being arranged in a circle on the surface of copper. As the iron atoms form a complete circle, a standing electron wave pattern appears inside the circular trap.

Key summary: standing waves in boxes

Waves on a string

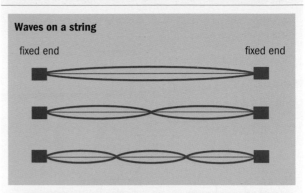

fixed end   fixed end

Waves on a circular diaphragm

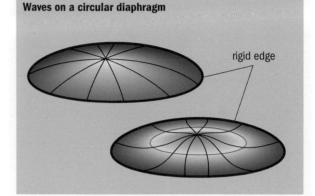

rigid edge

Only certain field patterns are possible because the waves must fit inside the box

Key summary: molecular guitar strings

Carotene molecule C$_{40}$H$_{56}$

electrons spread along the molecule

Analogy with guitar string

electrons make standing waves along the molecule

length L

electron wavelength proportional to L

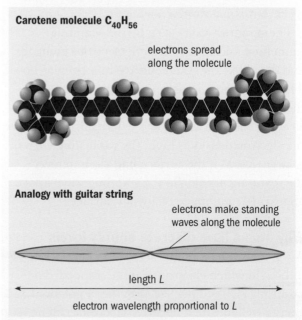

Long molecules absorb visible wavelengths of light

Key summary: waves and energy levels

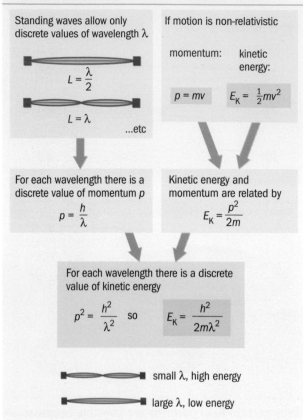

Standing waves allow only discrete values of wavelength λ

$$L = \frac{\lambda}{2}$$

$$L = \lambda$$

...etc

For each wavelength there is a discrete value of momentum p

$$p = \frac{h}{\lambda}$$

If motion is non-relativistic

momentum: kinetic energy:

$$p = mv \qquad E_K = \tfrac{1}{2}mv^2$$

Kinetic energy and momentum are related by

$$E_K = \frac{p^2}{2m}$$

For each wavelength there is a discrete value of kinetic energy

$$p^2 = \frac{h^2}{\lambda^2} \quad \text{so} \quad E_K = \frac{h^2}{2m\lambda^2}$$

small λ, high energy

large λ, low energy

Discrete wavelengths imply discrete energy levels

Key summary: trapping an electron

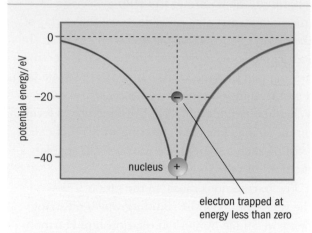

electron trapped at energy less than zero

nucleus +

Electrons are trapped in the box formed by the potential energy well of the nucleus

gravitational potential well (chapter 11).

If electrons are trapped in a box they must have definite discrete wavelengths, which depend on the size of the box. Given the wavelength λ you know the momentum p of the electron through the de Broglie relationship:

$$p = \frac{h}{\lambda}$$

The energy of an electron in an atom is quite small so you can use the non-relativistic expressions for momentum $p = mv$ and kinetic energy:

$$E_K = \tfrac{1}{2}mv^2 = \frac{p^2}{2m}$$

These give a relationship between electron kinetic energy and de Broglie wavelength:

$$E_K = \frac{h^2}{2m\lambda^2}$$

The smaller the box, the narrower the wavelength, and so the larger the momentum and the kinetic energy. The size and shape of the box also determine exactly which wavelengths are allowed. In turn this selects particular allowed values of the energy. The electrons in an atom can only have one of these allowed energy values. No in-between states exist, just as no in-between notes exist on a guitar. These permitted energies are the energy levels of electrons in atoms.

molecule absorbs blue light, so that it looks orange or yellow. The molecules that absorb light in your retina work in a similar way.

The standing waves seen by the scanning tunnelling microscope are not electrons going up and down like the water surface in a circular bowl. The tunnelling microscope detects places where the electron density – the probability to find an electron – is high and places where it is low. Where the electron density is high, the tunnelling current is large, just as if the surface had risen to meet the scanning needle. So the standing wave image appears like an image of a wavy surface.

Why atoms have discrete energy levels

An atom can be thought of as a kind of pocket, or box, in which electrons can be trapped. The positively charged nucleus provides a potential energy well in which negative electrons can be bound, unable to escape (chapter 16). It's similar to the way that you are trapped by Earth's

Key summary: spectral lines and energy levels

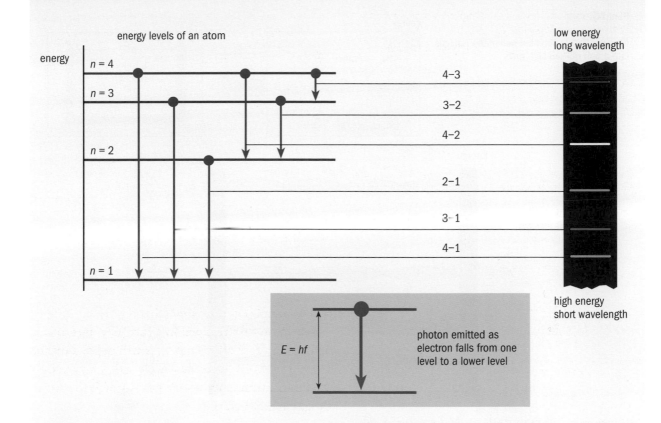

Spectral lines map energy levels. *E = hf* is the energy difference between two levels.

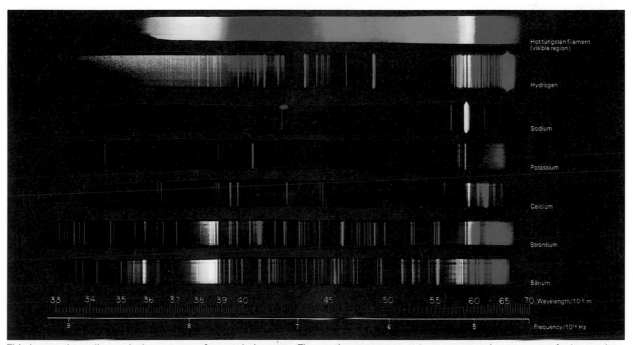

This image shows line emission spectra of several elements. The continuous spectrum from tungsten is one reason for its use in domestic light bulbs, whereas the orange light from sodium gives the familiar colour of many streetlights. The ultraviolet region (shown in white) starts at about 400 nm (40×10^{-8} m on the scale shown).

Key summary: a guitar-string atom

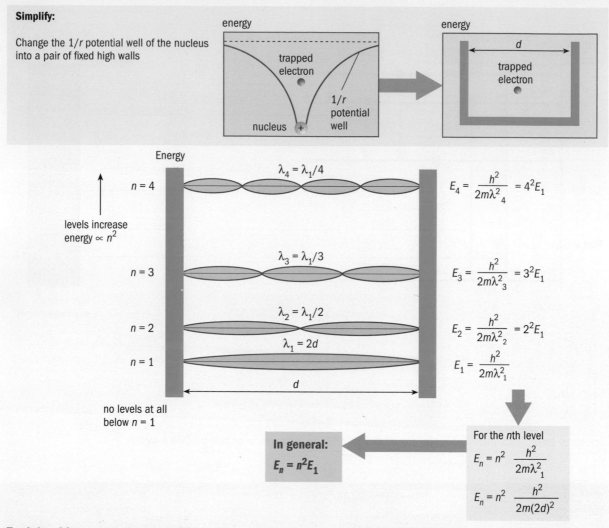

Simplify:

Change the $1/r$ potential well of the nucleus into a pair of fixed high walls

energy — trapped electron — $1/r$ potential well — nucleus +

energy — d — trapped electron

Energy — levels increase energy $\propto n^2$

$n = 4$ $\lambda_4 = \lambda_1/4$ $E_4 = \dfrac{h^2}{2m\lambda_4^2} = 4^2 E_1$

$n = 3$ $\lambda_3 = \lambda_1/3$ $E_3 = \dfrac{h^2}{2m\lambda_3^2} = 3^2 E_1$

$n = 2$ $\lambda_2 = \lambda_1/2$ $E_2 = \dfrac{h^2}{2m\lambda_2^2} = 2^2 E_1$

$\lambda_1 = 2d$

$n = 1$ $E_1 = \dfrac{h^2}{2m\lambda_1^2}$

d

no levels at all below $n = 1$

In general:
$E_n = n^2 E_1$

For the nth level
$E_n = n^2 \dfrac{h^2}{2m\lambda_1^2}$

$E_n = n^2 \dfrac{h^2}{2m(2d)^2}$

Each level has a quantum number n. The energy depends on the quantum number.

Energy levels and spectral lines

Every yellow sodium streetlight or red neon sign gives evidence that electrons in atoms have discrete energy levels. Their light comes in sharp discrete spectral lines. Each line is light carried by photons of discrete wavelength and frequency, and so of discrete energy $E = hf$. The energies of the photons are just the difference in energy between two of the possible energy levels of electrons in the atom.

The emission of radiation is a perfect experimental tool to map the energy levels of an atom. Excite the atoms, measure the spectrum and calculate energy differences from the frequencies. The spectrum thus provides a scrambled map of the energy levels inside the atom.

A guitar-string atom

To get an idea about how the music of the atoms works, involving simple numbers, we are going to describe the simplest imaginable model of an atom – a guitar-string atom. It's obviously wrong, because it changes the curving slopes of the potential energy well of the nucleus into hard fixed walls and it is only one-dimensional. But it has one correct, crucial feature – the electron waves are trapped in a definite space.

Because they are trapped, the electrons form standing waves. Because the walls are simply vertical, the standing waves are very simple – exactly like those on a guitar. Their music is almost as simple, with the energy depending just

Key summary: how small could a hydrogen atom be?

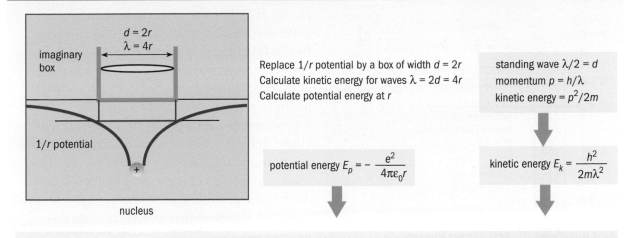

$d = 2r$
$\lambda = 4r$
imaginary box

$1/r$ potential

nucleus

Replace $1/r$ potential by a box of width $d = 2r$
Calculate kinetic energy for waves $\lambda = 2d = 4r$
Calculate potential energy at r

standing wave $\lambda/2 = d$
momentum $p = h/\lambda$
kinetic energy $p^2/2m$

potential energy $E_p = -\dfrac{e^2}{4\pi\varepsilon_0 r}$

kinetic energy $E_k = \dfrac{h^2}{2m\lambda^2}$

Find the minimum radius of an atom for total energy < 0

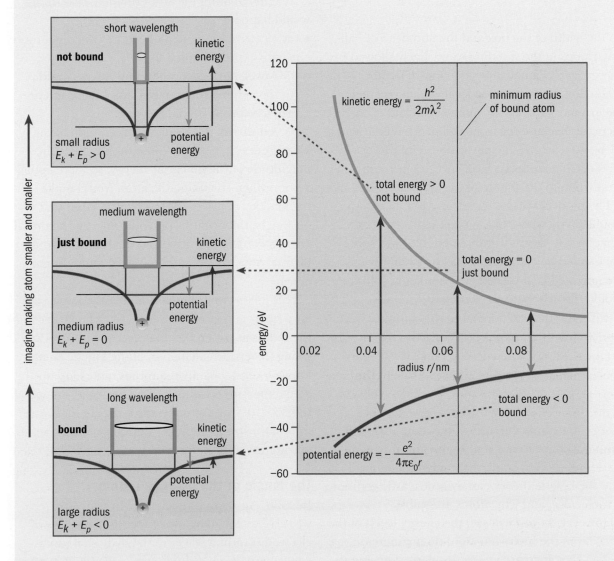

imagine making atom smaller and smaller

not bound
short wavelength
kinetic energy
potential energy
small radius $E_k + E_p > 0$

just bound
medium wavelength
kinetic energy
potential energy
medium radius $E_k + E_p = 0$

bound
long wavelength
kinetic energy
potential energy
large radius $E_k + E_p < 0$

kinetic energy $= \dfrac{h^2}{2m\lambda^2}$

minimum radius of bound atom

total energy > 0 not bound

total energy = 0 just bound

total energy < 0 bound

potential energy $= -\dfrac{e^2}{4\pi\varepsilon_0 r}$

energy/eV
120
100
80
60
40
20
0
−20
−40
−60

radius r/nm
0.02   0.04   0.06   0.08

If the size is too small, the kinetic energy is too large for electrical potential energy to bind the electron

Key summary: the energy levels of hydrogen

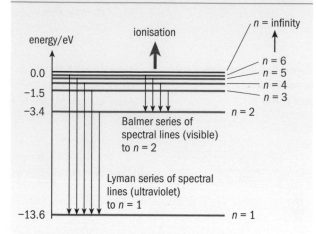

As *n* increases, E_n gets closer to zero

on the width of the box and the number *n* of half-wave loops in the standing wave. Each allowed energy level is labelled by a quantum number *n*, starting with the lowest level, which is called the ground state. So the momentum and energy increase. Because the energy depends on the square of the wavelength, the musical scale of this crude guitar-string atom has level energies increasing in proportion to the square of this quantum number *n*.

The model explains something that may have troubled you when, in the kinetic theory of gases (chapter 13), atoms were treated like hard spheres, unchanged by collisions. If electrons simply orbited the nucleus like a planetary system, any collision should disturb them (as two solar systems would be disturbed if they collided), but the standing wave model shows that atoms only have discrete possible values of energy with a lowest possible "ground floor" energy (*n* = 1). The collisions due to the thermal motion of atoms and molecules (energy of the order of kT) simply don't have enough energy to disturb the state of the atom unless the temperature becomes very high. This is why the matter all around you largely stays unchanged, why the kinetic theory can assume elastic collisions of molecules, and why atoms are stable.

However, as you will see, the energy levels of the electron in the hydrogen atom do not increase as n^2 (in fact they increase as $1/n^2$ instead). The guitar-string atom gets that part totally wrong. A much less crude approximation is needed.

Atoms can't be smaller than they are

The wave-in-a-box idea can explain why atoms are the size they are. The electron in (say) a hydrogen atom must be bound to the nucleus. Its total energy must therefore be negative. If the total energy is positive, the electron will escape. The attraction of the nucleus gives it negative potential energy, so that's all right. But – and here's the rub – if the electron is boxed into a closed space, the standing waves require it to have momentum, and therefore kinetic energy. The kinetic energy gets bigger the smaller the box. Adding the positive kinetic energy to the negative potential energy always brings the total energy up nearer towards zero.

Imagine making the atom smaller. Two things would happen:
- the electrostatic potential energy would get more negative (proportional to $1/r$);
- the wavelength of the standing waves would get shorter, making the electron kinetic energy increase (proportional to $1/r^2$).

As you shrink the atom the magnitude of the kinetic energy increases more rapidly than the magnitude of the potential energy, so there comes a point when shrinking the atom would make the kinetic energy of the electron so large that it outweighs the negative potential energy. The total energy would be positive and the electron would be free. Squeeze an atom too hard and it bursts apart.

This means that there's a smallest possible size for a hydrogen atom. Its radius must be larger than the radius at which the total energy becomes zero. Using the crude calculation of the kinetic energy for waves-in-a-box gives a radius quite close to the actual measured radius of a hydrogen atom, which is 0.53×10^{-10} m. Actually, the estimate makes the minimum size a bit bigger than the actual size, so beware the approximations here.

The music of the hydrogen atom

In 1926, Austrian physicist Erwin Schrödinger was able to calculate the shapes and energies of electron standing waves in the hydrogen atom. The energy levels of hydrogen came out as:

$$E_n = \frac{-13.6 \text{ eV}}{n^2}$$

The result is simple but the calculation is rather hard. Schrödinger had to work out standing wave patterns in three dimensions. And he had to do so for the soft box of the hydrogen atom with its sloping sided $1/r$ potential energy well. But this let him derive, just from the quantum behaviour of electrons and the shape of the electric potential, the energies and frequencies of the music of the hydrogen atom.

The principal quantum number n takes values 1, 2, 3, etc. with $n = 1$ the lowest allowed state, called the ground state. As n increases, the energies get closer and closer together, and the energy E_n gets closer and closer to zero. If $E > 0$ the electron has positive total energy and breaks free of the atom – the atom ionises. The ionisation energy is just the energy needed to get an electron from the lowest level at -13.6 eV up to zero energy, when it is free. So the ionisation energy of hydrogen is 13.6 eV.

If you are studying chemistry you may have used the notation 1s, 2p etc to denote electron orbitals. The principal quantum number n is just the number used in this notation (e.g., for 2s, $n = 2$).

The dependence of hydrogen energy levels on $1/n^2$ was by no means unexpected. Ten years earlier Danish physicist Niels Bohr had reached the same answer by a different argument. More than twenty years before that, Swiss school teacher Johann Balmer hit on a simple numerical formula for the frequencies of visible lines in the hydrogen spectrum. If Balmer took the numerical differences

$$\frac{1}{2^2} - \frac{1}{n^2}$$

for integer values of n greater than 2, and multiplied by a constant, he could get the line frequencies.

Looking back, knowing that $E = hf$, you can see that Balmer's formula was giving photon energies. It was saying that the energies of the photons are all differences between two terms. That's what you'd expect if photons are emitted when an electron goes from one energy level to a lower one.

The number 2 appears because the lines Balmer could see in the visible part of the spectrum came from transitions from higher levels down to the energy level $n = 2$. It wasn't long before American physicist Theodore Lyman found lines in the ultraviolet spectrum corresponding with

Key summary: standing waves in hydrogen

Energy	Principal quantum number	Standing wave patterns (two-dimensional representation)
$E = -13.6$ eV $= \dfrac{-13.6 \text{ eV}}{1^2}$	$n = 1$	0.5×10^{-10} m
$E = -3.4$ eV $= \dfrac{-13.6 \text{ eV}}{2^2}$	$n = 2$	
$E = -1.5$ eV $= \dfrac{-13.6 \text{ eV}}{3^2}$	$n = 3$	

Only spherically symmetric standing waves (s-states) are shown

density of shading shows probability of finding the electron

Some 3-dimensional standing waves in hydrogen

transitions to the lowest energy level, $n = 1$. In this way, the waves-in-a-box model of the atom explains how the music of spectral frequencies depends on integer numbers.

Achievements of quantum theory

We have outlined how quantum theory explains the simplest of all atoms, hydrogen. But it has achieved much more than this, explaining, for example:

- the structure and energy levels of more complex atoms, leading to an explanation of the periodic table of the elements;
- the origin and strengths of the different types of chemical bond;
- the behaviour of the free electrons in metals, which give them their high conductivity;
- the existence of superconductivity and superfluidity at low temperatures;

Quantum theory is also the basis of modern technologies including: scanning tunnelling microscopes; lasers; superconducting magnets; semiconductor chips in computers; magnetic resonance imaging; and digital memories.

For all but the simplest problems, computational models (chapter 10) are essential.

Quick check

Useful data: $h = 6.6 \times 10^{-34}\,\text{J}\,\text{Hz}^{-1}$, $c = 3.0 \times 10^8\,\text{m}\,\text{s}^{-1}$, $e = -1.6 \times 10^{-19}\,\text{C}$. Energy of a hydrogen atom is ground state $n = 1$ is $-13.6\,\text{eV}$

1. Why is the total energy of an electron negative when it is in an atomic energy level?

2. Show that the energy, frequency and wavelength of photons emitted by a hydrogen atom going from state $n = 4$ to $n = 2$ are about $2.5\,\text{eV}$, $6.2 \times 10^{14}\,\text{Hz}$ and $480\,\text{nm}$.

3. Show that the minimum energy needed to ionise a hydrogen atom when it is in the $n = 2$ excited state is about $3.4\,\text{eV}$.

4. Explain why nuclear spectra involve gamma-rays whereas atomic spectra involve visible (or near-visible) light.

5. Use the energy level equation $E_n = -\dfrac{13.6\ \text{eV}}{n^2}$ for hydrogen to derive Balmer's formula that he wrote in the form $\dfrac{1}{\lambda_n} = R\left(\dfrac{1}{2^2} - \dfrac{1}{n^2}\right)$

   where R is a constant. Show that $hcR = 13.6\,\text{eV}$.

6. The spectrum of light from sodium vapour has a bright yellow line at wavelength $590\,\text{nm}$. Show that sodium atoms have a pair of energy levels differing in energy by about $2.1\,\text{eV}$.

Links to the *Advancing Physics* CD-ROM

Practise with these questions:
150S Short answer *Spectra and energy levels*
160S Short answer *How small could a hydrogen atom be?*
170S Short answer *Carrots and guitar strings*
180S Short answer *The hydrogen spectrum*

Try out this activitiy:
150S Software-based *Sizing up a hydrogen atom*

Look up these key terms in the A–Z:
Atom; electric potential; electron; energy level; model of the atom; photon; quantum theory

Go further for interest by looking at:
40T   Text to read *Quantum theory in the twentieth century*

Revise using the revision checklist and:
1700 OHT *Standing waves in boxes*
1800 OHT *Colours from electron guitar strings*
1900 OHT *Energy levels*
2000 OHT *Standing waves in atoms*
2100 OHT *Size of the hydrogen atom*

17.4 Known and unknown

It is time to draw everything together.

- Matter is ultimately made from just two different classes of particles – quarks and leptons, which are both fermions.
- These particles interact by exchanging bosons (e.g. photons in electromagnetic interactions).
- Particles are created and annihilated in particle–antiparticle pairs.

Here we will summarise how these facts fit into what is called the standard model.

Matter

The view from particle physics about what the world is made of is breathtakingly simple. The ordinary matter around you is made from just one lepton (the electron) and the two different types of quark (up and down) in protons and neutrons. These three, together with the neutrino that goes with the electron, explain almost all the visible matter in the universe.

Almost, but not quite. For some reason nature repeats the pattern of four fermions twice more at higher energies and masses, in particles observable only at high energies. "Who ordered that?", the American physicist Isidor Rabi exclaimed, hearing about the first of the massive electrons, the muon.

The pattern gives three generations of particles. Are there more to be found? Bringing together astronomical and particle collision evidence, current opinion is that there are just these three generations of fundamental particles, no more.

Interactions

Most of what you see happening around you is due to the electromagnetic interaction, carried by photons. Virtual photon exchanges between electrons are at the bottom of chemical change, and of the physics of everyday things.

The idea of a single electromagnetic interaction comes from the 19th century, when James Clerk Maxwell used Michael Faraday's ideas to bring together electric and magnetic interactions in one unified scheme. In the 1960s, Steven

Matter

Generation	Leptons	Electric charge	Rest energy /MeV	Quarks	Electric charge	Rest energy /MeV
1 The world around you	e^- electron	−1	0.511	d down	−1/3	8
	ν_e electron–neutrino	0	very near 0	u up	+2/3	4
2	μ^- muon	−1	106	s strange	−1/3	150
	ν_μ muon–neutrino	0	very near 0	c charmed	+2/3	1500
3	τ^- tau	−1	1780	b bottom	−1/3	4600
	ν_τ tau–neutrino	0	very near 0	t top	+2/3	180 000?

Interactions

Interaction	Force carrier	Electric charge	Rest energy/GeV	Status
electromagnetism	photon	0	0	observed
weak interaction	Z^0	0	93	observed
	W^+	+1	81	observed
	W^-	−1	81	observed
strong interaction	8 different colour combinations of gluons	0	0	indirectly observed
gravity	graviton	0	0	expected theoretically

Left: The Hubble deep field, showing galaxies out to the farthest distances that telescopes can reach. Right: Particle tracks reconstructed from data about a collision in the DELPHI detector at CERN. There is still a problem in theoretical physics of combining theories of gravity, dominant on the largest scales in the universe, with quantum theories of fundamental particles. Can they be made compatible?

Weinberg and Abdus Salam, developing an idea from Sheldon Glashow, showed a way to unify electromagnetism and the weak interaction. The theory introduces three more bosons, in addition to the photon, to carry the interactions – a neutral Z particle (a kind of massive photon) and electrically charged W$^+$ and W$^-$ bosons.

The remaining subatomic interaction is the strong interaction, carried by gluons exchanged between quarks. Eight gluons are needed altogether. Many particle physicists believe that it will be possible to unify the strong and electroweak interactions, in a grand unified theory.

That leaves the oldest known field – gravity. Newton said he didn't understand its nature. It still has not been fitted into a unified picture with the other interactions.

The unknown

Here are some of the questions that particle physicists would like to be able to answer.

- What decides the masses of the various particles? At present, masses of particles have to be found experimentally as no theory predicts them from more basic principles.
- Where does mass come from, anyway? There is a theory (the Higgs field) of how particles acquire mass. An associated particle, the Higgs boson, is predicted.
- Why do fundamental particles come in pairs of two leptons and two quarks? Is there any relationship between the leptons and the quarks?

The energies required to test ideas about this could be so large that the theories might be effectively untestable.

- Why are there only three generations of fermions? Nobody knows.
- Can the strong interaction be unified successfully with the weak and electromagnetic interactions? Can there be a unified field theory?
- Can gravity be related to the other interactions? Can its exchange particle, the graviton, be detected?

Of these questions, perhaps the origin of mass and an explanation of the masses of the various particles are the most likely to be soon resolved. At the time of writing (late 2007) the LHC, designed to look for the Higgs boson that is conjectured to give particles their mass, is just about to start running.

Particle physics has helped to answer the big questions in astronomy. The energies in particle collisions approach the energy in the first fractions of a second after the Big Bang, so particle physics helps to describe the birth of the universe.

Finally, particle physics involves some mind-bending ideas. For example, antiparticles can be thought of as ordinary particles going backwards in time. The behaviour of matter may not be symmetrical with respect to reversing the direction of time. Emptiness is full of activity on the tiniest of scales, with particle–antiparticle pairs continually coming into and going out of existence. Nothingness may be the basis of everything.

Summary check-up

Particles and antiparticles ✓

- Particles can be created and annihilated in matter–antimatter pairs
- Electrons and positrons have opposite electric charge and lepton number
- Electrons are fermions, which obey the Pauli exclusion principle
- Leptons are particles such as the electron and neutrino and their antiparticles
- Quarks with fractional electric charges combine in threes to form neutrons, protons and other particles
- Quarks combine in particle–antiparticle pairs to form mesons

Interactions and conservation ✓

- Total relativistic energy and momentum, as well as charge, are conserved in interactions
- Electromagnetic interactions arise from the exchange of photons
- The strong colour force between quarks is carried by gluons, which are bosons, like photons. Particles that feel this force are called hadrons.

Scattering ✓

- Scattering experiments reveal the structures of atoms, nuclei and nucleons; the smaller the scale the greater the energy needed
- Atoms have tiny, positively charged nuclei made of protons and neutrons packed together at high density
- At high energies, scattering experiments create a large number of new kinds of particle

Atoms ✓

- Electrons confined in a region of space can be modelled as standing waves, with wavelengths determined by the size and shape of the confining region
- The de Broglie wavelength is given by $\lambda = \frac{h}{p}$
- Discrete atomic energy levels correspond to discrete electron standing waves in an atom
- Electrons can make quantum jumps between allowed energy levels. A downwards jump emits a photon whose energy is given by $E = hf = E_{initial} - E_{final}$.
- The energy levels in hydrogen are given by $E_n = \frac{13.6 \text{ eV}}{n^2}$, where n is the principal quantum number

Questions

Useful data: $e = -1.6 \times 10^{-19}$ C, $\varepsilon_0 = 8.9 \times 10^{-12}$ C^2 N^{-1} m^{-2}, $c = 3.0 \times 10^8$ m s^{-1}, $h = 6.6 \times 10^{-34}$ J Hz^{-1}, $u = 1.7 \times 10^{-27}$ kg; $G = 6.7 \times 10^{-11}$ N m^2 kg^{-2}

1. **Data on the radii of nuclei and the number of nucleons in each nucleus are shown in the table below.**

Nucleus	Atomic number Z	Atomic mass number A	Radius/ 10^{-15} m
^{1}H	1	1	1.00
^{4}He	2	4	2.08
^{12}C	6	12	3.04
^{16}O	8	16	3.41
^{28}Si	14	28	3.92
^{59}Co	27	59	4.94
^{88}Sr	38	88	5.34
^{122}Sb	51	122	5.97
^{197}Au	79	197	6.87

(a) Plot a graph to test the idea that the volume per nucleon is approximately constant.
(b) Obtain from your graph an estimate of the density of nuclear matter.
(c) Estimate the mass of a teaspoonful of nuclear matter.
(d) Estimate the gravitational field at the surface of a neutron star of radius 1 km.

2. **Outline experimental evidence for each of the following:**
(a) Atoms contain a tiny nucleus.
(b) The atomic nucleus is positively charged.
(c) The atomic nucleus contains most of the mass of the atom.
(d) Radii of nuclei are of the order of 10^{-15} m.
(e) Protons and neutrons contain smaller particles, called quarks.

3. **What is the de Broglie wavelength of:**
(a) an electron accelerated through 1000 V?
(b) an alpha particle with 5 MeV of kinetic energy?
(c) an electron with energy 10 GeV? Why must you use the relativistic approximation $p = E/c$ here?

4. **How, approximately, do the following properties of lead-206 ($Z = 82$) and strontium-88 ($Z = 38$) nuclei compare? Give answers as simple numerical ratios.**
(a) mass;
(b) charge;
(c) density;
(d) neutron number;
(e) radius.

5. **Give an example of a scattering experiment. Take care to explain:**
(a) what the target particles are;
(b) what the probe particles are;
(c) what energy is involved;
(d) the scale on which the experiment gives information;
(e) a result inferred from the experiment.

6. **Describe how up quarks of charge $+\frac{2}{3}$ and down quarks of charge $-\frac{1}{3}$ can be used to construct the following:**
(a) proton, charge +1;
(b) neutron, charge 0;
(d) Δ^{++} particle, charge +2;
(e) Δ^- particle, charge −1.

7. **State the charge and mass for each of the following antiparticles:**
(a) antiproton;
(b) antielectron;
(c) antineutrino.

8. **What is the minimum energy of a gamma-ray required to create:**
(a) an electron–positron pair?
(b) a muon–antimuon pair?
(c) a proton–antiproton pair
Mass of electron = 9.1×10^{-31} kg; mass of muon = $207 \times$ mass of electron; mass of proton = 1.67×10^{-27} kg.

9. **Carotene is a straight molecule about 2 nm long. Certain delocalised electrons effectively occupy the whole length of the molecule. Treating their standing waves as similar to those on a stretched string, calculate:**
(a) the electron wavelengths for the lowest energy state ($n = 1$) and the next state ($n = 2$);
(b) the difference in kinetic energy of an electron in the $n = 1$ and $n = 2$ states;
(c) the frequency of a photon emitted in a transition from states with $n = 2$ to $n = 1$;
(d) the wavelength of the photon in (c).
(e) Give a reason why carotene molecules can absorb light in the visible part of the spectrum.

18 Ionising radiation and risk

Ionising radiation is used to treat cancers but it can also cause them. In this chapter we discuss the risks and benefits of using ionising radiations and the processes that produce them. We will give examples of:

- ionising radiations in use in industry, at home and in medicine
- ways of calculating and comparing risks
- nuclear transformations producing radiations
- fission and fusion in nuclear power production

18.1 Radiation put to use

Ionising radiations are put to good use in the home, in industry and in medicine.

- Ionising radiations are easily detected, which is why they can be used as tracers, for example in medicine and the oil industry.
- As these radiations ionise matter they lose energy and so they are absorbed as they pass through matter. The fire alarm in your home sounds when alpha particles in it are absorbed by smoke particles in the air.
- When absorbed, the energy from the radiations warms things up. This is how radioactive sources help power the *Cassini* spacecraft.
- The energy carried by the radiations means that

Ionising radiation: some benefits...

A rough day in the North Sea on an oil exploration platform. The flow patterns of a new oil deposit are investigated using radioactive tracers that emit gamma-radiation.

The *Cassini* probe sets out across space to Saturn. Its power comes from thermoelectric generators. The temperature difference required to make these work is produced when plutonium decays and emits alpha radiation, the energy from which is used to generate electrical power.

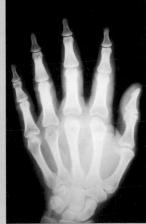

Each day these machines sterilise about 300 m³ of medical supplies, enough to completely fill a large room. They use gamma-radiation to irradiate the syringes, bandages and dressings.

An X-ray image is made by X-ray photons ionising silver atoms in a photographic film. Grains of silver are formed around the silver atoms affected by radiation.

On 19 September 1991 a frozen corpse was found by two German hikers high on the Similaun glacier in the Ötztal Alps, at the border between Italy and Austria. Carbon dating techniques using the amount of beta radiation emitted confirmed that this was an unfortunate Neolithic traveller.

Ionising radiation: some risks...

The beam from a particle accelerator is a powerful and dangerous form of ionising radiation. Here you see a blue glow as the beam ionises the air in its path.

Cosmic rays are a matter of concern to airline crew and passengers flying high in the atmosphere.

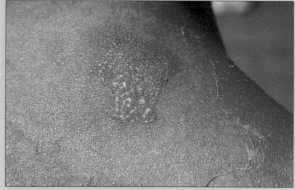

Ultraviolet light can ionise molecules in tissue, producing chemically active free radicals that can cause tissue damage. This is what makes sunburn dangerous.

they can damage living matter. This is put to good use in sterilising medical equipment.

● The amount of radiation from a radioactive source decreases with time in a predictable way. For example, carbon-14 dates the iceman pictured on p221 to 5300 years old (chapter 10). Nothing is completely safe – there are always risks as well as benefits. Ionising radiations must be treated with particular respect. The radioactivity

that warms the thermocouple in the *Cassini* space probe is essentially the same as that released by nuclear fission in a nuclear power station. Replacing fossil fuel power stations by nuclear power would help reduce greenhouse gases. But then what should be done with their radioactive waste? Ionising radiation and living things must be mixed with care.

Exposure to radiation is one way that the mutations that drive evolution can occur, but exposure to too much radiation kills. The radiation that kills cancerous cells must not harm healthy cells so much that they cannot repair themselves. The same mechanism of energy release that gives the potential for mass destruction in nuclear bombs also drives the Sun.

No use of ionising radiation can have zero risk. The only sensible answer to the question: "Is it totally safe?" is "No", but you can ask how the risk can be reduced, and by how much. You can ask whether the benefits are worth the risk; and you also have to consider the consequences of not using radiation.

Ionising atoms

Ionising radiations are simply radiations that ionise the atoms of the material they are passing through. There are many kinds of ionising radiations including cosmic rays, beams of protons or neutrons, X-rays and ultraviolet light, and the radiations from radioactive decay.

To be able to ionise an atom by knocking an electron out of it requires an energetic form of radiation. Radio waves, infrared and visible light can't do it. The air in your toaster doesn't begin to conduct when the heating elements glow (chapter 14). To remove an electron from nitrogen needs about 14 eV (chapter 17).

A visible photon has only about 2 eV, but an ultraviolet photon has sufficient energy to break molecular bonds. A burnt red nose after a day in the sunshine is uncomfortable, but also increases the risk of a skin cancer. X-ray and gamma-ray photons both have large energies, from thousands to millions of electron volts. The only difference is in their origin – X-rays are photons produced by accelerated electrons, gamma-rays are photons emitted by nuclei.

A cloud chamber picture of alpha particle tracks. The tracks are straight and have a definite range. Particles of different energies show different ranges.

In a cloud chamber, electrons (β^-) make rather faint and erratic tracks. They do not ionise air very strongly and are easily deflected by collisions with electrons in atoms.

The nuclear radiations from radioactive decay can have energies of a few MeV per particle, so all of them – whether alpha (He nuclei), beta (electrons or positrons) or gamma (photons) – can have enough energy to ionise 100 000 or so atoms.

Ionisation and radiation protection

Ionising radiations lose energy as they pass through matter. A thick enough shield of matter will protect you from a source of ionising radiation. The more readily the radiations ionise atoms in the shield material, the less the thickness required.

At one extreme are alpha particles, strongly ionising and consequently easily stopped. At the 5 MeV typical of radioactive sources they are stopped by a sheet of paper, aluminium foil or even a tenth of a millimetre of water. The smoke detector in your home is triggered by the extra absorption of alpha particles by smoke in air. Even in air, they typically travel only 50 mm or so before stopping so alpha particles outside your body present little danger. By contrast, if you breathe

in or ingest a substance that emits alpha particles you can be sure that they will be stopped in nearby tissue, damaging cells and DNA. Radon gas from bricks, concrete and the ground is the most common source of this kind of danger (p227).

At the opposite extreme are neutrinos, which are not absorbed significantly by passing through the whole Earth, and equally present negligible danger as they interact even less with tissues in your body.

It is between these extremes that radiation protection becomes important. Lead-lined canisters are used to store radioactive sources. Thick steel and concrete walls shield nuclear reactors. As a rough rule, the more dense an absorber, the less the thickness needed to stop the radiation. Combining density and thickness gives the useful measure mass per unit area ($\text{kg m}^{-3} \times \text{m} = \text{kg m}^{-2}$). Within a factor of two or three, the mass per unit area needed to stop radiation of a given energy is similar across various absorbing materials. For a given material

Key summary: shielding from X-rays and gamma-rays

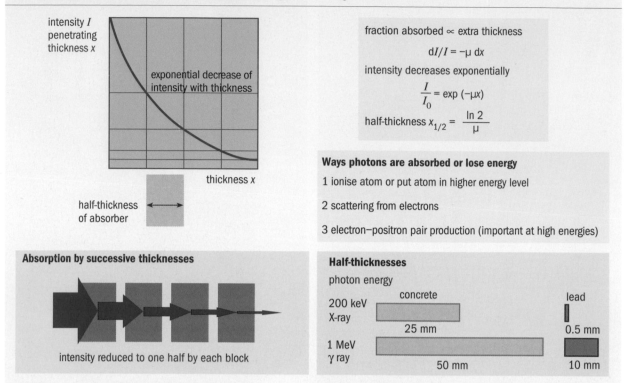

intensity I penetrating thickness x

exponential decrease of intensity with thickness

thickness x

half-thickness of absorber

fraction absorbed ∝ extra thickness

$$dI/I = -\mu \, dx$$

intensity decreases exponentially

$$\frac{I}{I_0} = \exp(-\mu x)$$

half-thickness $x_{1/2} = \dfrac{\ln 2}{\mu}$

Ways photons are absorbed or lose energy

1 ionise atom or put atom in higher energy level

2 scattering from electrons

3 electron–positron pair production (important at high energies)

Absorption by successive thicknesses

intensity reduced to one half by each block

Half-thicknesses

photon energy

200 keV X-ray — concrete — 25 mm — lead — 0.5 mm

1 MeV γ ray — 50 mm — 10 mm

The intensity of radiation let through decreases exponentially with thickness of the shield

the thickness of the shielding needed increases dramatically as the energy of the radiation increases. Take two examples:

Beta radiation The electrons (β-rays) emitted by radioactive substances typically have energies in the range 1–5 MeV. A few millimetres of solid material is needed to stop them, for example between 1 and 10 mm of aluminium. So they go right through your skin and are absorbed in the tissue a few millimetres beneath.

High-energy protons Astronauts may be exposed to high-energy cosmic rays (mostly protons), especially when there are solar storms on the Sun. At an energy of 5 MeV, protons, like alpha particles, are easily absorbed. A shield needs to provide roughly $1 \, \mathrm{kg \, m^{-2}}$ of matter in their path; that's 1 mm of water and less of aluminium or lead. But at 100 MeV, which protons from the Sun can reach, much more shielding is required, approximately $100 \, \mathrm{kg \, m^{-2}}$.

So the astronauts might shelter behind their water tanks if the water was around 100 mm thick but they can't expect the metal walls of the

spacecraft to protect them. Sheer weight prevents the capsule walls being made 30 mm thick.

However, if the proton energies reach 1 GeV, the shielding required increases to as much as several thousand $\mathrm{kg \, m^{-2}}$. Such energies are common in particle accelerators. You can see why several metres thick of shielding may be needed in the path of the beam. The whole thickness of Earth's atmosphere is needed to protect against 1 GeV or higher-energy particles in cosmic radiation.

Shielding from X-rays and gamma-rays

A doctor or dentist sees bone or teeth as white areas on an X-ray film because these materials absorb X-rays. Flesh or dental cavities absorb less well, the X-rays get through and the exposed film looks black. Both work because X-rays are not very strongly absorbed by matter so they can penetrate right through your body. For the same reason, thick dense shielding is needed to provide protection against X-rays and gamma-rays.

To a good approximation a given thickness of material reduces the number of X-ray or gamma-

Historical uses of X-rays

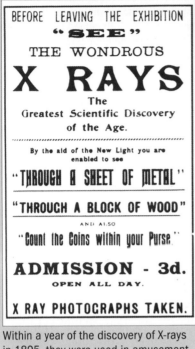

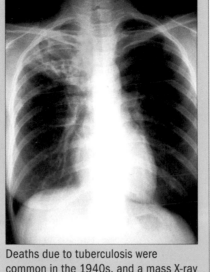

Within a year of the discovery of X-rays in 1895, they were used in amusement parks for entertainment. Today, taking the risk just for fun is unacceptable.

Deaths due to tuberculosis were common in the 1940s, and a mass X-ray programme to diagnose the disease was started. Now, with just two deaths per 100 000 population in the UK each year, the risk of the disease does not make the risk of the X-ray worthwhile. Here tubercular patches in the lung are picked out in false colour.

As late as the 1950s X-rays were used to ensure well fitting shoes. Comfort considerations are now considered to be not worth the risk.

ray photons by a constant fraction that depends on the photon energy. This means that the intensity of the radiation decreases exponentially with thickness. Just as for exponential radioactive decay where you can define the half-life, so for exponential decrease of intensity with thickness you can define the half-thickness. This is simply the thickness of absorber needed to halve the number of photons that on average get through.

At an energy of 200 keV, typical of medical X-rays, the half-thickness of a lead shield is about 0.5 mm. Ten such layers – 5 mm thick – would reduce the intensity by about 1000 times (2^{10}). This is why your dentist's X-ray machine is built into a lead shielded canister. A 250 mm thick concrete wall does the same, which makes it safe for other people outside the dental surgery.

However, gamma-rays in the region of 1 MeV, typical of radioactive substances, are much more penetrating. Half get through 10 mm of lead or 50 mm of concrete. For this reason medical gamma-ray sources such as cobalt-60 need to be given very substantial shielding.

"I'll just take an X-ray of that"

Your dentist takes an X-ray picture of the root of your aching tooth. You notice that the dentist and nurse leave the room or keep well away from the X-ray machine as the picture is taken. They are reducing the dose of X-rays that they get to minimise the risk to them of cell damage caused by ionising radiations. But what about you? Should you take a risk that they are careful to avoid?

The difference between you and the dentist is that the dentist may use the X-ray machine several times a day but you may only have a dental X-ray once a year. So the total dose the dentist gets is much larger than yours. By staying close to the X-ray machine, a dentist has to accept a much larger risk than each patient has to accept.

Still, there is some risk to you. The benefit to you is that, by seeing just what the problem is, the dentist can decide better what to do. A tooth may be saved that you otherwise might have lost. But the risk to you has to be pretty small for it to be worth taking for this much benefit.

Changes in public opinion influence perceptions

Key summary: radiation dose measurements

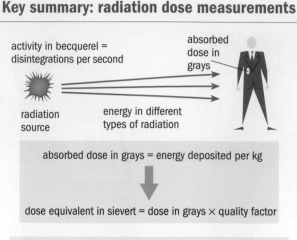

activity in becquerel = disintegrations per second

radiation source

energy in different types of radiation

absorbed dose in grays

absorbed dose in grays = energy deposited per kg

dose equivalent in sievert = dose in grays × quality factor

Radiation quality factors		
radiation	factor	dose equivalent of 1 gray/Sv
alpha	20	20
beta	1	1
gamma	1	1
X-rays	1	1
neutrons	10	10

The round numbers for quality factors show that scientific understanding of them is not precise

Hal Gray, after whom the unit of absorbed dose is named, came from a poor London family. His father taught him to do mental arithmetic and his mother taught him useful crafts. Both helped him become a first rate experimental physicist.

of how much benefit is needed to make such a risk acceptable; so too do new medical data and changes in medical practice. Trust is involved – you have to trust that medical practice is soundly based on fact and calculation, that your dentist is doing the best thing for you. Medical physics is a rewarding career for many physicists.

Ionising radiation and living matter

A little knowledge is a dangerous thing. For much of the first half of the 20th century, a little knowledge is just what physicists and doctors had about the effects of ionising radiations on living matter. Louis Harold "Hal" Gray was among the first hospital physicists to be appointed without medical training. He joined the Mount Vernon Hospital in London in 1933 after working on the absorption of gamma-rays by matter as a physicist at the Cavendish Laboratory in Cambridge. He developed quantitative methods of testing the action of radiation on living materials through painstaking experiments with bean roots.

Physicists knew how to count particles and

so how to measure radiation levels in particles per second. The becquerel is the unit of activity in disintegrations per second, or of intensity in particles arriving per second (chapter 10). But Gray realised that what matters is the energy absorbed by the tissue as well as the amount of radiation. It's the energy of the radiation and the ionisation damage it inflicts on cells that does harm. The absorbed dose – the number of joules absorbed per kilogram of tissue – is now measured in grays (Gy).

However, the energy absorbed isn't everything. It can do more harm or less. How much harm depends on the type of radiation and on the type of tissue exposed. Alpha radiation is highly ionising, beta and gamma-radiation less so. It was known as early as 1906 that the sensitivity of body cells to radiation varies, so a tissue weighting factor is also used. Rapidly dividing cells are most at risk – the testes and ovaries are more affected than skin and bone surfaces, by a factor of about 10.

For these reasons, the gray – simply the energy absorbed – is not a good enough unit for predicting possible consequences of exposure to radiation. So the sievert (Sv), named after the Swedish radiologist Rolf Sievert, is used. The various components of the absorbed dose in grays are multiplied by quality factors depending on the type of radiation and the type of tissue to give the dose equivalent in sieverts.

What is a "big" dose?

In Europe people typically get a dose equivalent to nearly 2000 microsievert (μSv) each year from natural background sources, including the food they eat and the air they breathe. When a government minister mentioned this in a TV broadcast in Hungary soon after the Chernobyl nuclear accident, people thought that it sounded a lot. Would they have thought so if he had said it

Key summary: average whole body dose equivalent per year from various sources (Europe)

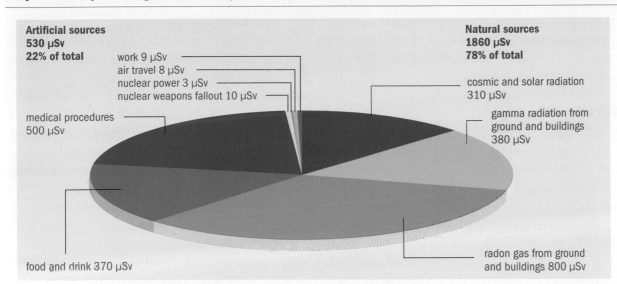

Artificial sources
530 µSv
22% of total

work 9 µSv
air travel 8 µSv
nuclear power 3 µSv
nuclear weapons fallout 10 µSv

medical procedures
500 µSv

Natural sources
1860 µSv
78% of total

cosmic and solar radiation
310 µSv

gamma radiation from
ground and buildings
380 µSv

food and drink 370 µSv

radon gas from ground
and buildings 800 µSv

Natural sources of radiation account for more than three-quarters of the average annual radiation dose

was 2 mSv or 0.002 sievert? It can be important to choose the appropriate units with which to communicate information to the public.

You might say that a big dose is one that adds appreciably to the natural background dose people receive. Take a chest X-ray, for example. The dose equivalent is about 200 mSv, or less when tissue factors are taken into account. That's up to a 10% addition to background, so you might decide to think of it as being appreciable. But before deciding you would want to know the risks incurred by such doses.

Assessing the risk of exposure to radiation starts with the dose equivalent in sieverts. It makes sense to ask: "Given such and such a dose equivalent, what is the probability of this or that consequence?" – for example, a genetic defect or developing a cancer. Over the years, the International Commission on Radiological Protection has collected data on the risks of radiation, starting with the effects on survivors of the atomic bombs dropped on Hiroshima and Nagasaki in 1945. Estimates of probability have changed, and are still very approximate. At the time of writing, the best estimate of the probability of radiation-induced cancer was 3% per sievert. That is, three out of a hundred people on average develop a cancer from exposure to a dose

Reduce the risk in proportion

Dose equivalent	Risk	
1 Sv	3%	or 3 in 100
200 µSv	$3 \times 200 \times 10^{-6}$%	or 6 in 1 million

equivalent of one sievert.

But 1 Sv is a very large dose – 500 times background. For smaller doses one approach is simply to reduce the risk in proportion. The table above shows an example of making the calculation this way for the risk from a chest X-ray of 200 µSv.

As a rough, widely used rule of thumb the probability is often rounded up to 5% per sievert. On this basis, the cancer risk from an amount of radiation equal to the natural background is of the order of 0.01% or 1 in 10 000. That's roughly the same as the risk of a fatal accident from driving a car for a year. You can see that the risk from chest X-rays is relatively small. Even so, mass programmes of routine chest X-rays to detect tuberculosis are no longer carried out (p225).

There is evidence that the simple linear relation between dose equivalent and the probability of developing a cancer (or other consequence) is not adequate. In some cases there appears to be a threshold, that only a dose over a certain limit has effects. At less than the threshold, the dose has no observable effect. It may even be the case that very low regular exposure has benefits. Radiologists are continually extending knowledge of the effects of ionising radiations on tissue.

Will I risk it?

"Shall I risk not doing this assignment?" A student is weighing the risk that material he or she doesn't study will subsequently turn up in the exam, against the benefits of the extra free time. The probability, the costs and the benefits must all be taken into account. The risk is calculated by analysing how likely it is that the topic will come up on the exam and assessing the consequences of a poorer performance if it does. The benefits hinge on what can be done with the extra free time. In formal calculations the seriousness of a risk is calculated as

$$\text{risk} = \text{probability (of the event occurring)} \times \text{consequence (if it does)}$$

The probability is to do with how likely the event is to occur. Some events are very likely to occur and others very unlikely. A probability of 0.001 means that there is a one in a thousand chance of it occurring. A probability of 0.5 means that there is an even chance of an event happening or not. On this scale, probabilities of 1 and 0 correspond to events that are certain to happen and certain not to happen, respectively.

It is easy to be misled by statements of probability. Suppose an individual reactor is designed to have a probability of serious failure once in 5000 years. That sounds very safe. But it means a probability of 1/5000 per reactor for an accident per year. There are several hundred reactors in the world, so one should expect an accident once every decade or so. That sounds much more serious.

The consequence is what happens when the event occurs. The consequence of a minor accident might be a few days off work with loss of earnings. The probability may not be small but the consequence is not very serious. The cost of a hydroelectric dam failure could be thousands of lives as well as the expense of rebuilding the dam. Now the probability is very small but the consequences are large. It may or may not be appropriate to try to put a cash value on consequences but, if risks are to be compared, it is necessary to find some common basis.

People accept risks that they have chosen to take much more easily than risks imposed on them. For example, mountain climbing is relatively dangerous, with several fatalities every year among the modest number of people who choose do it. Skiing accidents resulting in injury are also common. The same people may find a similar risk from industrial accidents intolerable and scandalous.

People also get used to risks. In Britain more than 3000 people are killed in car accidents each year and more than 500 die from house fires. But new kinds of risk, which people do not understand, are particularly worrying. This may be one reason for considerable public anxiety about risks from radiation. Understanding the risks can be important.

An example is the 800 µSv background radiation from radon in buildings as a result of the decay of natural radioactive materials in the ground and in brick or concrete walls (p227). You can calculate (using 5% per sievert) that this could be responsible for about 2000 cancer deaths in 50 million people exposed. That risk can be very much reduced by good ventilation, though this does increase heating costs. For one person, the risk is around one in 25 000. If you were willing to put a price on a human life, you could calculate how much it is worth spending on ventilation to reduce the risk. Risk analysts make these kinds of calculation based partly on fact, and partly on intuition and guesswork.

Allowed doses

The maximum allowed radiation exposure for a member of the public from artificial sources is 1000 µSv per year for frequent exposure. This means that the authorities accept a risk of 50 deaths per million. Is that too much? To compare, a similar risk would be associated with smoking 75 cigarettes a year or cycling 2 km to work each day. Farming brings with it a risk of 80 deaths per million. What level of risk do *you* find acceptable?

Quick check

Useful data: $e = -1.6 \times 10^{-19}$ C, $c = 3.0 \times 10^{8}$ m s^{-1}, $h = 6.6 \times 10^{-34}$ J Hz^{-1}

1. Find out how one of the following works: home smoke detectors using alpha radiation; sterilisation of medical equipment using gamma-radiation; sterilisation of foodstuffs using gamma-radiation; power generation in spacecraft using radioactive decay; use of radioactive tracers in medical diagnosis; a medical use of X-rays.

2. It takes 14 eV to ionise nitrogen. Show that the maximum number of ions that an alpha particle of energy 3 MeV could produce is about 200 000.

3. Show that the longest wavelength of an ultraviolet-light photon that could ionise an atom with an ionisation energy of 10 eV is 120 nm.

4. A thickness of 0.5 m of concrete reduces the intensity of a certain gamma-radiation by a half. Show that 94% of the radiation is absorbed by 2 m of concrete and that the absorption coefficient µ is 1.4 m^{-1}.

5. Show that over a billion 5 MeV particles need to be absorbed in 1 kg of tissue to give an absorbed dose of 1 mGy.

6. Show that the dose equivalent in sievert of a combination of doses of 1 mGy, of which 20% is from alpha radiation and the rest from gamma-radiation, is 4.8 mSv. (See p226 for the quality factors.)

Links to the *Advancing Physics* CD-ROM

Practise with these questions:

10D   Data handling *The cost of taking a chance*

20X   Explanation–exposition *Telling people about risk*

50C   Comprehension *How safe are X-rays?*

130C Comprehension *Radiation protection and dosimetry*

120S Short answer *Summary questions for 18.1*

Try out these activities:

10E   Experiment *Radon in the home*

20H   Home experiment *Radiation in dust*

Look up these key terms in the A–Z:
Alpha radiation; background radiation; beta radiation; biological effects of ionising radiation; gamma radiation; ionising radiation; range of radiation; risk; uses of ionising radiation

Revise using the revision checklist and:

100   OHT *Absorbing radiations*

200   OHT *Doses*

300   OHT *Whole body dose equivalents*

18.2 The nuclear valley

What makes a nucleus radioactive? What makes a nucleus stable? The answer lies in the balance between the numbers of protons and neutrons.

The nucleus of carbon-12 ($^{12}_{6}C$) is stable and has equal numbers of protons and neutrons. But the nucleus of carbon-14 ($^{14}_{6}C$), with just two more neutrons, is radioactive. It emits an electron (and an antineutrino, chapter 17) to give the stable isotope nitrogen-14 ($^{14}_{7}N$), which again has equal numbers of protons and neutrons.

A plot of proton number against neutron number for stable isotopes shows that for light nuclei an equal number rule holds up quite well. There's a good reason. A strong nuclear attraction acts equally between neutrons and protons, neutrons and neutrons, and protons and protons. Also, because protons and neutrons obey the Pauli exclusion principle (chapter 17), a proton–neutron pair can occupy the same quantum state though two neutrons or two protons can't.

More massive nuclei have progressively more neutrons than protons. Stable iron-56 ($^{56}_{26}Fe$) has 30 neutrons for 26 protons; stable lead-206 ($^{206}_{82}Pb$) has more – 124 neutrons for 82 protons. As the number of protons increases, so does the potential energy of their mutual electrical repulsion. The extra neutrons "dilute" the protons and provide further nuclear attraction to overcome the electrical repulsion.

Stability must be a question of energy. If energy is needed to pull the neutrons and protons apart, the combination can be stable. Compare gravitational and electrical examples. Gravitational forces hold planets, stars and galaxies together. Electrical forces hold atoms together. That is, it takes energy to pull them apart. In each case, the total energy of the bound system is less than the energy of all the pieces if pulled apart. Thus the energy of a bound nucleus must be less than the energy of its protons and neutrons taken separately.

Stable and unstable nuclei: the balance of numbers of protons and neutrons

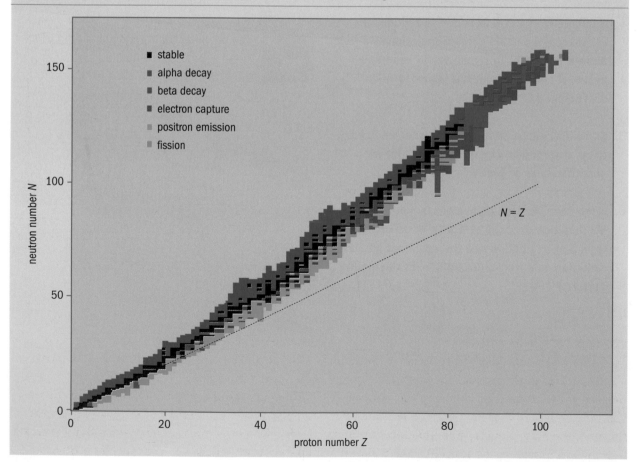

Key summary: the binding energy of a carbon-12 nucleus

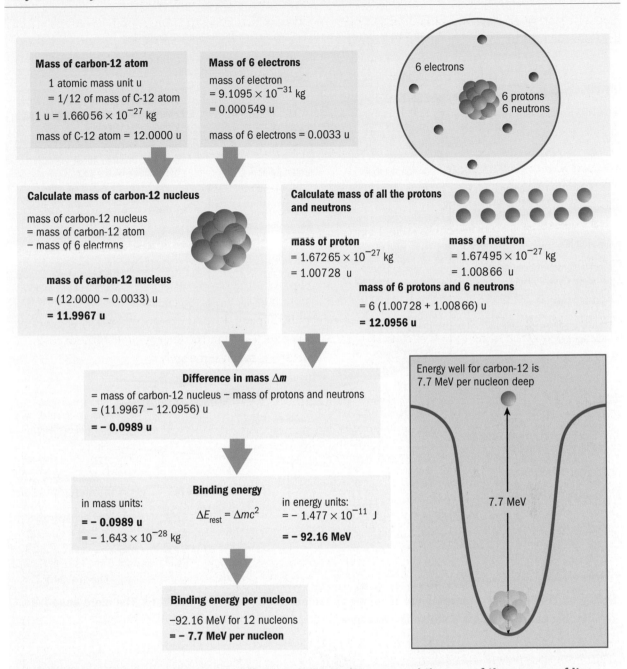

Mass of carbon-12 atom

1 atomic mass unit u

= 1/12 of mass of C-12 atom

$1\,u = 1.660\,56 \times 10^{-27}$ kg

mass of C-12 atom = 12.0000 u

Mass of 6 electrons

mass of electron

$= 9.1095 \times 10^{-31}$ kg

= 0.000549 u

mass of 6 electrons = 0.0033 u

6 electrons

6 protons
6 neutrons

Calculate mass of carbon-12 nucleus

mass of carbon-12 nucleus
= mass of carbon-12 atom
− mass of 6 electrons

mass of carbon-12 nucleus

= (12.0000 − 0.0033) u

= 11.9967 u

Calculate mass of all the protons and neutrons

mass of proton
$= 1.672\,65 \times 10^{-27}$ kg
= 1.00728 u

mass of neutron
$= 1.674\,95 \times 10^{-27}$ kg
= 1.008 66 u

mass of 6 protons and 6 neutrons

= 6 (1.007 28 + 1.008 66) u

= 12.0956 u

Difference in mass Δm

= mass of carbon-12 nucleus − mass of protons and neutrons

= (11.9967 − 12.0956) u

= − 0.0989 u

Binding energy

in mass units:

= − 0.0989 u

= − 1.643 × 10^{-28} kg

$\Delta E_{\text{rest}} = \Delta m c^2$

in energy units:

= − 1.477 × 10^{-11} J

= − 92.16 MeV

Energy well for carbon-12 is
7.7 MeV per nucleon deep

7.7 MeV

Binding energy per nucleon

−92.16 MeV for 12 nucleons

= − 7.7 MeV per nucleon

The binding energy of a nucleus is the difference between its mass and the sum of the masses of its neutrons and protons

Binding energy of a nucleus

How do you find the energy of a nucleus? Easy: just find its mass (the mass of the atom less the mass of its electrons) then calculate the rest energy from $E_{\text{rest}} = mc^2$. That is its energy. Then compare this with the masses of the individual protons and neutrons all added together, finding their total rest energy in the same way. If the nucleus is bound, its mass and so its rest energy will be less than that of the sum of its parts.

Take the stable nucleus $^{12}_{6}\text{C}$. Because of the way atomic masses are defined, the mass of a carbon *atom* is 12 atomic mass units exactly. But if you add up the masses of the six neutrons, six protons and

Key summary: nuclear landscape: the nuclear valley of stability

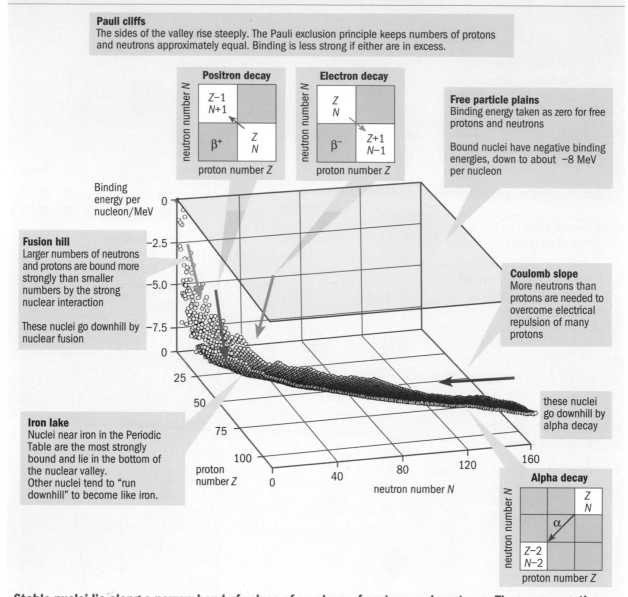

Pauli cliffs
The sides of the valley rise steeply. The Pauli exclusion principle keeps numbers of protons and neutrons approximately equal. Binding is less strong if either are in excess.

Positron decay

Electron decay

Free particle plains
Binding energy taken as zero for free protons and neutrons

Bound nuclei have negative binding energies, down to about −8 MeV per nucleon

Fusion hill
Larger numbers of neutrons and protons are bound more strongly than smaller numbers by the strong nuclear interaction

These nuclei go downhill by nuclear fusion

Coulomb slope
More neutrons than protons are needed to overcome electrical repulsion of many protons

Iron lake
Nuclei near iron in the Periodic Table are the most strongly bound and lie in the bottom of the nuclear valley.
Other nuclei tend to "run downhill" to become like iron.

these nuclei go downhill by alpha decay

Alpha decay

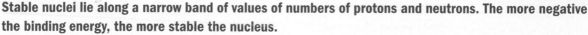

Stable nuclei lie along a narrow band of values of numbers of protons and neutrons. The more negative the binding energy, the more stable the nucleus.

six electrons you get a larger mass. The difference comes to −0.0989 atomic mass units. That's almost 1% of the total mass and nearly a tenth of the mass of one nucleon.

The rest energy $E_{\text{rest}} = mc^2$ associated with one atomic mass unit (roughly the mass of a proton or neutron) is 931 MeV so the stable carbon nucleus lies at an energy of −92 MeV compared to its parts. It would take 92 MeV to pull it all apart.

With 12 nucleons altogether, that's nearly −7.7 MeV for each proton or neutron. This measures how strongly each nucleon is bound. It is called the binding energy per nucleon. So if these 12 nucleons come together and form a nucleus they lower their energy by 7.7 MeV per nucleon. The energy well in which a nucleon lies is 7.7 MeV deep. Unstable nuclei, not surprisingly, are less strongly bound. Spontaneous radioactive decay happens because nucleons tend towards possible lower-energy states.

Although the energy differences are large, and the corresponding mass differences are quite measurable, accurate knowledge of the binding energy of nuclei depends on careful precision measurements of atomic masses. A precision of six or more significant figures is needed, as you can see from the sample calculation for carbon-12 (p231). The data for thousands of nuclei, stable and unstable, show how the binding energy depends on the numbers of protons and neutrons in a nucleus.

The valley of stability

A plot of the way binding energy varies with proton and neutron number, shown in "Key summary: nuclear landscape: the nuclear valley of stability", looks like a deep narrow valley. The binding energy is negative everywhere in the valley. The energy is zero up on the high flat plains around the valley, where the nucleus has been taken apart into free protons and neutrons.

The valley descends steeply to begin with, as extra neutrons and protons are added and the strong nuclear force takes effect. We have called this the "fusion hill" because it is the hill down which nuclei in stars go as heavier elements are made in nuclear fusion reactions (p241). The lowest point and strongest binding is around the element iron $^{56}_{26}$Fe, with binding energy -8.8 MeV per nucleon. We have called this region the "iron lake", since lakes rest at the bottom of valleys. Beyond the iron lake, the valley floor rises gently and the binding per nucleon becomes weaker. This rise we have called the "coulomb slope" because the reduction in strength of binding is caused by the growing electrical coulomb repulsion of the protons. Nuclei can fall down this slope, back towards iron. Many do so by alpha decay, losing two protons and two neutrons in one go. Some suffer fission, breaking into smaller pieces (p237).

The valley has steep side walls, like a canyon. The valley walls are where there are unstable nuclei with a slight excess of protons or neutrons. We have called these steep walls the "Pauli cliffs". The Pauli exclusion principle means that extra neutrons or protons have to go in states of higher energy because both are fermions (chapter 17), so the energy cliffs rise on both sides of the valley.

Nuclei fall down the sides of the valley by emitting electrons or positrons. Emitting an

electron changes a neutron to a proton; emitting a positron changes a proton to a neutron. A proton can be changed to a neutron in another way, too – the nucleus can capture an electron from those close to it in the atom.

Nuclei, like atoms, can exist in excited states of higher energy – the nucleus has been lifted above the floor of its part of the valley. Like atoms, nuclei emit photons when they fall from an excited state to a lower one, when they drop to the valley floor again. But these are very high-energy photons – gamma-rays with energy in the range 0.1 to more than 5 MeV. This is the origin of the gamma-rays emitted by radioactive materials.

Designer isotopes

Both naturally occurring and artificially created, designer isotopes are put to use in a variety of applications (p221).

Naturally occurring radioactive isotopes such as uranium-238 and thorium-232 are isotopes that were present at the formation of Earth and the solar system. With half-lives of 4.5×10^9 years and 1.4×10^{10} years respectively, they are still around.

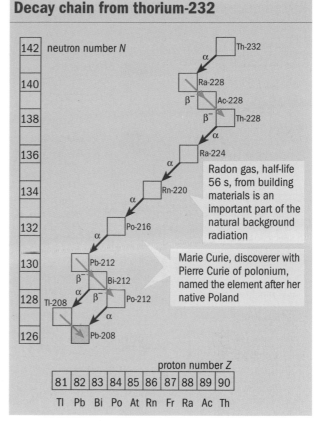

Decay chain from thorium-232

Radon gas, half-life 56 s, from building materials is an important part of the natural background radiation

Marie Curie, discoverer with Pierre Curie of polonium, named the element after her native Poland

Key summary: radioactive decay processes

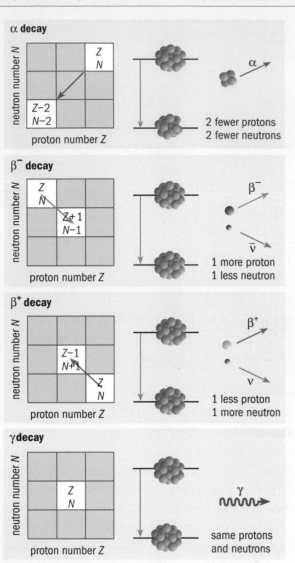

N and Z are affected differently by each type of decay process

But their nucleons are slowly trickling down the coulomb slope of the nuclear valley. Each decays to another isotope, which in turn decays to others farther down.

Some way down the chain from thorium-232 is radon-220. It decays with a half-life of 56 seconds to polonium-216, taking the chain still farther downhill by two protons and two neutrons. The decay reaction is:

$$^{220}_{86}\text{Rn} \longrightarrow ^{216}_{84}\text{Po} + ^{4}_{2}\text{He}$$

Thorium exists in many building materials (and in the ground), so bricks and concrete can leak radon gas into rooms. With poor ventilation you will breathe in a little radon; the danger comes from the alpha particles it emits inside you (p226). Radon concentrations are quite low in most parts of the UK, but there are places in the world where they are very much higher.

Alpha-emitter americium-241, used in smoke alarms (p223), is an example of a designer isotope. It comes from the decay of plutonium, obtained by bombarding uranium-238 with neutrons. The 432-year half-life of americium-241 outlasts the lifetime of the alarm but is short enough to give a reasonable activity from a small amount of material.

Positron (β^{+}) emitters are used in positron emission tomography scans (p189). An example is oxygen-15:

$$^{15}_{8}\text{O} \longrightarrow ^{15}_{7}\text{N} + ^{0}_{+1}\text{e} + ^{0}_{0}\bar{\nu}$$

A short half-life is essential in such medical tracer applications so that the tracer lasts as little time as possible in the body.

An example of a gamma-emitting tracer used in medicine is technetium-99$^\text{m}$. Its name reflects the fact that it is purely artificial. It is formed in an excited but unusually stable ("metastable") state from another decay. Its nucleons simply fall to a lower energy level, emitting gamma-ray photons, with a half-life of six hours.

The gamma-ray photons from cobalt-60 find a very different use. Cobalt-60 is used both medically to kill tumours and industrially to detect flaws in, for example, steel bridge girders. Its half-life of 5.3 years makes it very active, but long-lasting enough for the uses made of it by doctors and engineers. Such a source needs thick lead shielding and has to be treated with great care.

Many of these medically useful isotopes have to be made to order in hospitals. As a consequence, hospitals employ physicists to, for example, run cyclotrons to irradiate materials to make them on demand. And of course the physicists' knowledge is essential if the risks of using them are to be minimised.

Quick check

Useful data: $m_p = 1.672649 \times 10^{-27}$ kg,
$m_n = 1.674954 \times 10^{-27}$ kg, $c = 3.000 \times 10^8$ m s^{-1},
$e = 1.602 \times 10^{-19}$ C

1. Here are four radioactive isotopes of zinc: $^{57}_{30}$Zn, $^{60}_{30}$Zn, $^{75}_{30}$Zn, $^{76}_{30}$Zn. Two emit positrons and two emit electrons. Which must be the positron emitters?

2. Bismuth-212 ($^{212}_{83}$Bi) decays in two ways: by emitting an alpha particle or by emitting an electron. Write symbols for the resulting nuclei. The nearby elements are thallium (Tl) $Z = 81$, lead (Pb) $Z = 82$, polonium (Po) $Z = 84$; astatine (At) $Z = 85$.

3. There are four successive alpha decays in a chain from thorium-227 ($^{227}_{90}$Th). Show that the result is an isotope of lead, $Z = 82$. How many nucleons does this isotope have?

4. The mass of the helium ($^4_2$He) nucleus is 6.645×10^{-27} kg. Use the data on the mass of a proton and a neutron given above to show that the mass of the nucleus is 0.75% less than the sum of the masses of its component nucleons.

5. Using the data in question 4 show that the binding energy per nucleon of an alpha particle is −7.07 MeV.

6. The mass of the deuterium nucleus $^2_1$H, consisting of one proton and one neutron, is 3.344×10^{-27} kg. Show that the binding energy per nucleon is −1.11 MeV. Compare this with the answer to question 5 and state whether the deuteron or the alpha particle is the more strongly bound.

Go to the *Advancing Physics* CD-ROM

Practise with these questions:

140D Data handling *Radioactive decay series*

170D Data handling *Binding energy and mass defect*

200S Short answer *Change in energy: Change in mass*

210S Short answer *Summary questions for 18.2*

Try out these activities:

140S Software-based *Binding energy of nuclei*

150S Software-based *Exploring the nuclear valley*

Look up these key terms in the A–Z:

Binding energy; nuclear stability; nuclear transformations; nucleon

Go further for interest by looking at:

20T Text to read *What holds nuclei together*

Revise using the revision checklist and:

500 OHT *Stability: Balanced numbers of neutrons and protons*

60P Poster *Finding binding energy*

90O OHT *Decay processes*

100S Computer screen *Views of the nuclear landscape*

18.3 Fission and fusion

Fission in war and peace

In 1939, in exile in Copenhagen from Nazi Germany, the Austrian physicist Lise Meitner and her nephew Otto Frisch had an idea that changed the world. Meitner and Frisch realised that results obtained by the German chemist Otto Hahn, with whom Meitner had worked in Germany for many years, meant that the uranium-235 nucleus could split in two. This explained how Hahn found traces of lighter elements when the uranium was bombarded by neutrons. The energy released in this nuclear fission was much larger than that obtained from simple radioactive decay.

Events moved fast. Calculations suggested the possibility of a nuclear bomb. The US government started the Manhattan Project to develop it and physicist Enrico Fermi – another exile, this time from fascist Italy – set about building the world's first man-made nuclear reactor.

Today's nuclear power stations descend from that first reactor, and from the insight of Meitner and Frisch. So did the bombs that devastated Hiroshima and Nagasaki in 1945.

Deciding the future

Like it or not, the world's economy has become crucially dependent on oil and natural gas to satisfy the demand for heat and power, but the price of oil, a finite resource, is unstable. Cheap oil has allowed periods of economic expansion; rises in oil prices can push national economies into recession. Sooner or later, a replacement for oil has to be found. Decisions have to be made in your lifetime about whether nuclear power has a part to play in this. It's a question of whether what can be done, should be done.

Energy from fission

Large nuclei sit up at the high end of the coulomb slope in the nuclear valley of stability (p232). The drop to the iron lake at the bottom is a little more than 1 MeV per nucleon. Fission exploits this energy difference by making a large nucleus break into two lower energy pieces. Where alpha decay chains walk down the slope step by step, fission jumps down the hill in one go. That's why Meitner

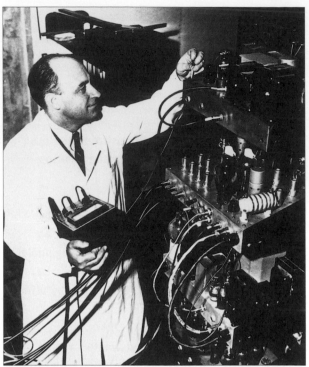

On 2 December 1942 the world's first fission reactor, built by Italian Enrico Fermi (above) and his team on a Chicago squash court, worked exactly as theory had predicted.

Fermi builds the first nuclear reactor

"The fifty-first layer would make the reactor critical. Everybody was growing tense as the moment approached when the reaction would become self sustaining. Then, unexpectedly and very calmly, Fermi announced, 'I'm hungry, let's go to lunch.'"

and Frisch's idea has such huge consequences.

Fission works because the nucleus behaves rather like a liquid drop (p202). A neutron captured by a uranium-235 nucleus can set the drop-like nucleus oscillating, sometimes into a dumbbell-like mode. If its shape gets close to that of two drops, it may split in two. The reason is that the two smaller droplets together have less energy than the original larger nucleus – they are nearer the valley bottom.

Nuclear fission releases energy of the order of 1 MeV per particle. Using $\varepsilon = kT$ (chapter 14) this corresponds to a temperature of 6 billion kelvin or so. This enormous temperature reflects the extreme conditions inside supernovae, where heavy elements were first made. Such power has to be treated with respect.

Key summary: nuclear fission of uranium-235

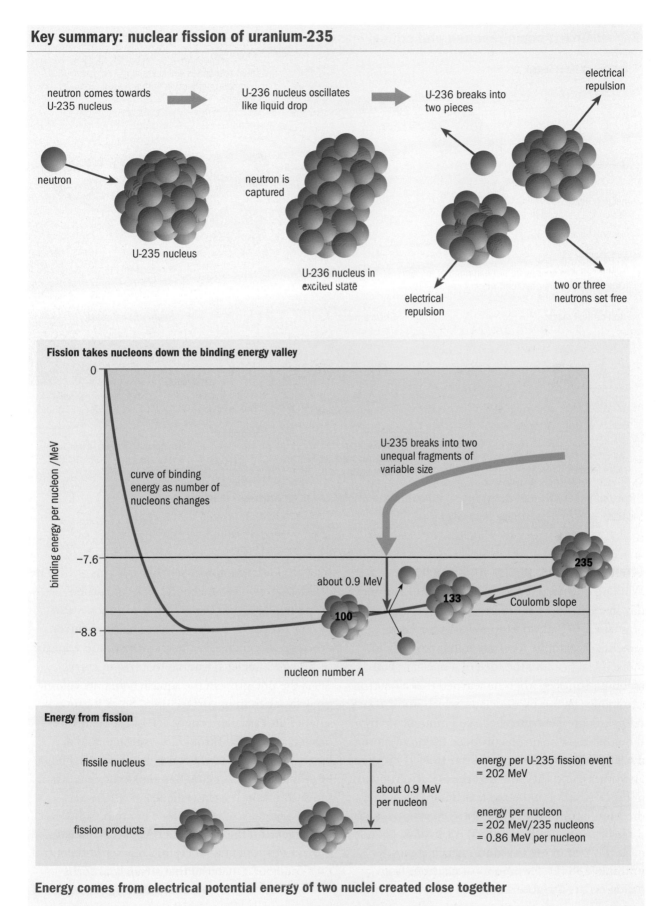

neutron comes towards U-235 nucleus

neutron

U-235 nucleus

U-236 nucleus oscillates like liquid drop

neutron is captured

U-236 nucleus in excited state

U-236 breaks into two pieces

electrical repulsion

electrical repulsion

two or three neutrons set free

Fission takes nucleons down the binding energy valley

binding energy per nucleon /MeV

0

−7.6

−8.8

curve of binding energy as number of nucleons changes

U-235 breaks into two unequal fragments of variable size

about 0.9 MeV

100

133

235

Coulomb slope

nucleon number A

Energy from fission

fissile nucleus

fission products

about 0.9 MeV per nucleon

energy per U-235 fission event = 202 MeV

energy per nucleon = 202 MeV/235 nucleons = 0.86 MeV per nucleon

Energy comes from electrical potential energy of two nuclei created close together

Key summary: chain reaction and critical mass

Critical chain reaction

The chain reaction is self-sustaining at a steady rate if on average one neutron from a fission produces a further fission

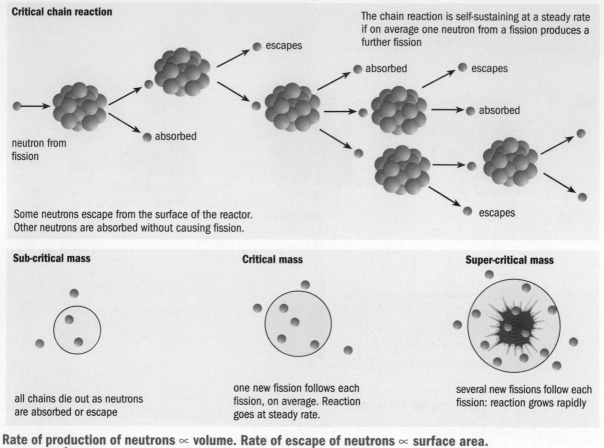

escapes

absorbed

escapes

absorbed

neutron from fission

absorbed

escapes

Some neutrons escape from the surface of the reactor. Other neutrons are absorbed without causing fission.

Sub-critical mass

all chains die out as neutrons are absorbed or escape

Critical mass

one new fission follows each fission, on average. Reaction goes at steady rate.

Super-critical mass

several new fissions follow each fission: reaction grows rapidly

Rate of production of neutrons ∝ volume. Rate of escape of neutrons ∝ surface area.

Ratio $\dfrac{\text{volume}}{\text{surface area}}$ increases with size.

Chain reactions: power and bombs

Walking down London's Southampton Row one day in 1933, another emigré, the Hungarian physicist Leo Szilard, stopped abruptly at a crossing. The lights went green but Szilard didn't move. He'd just had the idea of a nuclear chain reaction. In 1939, when he learned that a neutron can induce fission in a nucleus, which in turn releases more neutrons, he asked himself: "Why shouldn't these neutrons produce fission in more nuclei, and so on and so on?" He briefed Einstein and others, and they warned the US government of the possibility of the nuclear bomb. So began the Manhattan Project (p236).

Fissile uranium-235 (U-235) is less than 1% of naturally occurring uranium, which is mostly uranium-238 (U-238). So most neutrons from fission events are absorbed by U-238 nuclei without causing fission, or escape from the surface of the

reactor. The reason why Fermi's pile was a mixture of bricks of nuclear fuel and bricks of carbon graphite was to use the graphite as a moderator to slow down the neutrons. Slowing the neutrons down greatly increases the chance of their capture by a U-235 nucleus, leading to a fission event.

As the pile grew in the squash court, its volume grew faster than its surface area. So as it grew, a smaller and smaller proportion of the neutrons produced escaped through the surface, and a larger proportion could cause new fissions. The fifty-first layer was enough to make the reaction self-sustaining. It went critical.

The critical size is a fine line to cross. A little bit bigger and the chain reaction can run away, growing exponentially. Fermi's reactor had control rods made of a material that strongly absorbs neutrons so that he could stop the chain reaction if it grew too big. Fermi used three sets of cadmium

Key summary: the design of a pressurised water reactor

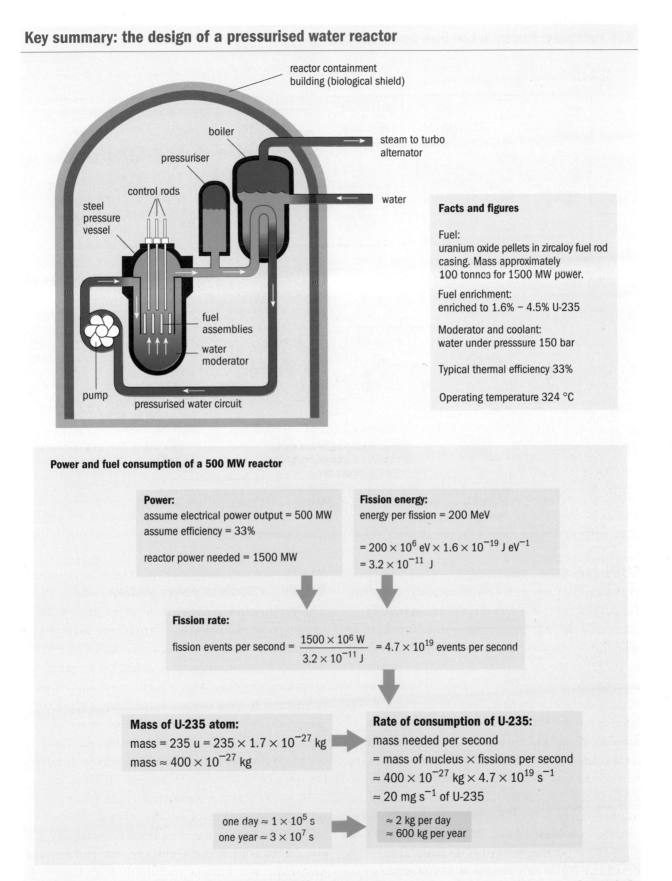

reactor containment
building (biological shield)

boiler

steam to turbo
alternator

pressuriser

control rods

water

steel
pressure
vessel

fuel
assemblies

water
moderator

pump

pressurised water circuit

Facts and figures

Fuel:
uranium oxide pellets in zircaloy fuel rod
casing. Mass approximately
100 tonnes for 1500 MW power.

Fuel enrichment:
enriched to 1.6% – 4.5% U-235

Moderator and coolant:
water under presssure 150 bar

Typical thermal efficiency 33%

Operating temperature 324 °C

Power and fuel consumption of a 500 MW reactor

Power:
assume electrical power output = 500 MW
assume efficiency = 33%

reactor power needed = 1500 MW

Fission energy:
energy per fission = 200 MeV

$= 200 \times 10^6 \text{ eV} \times 1.6 \times 10^{-19} \text{ J eV}^{-1}$
$= 3.2 \times 10^{-11} \text{ J}$

Fission rate:

fission events per second $= \dfrac{1500 \times 10^6 \text{ W}}{3.2 \times 10^{-11} \text{ J}} = 4.7 \times 10^{19}$ events per second

Mass of U-235 atom:
mass = 235 u = $235 \times 1.7 \times 10^{-27}$ kg
mass ≈ 400×10^{-27} kg

Rate of consumption of U-235:
mass needed per second
= mass of nucleus × fissions per second
≈ $400 \times 10^{-27} \text{ kg} \times 4.7 \times 10^{19} \text{ s}^{-1}$
≈ 20 mg s^{-1} of U-235

one day ≈ 1×10^5 s
one year ≈ 3×10^7 s

≈ 2 kg per day
≈ 600 kg per year

If uranium fuel is enriched to roughly 3% U-235 then about 20 tonnes of enriched uranium a year is needed

Key summary: fusion in the Sun and on Earth

Fusion in the Sun: three-stage process

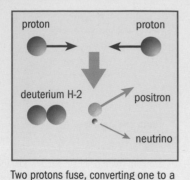

Two protons fuse, converting one to a neutron, to form deuterium H-2.

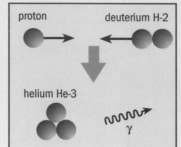

The deuterium H-2 captures another proton to form He-3.

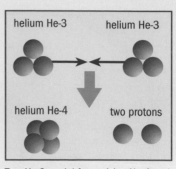

Two He-3 nuclei fuse, giving He-4 and freeing two protons.

Fusion on Earth: two-stage process

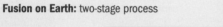

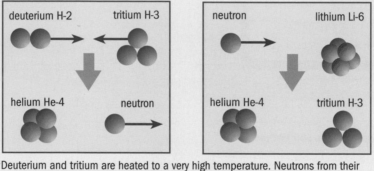

Deuterium and tritium are heated to a very high temperature. Neutrons from their fusion then fuse with lithium in a "blanket" around the hot gases. Tritium is renewed.

The solar fusion reaction is far too slow for use in a fusion reactor on Earth

control rods. One set was automatically operated from a distance; one set was pulled out little by little until the reaction began to give fine control. The third emergency set of control rods has given us the jargon word "scram" for shutting down a reactor in a hurry. The letters stand for safety control rod axe man, recalling the fact that Fermi had his emergency rods slung on a rope with a man standing by with an axe to cut the rope and let them fall into the reactor. Being very cautious, Fermi also had a three-person "suicide squad" stood above the pile with buckets of cadmium solution ready to flood the pile.

A bomb is made by creating a supercritical mass of U-235 or plutonium. A U-235 bomb is made by creating a supercritical mass of U-235 enriched uranium. This is done by firing two subcritical masses together using chemical explosives.

Building a nuclear power station

The Westinghouse pressurised water design has proved to be the most popular type of nuclear reactor. The design, originally developed for nuclear ships and submarines, has been scaled up for power stations. Where Fermi used carbon as a moderator to slow down the neutrons, this design uses water, which also carries energy away from the core of the reactor to generate steam. Thus the moderator is also the coolant, which simplifies the design.

About 80% of the energy from each fission is carried by the kinetic energy of the fission fragments. As they are stopped, this energy is spread out as random thermal motion and the core heats the water coolant pumped through it.

Coming as they do from a large neutron-rich nucleus, the fission fragments are neutron-rich for

Inside the Joint European Torus' (JET) toroidal vacuum vessel, the world's largest Tokamak, showing the remote handling boom and manipulator on the right-hand side. The height of the inside of the vessel is 4.2 m.

their size – they are mostly radioactive, emitting electrons (β^-) and thus increasing the proportion of protons to neutrons (p234). Antineutrinos emitted in these decays (p196) flood out of the reactor. The decays, and the stopping of neutrons in the reactor material, add to the energy produced.

Neutrons are also absorbed by the non-fissile U-238, breeding plutonium. The plutonium is extracted by reprocessing the fuel rods. It was used to build up the world's arsenal of nuclear weapons and it can also be used as a reactor fuel, thus putting the original non-fissile U-238 to productive use.

The crucial problem is disposing safely of the radioactive waste, some of which is long-lived. In thinking about environmental impact, don't forget that conventional fossil fuel stations also produce waste, including carbon dioxide. A 1500 MW conventional power station consumes about 2 million tonnes of coal or oil a year.

Fusion: in the future?

Fusion is the reaction that powers the stars. At very high temperatures small nuclei collide and fuse. As they do so energy is released – they fall down the initial steep slope of the valley of nuclear stability (p232). But a long way downhill though it is, it isn't easy to fuse protons. Very high energies are needed to get them close enough to each other against the electrical potential barrier. Also, at the very moment of fusion, one proton

must decay to a neutron so that together they can form deuterium. This is why nuclear fusion in the Sun is a very slow process.

Immediately after the Second World War, the technology to build a fusion bomb (hydrogen bomb) was rapidly developed and horrifically successful. The technology to produce power from controlled fusion has not yet succeeded. It was Andrei Sakharov – who also designed the Russian hydrogen bomb – who proposed one of the most promising devices, the Tokamak. The idea is to use a magnetic field to confine a hot plasma of deuterium and tritium in a ring. An electrical discharge heats the plasma to the necessary millions of degrees. The big problem is keeping the plasma hot enough and compressed enough for long enough for fusion to happen. Although fusion has been achieved, it has not yet been possible to get more power out than was put in.

The current plan is to surround the plasma by a blanket of lithium. The lithium captures neutrons from the fusion of deuterium and tritium. This produces helium and regenerates the tritium. The lithium blanket also carries the cooling water, which takes energy away to raise steam. Another possible route to sustained fusion uses laser pulses to compress pellets of deuterium and tritium.

Power from fusion, if it could be achieved, would postpone the depletion of fossil fuels for a very long time. But fusion machines would also produce radioactive waste because the neutrons produced irradiate the machine.

Can do – should do?

Plainly, you should not do everything that you can do. Yet it is clearly arguable that physicists have exploited their knowledge in unacceptable ways, leading some people to be suspicious of them. Similar issues now face biologists and genetic engineers.

Whether to pursue nuclear power is just such a question. There are relatively few nuclear reactors in the UK and the development of new ones is under debate because of public concern. By contrast, France and Japan, both countries lacking large fossil fuel reserves, generate the majority of their electricity in this way. Different judgements have been made about the risks and benefits.

Energy sources and effects

A wind farm

A dam providing hydroelectric power

A coal-fired power station in Kentucky, USA

Miners coming off shift

The sarcophagus built around the Chernobyl nuclear plant after the explosion in 1986

Trees in the Krusne Hory mountains in the Czech Republic that have been poisoned by acid rain

Quick check

Useful data: $m_p = 1.672649 \times 10^{-27}$ kg,
$m_n = 1.674954 \times 10^{-27}$ kg, $c = 3.000 \times 10^8$ m s^{-1},
$e = 1.602 \times 10^{-19}$ C

1. The mass of the $^{235}_{92}$U nucleus is
   3.902×10^{-25} kg. Show that the binding energy
   per nucleon is -7.6 MeV. What is the importance
   of the difference between this and the -8.8 MeV
   binding energy of $^{56}_{26}$Fe?

2. Show that niobium $^{100}_{41}$Nb and tin $^{132}_{51}$Sb are a
   pair of possible fission products of $^{235}_{92}$U if three
   neutrons are set free in the fission.

3. Both $^{100}_{41}$Nb and $^{132}_{51}$Sb (question 2) are neutron
   rich. Suggest a beta decay equation for each.

4. Why does a nuclear reactor require a moderator
   to slow down neutrons?

5. The fission energy of U-235 is about 1 MeV per
   nucleon. Show that about 5×10^{12} J are released
   if all the U-235 nuclei in 1 kg of nuclear fuel
   containing 5% U-235 undergo fission.

6. Suppose that the average number of neutrons
   produced by a fission event is 2.5. Show that
   for a chain reaction that is just critical the
   probability that a neutron will cause a fission
   is 0.4. What happens to neutrons that do not
   cause fission events?

Links to the *Advancing Physics* CD-ROM

Practise with these questions:
240C Comprehension *Power in space*
230D Data handling *The disappearing Sun*
250S Short answer *Fission and fusion –
      practice questions*

Try out this activity:
40S Software-based *Sources of
     radiation in the UK: Some facts and
     figures*

Look up these key terms in the A–Z:
Chain reaction; fission; fusion; nuclear
power; radioactive waste; renewable
energy

Go further for interest by looking at:
30T Text to read *A walk in the snow*

Revise using the revision checklist and:
1100 OHT *Nuclear fission*
1200 OHT *Chain reactions*
130P Poster *A pressurised water reactor*
1400 OHT *Fusion*

Summary check-up

Ionising radiation ✓

- Ionising radiations have a wide range of uses in medicine, technology and everyday life
- Alpha radiation is strongly ionising; beta and gamma-radiation less so
- Alpha particles have a definite range in air. Beta particles have a variable range. Gamma-radiation is attenuated exponentially in absorbing material, with $I = I_0 e^{-\mu x}$.
- Absorbed dose is the energy in joules absorbed per kilogram of material, unit gray (Gy). The unit of dose equivalent is the sievert (Sv), the absorbed dose in gray multiplied by numerical factors to allow for the effects of different types of radiation on various tissues.
- The concept of risk combines the probability of an event with the consequences of that event occurring − risk = probability × consequences

Stability and binding of nuclei ✓

- The relative numbers of protons and neutrons are roughly equal in small stable nuclei. The larger the nucleus, the more the neutrons predominate.
- The binding energy of a nucleus is the difference between the rest energy of the nucleus and the total rest energy of its individual nucleons. Rest energies are given by $E_{rest} = mc^2$.
- The binding energy per nucleon forms a valley with a minimum at around $^{56}_{26}Fe$
- Alpha emission reduces proton and neutron numbers each by two. Several chains of such decays are known.
- Electron emission changes one neutron into a proton. Positron emission changes one proton into a neutron.
- Gamma-ray emission takes energy from the nucleus without change in numbers of protons and neutrons

Fission and fusion ✓

- Fissile nuclei such as U-235 break into two parts, releasing energy of the order 1 MeV per nucleon
- A fission chain reaction can occur, in which neutrons from one fission event cause other nuclei to undergo fission. There is a critical mass for fissile material at which the chain reaction becomes self-sustaining.
- Slow neutrons are captured by U-235 more efficiently than fast neutrons. Nuclear reactors use a moderator to slow the neutrons and a coolant to carry away energy.
- Low mass nuclei can fuse to form more massive nuclei, releasing energy of several MeV per nucleon
- Fusion of hydrogen to helium occurs in the Sun. On Earth, fusion has been achieved but fusion as a source of power remains only a possibility.

Questions

Useful data: $c = 3.0 \times 10^8 \, \mathrm{m \, s^{-1}}$.

1. Describe the construction of any one form of fission reactor. In your account, be sure to:
   (a) identify the fissile material used;
   (b) say whether the neutrons are moderated and, if so, how;
   (c) say how the reactor is kept just critical;
   (d) explain how energy from the reaction is used to generate electrical power;
   (e) describe the necessary safety features of the design.

2. The rest energy associated with mass m is given by $E_{rest} = mc^2$. The table below gives masses of some nuclei and the total mass of their neutrons and protons.

Nucleus	Mass of nucleus/ 10^{-27} kg	Mass of neutrons and protons /10^{-27} kg
$^2_1$H	3.34364	3.34760
$^4_2$He	6.64477	6.69521
$^3_1$H	5.00744	5.02256
$^6_3$Li	9.98577	10.04280

   (a) Explain the significance of the fact that the values in the second column are larger than those in the first.
   (b) Show that the rest energy of a mass of 1×10^{-27} kg is 562 MeV.
   (c) Calculate the rest energy of the differences in mass between the first and second columns.
   (d) Calculate the energy differences per nucleon and arrange the four nuclei in order of stability.

3. Write equations for the six decays shown below. Indicate in each case the nature of the decay.

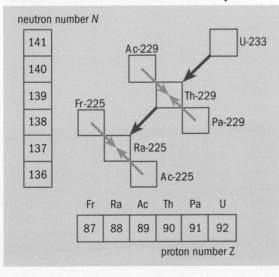

4. Draw some inferences about risks of ionising radiations from the following data, which show a typical European's whole body dose equivalent a year from various sources.

	Dose equivalent/ mSv	% of total
Natural radiation		
Cosmic and solar radiation	310	13
Gamma-radiation from soil, rocks, water	380	16
Radon decay products (in buildings)	800	33
Radiation from inside body (potassium-40, carbon-14 etc)	370	15
Total from natural sources	**1860**	**78**
Man-made radiation		
Medical and dental procedures	500	21
Weapons fall-out	10	0.4
Nuclear power	3	0.15
Occupational	9	0.36
Air travel	8	0.34
Total from man-made sources	**530**	**22**
Total from all sources	**2390**	**100**

Risk of developing a cancer approximately 5% per sievert; population of Britain approximately 60 million
Radiation data from Dowsett D J, Kenny P A and Johnstone R E 1998 *The Physics of Diagnostic Imaging* (Chapman and Hall Medical)

5. Give your own example of a practical use of ionising radiation. In your account, be sure to:
   (a) state how the ionising radiation is used in this case;
   (b) explain the choice of ionising radiation;
   (c) describe the risks involved and how they are dealt with;
   (d) suggest any alternatives to the use of ionising radiation.

6. A $^{235}_{92}$U nucleus captures a neutron, giving $^{236}_{92}$U, which splits into two equal parts, 10^{-14} m apart.
   (a) Show that the electrical potential energy of two protons 10^{-14} m apart is 0.14 MeV.
   (b) Calculate the electrical potential energy of the two fragments.
   (c) How might the existence of a strong nuclear attraction between the fragments explain why the answer to (b) is larger than the 200 MeV obtained from the fission of $^{235}_{92}$U?

Case studies: advances in physics

This chapter brings together, in this book and on the CD-ROM, case studies of physics in action. They give examples of:

- ideas in physics being used in new ways
- physics being used to solve a variety of problems
- the human side of advances in physics and technology

Questions about all the case studies are on the CD-ROM. A selection of these questions are in this chapter.

The case studies in the book

Energy efficient offices

describing how clever design can heat and ventilate a building with minimal use of energy

Profiles from space

explaining how lasers in space are used to map the surface of Mars, and to help foresters on Earth

Dark matter

telling the story of how it was found that 90% of matter in the universe seems to be invisible

Hybrid vehicles

looking to the future design of more efficient and less-polluting cars

The case studies on the CD-ROM

The London Eye

explaining how this remarkable structure was built, tested and lifted into position

Black gold

describing how oil is prospected from deep under the ground

Food for thought

showing that physics has a lot to say about the how and why of cooking

Magnifying with magnets

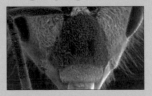

telling the story of how the scanning electron microscope was invented almost by accident

Case study: energy efficient offices

People in the UK use more energy for keeping warm than for anything else. Almost 90% of our energy comes from fossil fuels (coal, oil and gas). Apart from their finite nature, the carbon-dioxide emissions from the large-scale burning of these fuels are a major contributor to climate change.

It has become obvious that the best way forward is not to rely on dramatic technological innovations in energy supply (for example, cheap 50% efficient photocells or nuclear fusion) but to make better use of the energy-supply systems that we already have. Energy efficiency is an imperative for all of our buildings. Controlling the environment inside buildings is the largest contributor to the annual demand for fuel and power in the UK.

Architects can introduce new materials and construction processes, but generally they are constrained by the commercial aims of property developers. If it is not in the interests of a property developer to take unnecessary financial risks, then how is the building industry to experiment and achieve more environmentally friendly buildings?

The Building Research Establishment (BRE) provides a testing ground for new ideas in building in the UK. In 1991 it designed and constructed an Energy Efficient Office of the Future (EOF), a building that sets an excellent example for others to follow. It demonstrates principles of design, construction and management of office buildings that require significantly less energy than is normal today. The BRE designed an office that improves on typical buildings by a factor of about four.

The design brief

The EOF provides a floor area of $2000\,\mathrm{m}^2$ over three storeys – space for 100 people in an open-plan environment. It cost £2.5 m to construct (about the same as a normal office building of this size) and was finished in 1997. As part of the minimum-energy strategy, recycled materials were

In this photo of EOF you can see the ventilation stacks rising above the building, the glazed south-facing wall and the tilted shutters to control light in the building.

Operating energy targets for EOF

Energy performance	Gas/ $\mathrm{kWh\,m^{-2}\,y^{-1}}$	Electricity/ $\mathrm{kWh\,m^{-2}\,y^{-1}}$	Carbon/ $\mathrm{kg\,y^{-1}\,m^{-2}}$
typical	165	156	31.1
good practice	87	91	17.7
EOF specification	47	36	7.7

The BRE set itself the above operating energy targets, including the amount of carbon emitted as carbon dioxide in generating the electricity and in burning the gas.

used frequently. This included 80 000 bricks (at the low price of 48 p per brick), recycled flooring and timber from sustainable sources. Concrete aggregate was taken from the crushed panels of a 1970s building, instead of using fresh gravel.

Buildings are designed for people. They need a comfortable range of temperatures (around 19 °C is optimal) and fresh air. (A replacement time of one hour is usually enough to clear away exhaled breath and unpleasant smells.) They need to be well lit, avoiding both glare and dark areas, with lighting that is easily controlled. Since heating is the main energy use for buildings in the UK, much of the design effort of the EOF focused on keeping the temperature between 19 °C and 25 °C for most of the year without recourse to air-conditioning. At the same time, the use of electric lighting was kept to a minimum, relying on daylight as far as possible.

Model calculation of the thermal resistance of a cavity wall

Data: double-skin brick wall, 100 mm bricks, with 100 mm cavity filled with rock-wool thermal insulation. Area of wall = 100 m².

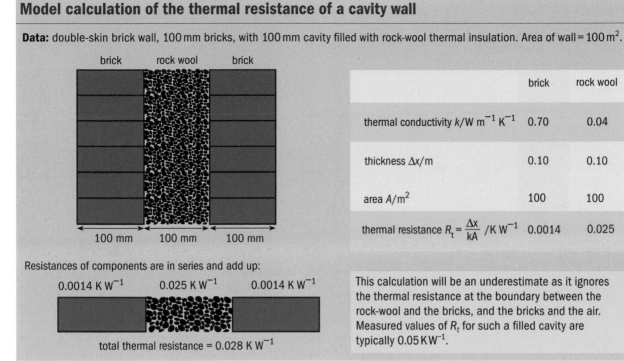

	brick	rock wool
thermal conductivity k/W m^{-1} K^{-1}	0.70	0.04
thickness Δx/m	0.10	0.10
area A/m²	100	100
thermal resistance $R_t = \dfrac{\Delta x}{kA}$ /K W^{-1}	0.0014	0.025

Resistances of components are in series and add up:

0.0014 K W^{-1} 0.025 K W^{-1} 0.0014 K W^{-1}

total thermal resistance = 0.028 K W^{-1}

This calculation will be an underestimate as it ignores the thermal resistance at the boundary between the rock-wool and the bricks, and the bricks and the air. Measured values of R_t for such a filled cavity are typically 0.05 K W^{-1}.

Passive heating

The EOF has two sources of energy that don't rely on fossil fuels. The first is the people who occupy the office. Even when sitting down, each person's warm body radiates and convects away energy at a rate of about 100 W, keeping the office appreciably warmer than the outside world. A calculation shows that in a well insulated building the body warmth of just one person can keep a good sized room 3 °C warmer than the outside. So no extra heating is needed to maintain a comfortable 19 °C when the outdoor temperature is 16 °C. But to achieve this, lots of insulation in the walls, roof and ceiling is necessary, as is double-glazing on the windows.

The designers need to be able to calculate the rate of loss of energy through the walls, floors and ceilings. The equation

$$P = kA\frac{\Delta T}{\Delta x}$$

says that the rate of thermal conduction of energy P is proportional to the temperature gradient. It is similar to the equation that says the rate of flow of charge is proportional to the potential gradient:

$$I = \sigma A\frac{\Delta V}{\Delta x}$$

Key summary: thermal conductivity

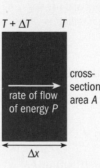

Energy leaks spontaneously through a slab of material when there is a temperature difference across it. Particles on the hotter side have greater average energy than particles on the cooler side. Random exchanges of energy between particles send energy on average from the hotter to the cooler end. The rate of thermal flow of energy is proportional to the temperature gradient $\Delta T/\Delta x$ across the material. This thermal flow of energy through a material is rather like the diffusion of particles from high to low concentration.

rate of thermal flow of energy

$$P = kA\frac{\Delta T}{\Delta x}$$

$\dfrac{\Delta T}{\Delta x}$ is the temperature gradient, units K m^{-1}

k is the thermal conductivity, units W m^{-1} K^{-1}

Thermal conductivity is analogous to electrical conductivity

This is useful. It means that many of the ideas acquired through the study of electricity can be used immediately in the field of thermal conduction. (Historically, the analogy was used the

Sunlight availability and heating demand

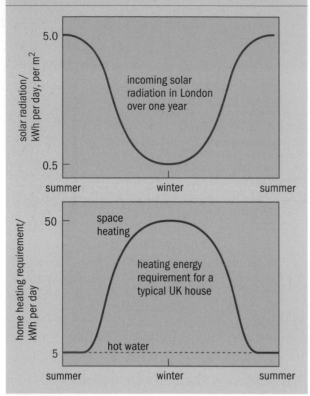

These graphs plot the availability of sunlight and heating demand in the UK over a yearly cycle. Solar radiation varies by a factor of almost 10 over a year.

other way round.) For example, consider the useful idea of electrical resistance R_e:

$$R_e = \frac{\Delta x}{\sigma A}$$

It suggests that thermal resistance R_t might be worth considering:

$$R_t = \frac{\Delta x}{kA}$$

In terms of R_t, P may be written as:

$$P = \frac{\Delta T}{R_t}$$

An example calculation using thermal resistances is given in the illustration on p249.

Solar heating

The second source of energy, delivered for free, is the Sun. Unfortunately, as you can see from the graph above, solar energy is least available when it is most needed. At only $0.5\,\mathrm{kW\,h\,m^{-2}\,day^{-1}}$,

The "storage cooler" ceiling

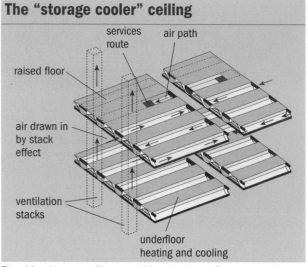

The ridged, wavy ceilings provide a larger surface area and channels for air flow across the ceiling. Water pipes for heating or cooling are also laid under the floors.

the weak January sunshine in the UK will not provide much heat. On the other hand, the $5\,\mathrm{kW\,h\,m^{-2}\,day^{-1}}$ available in June is not needed at all for heating purposes at that time of the year. So some of it is converted to electricity by the $47\,\mathrm{m^2}$ of photovoltaic cells that have been used as cladding on the south-facing wall instead of granite slabs. This provides about $20\,\mathrm{kW\,h}$ a day in the summer.

Summer cooling

In the EOF, the building becomes part of the temperature-control system. It contains a new type of wavy-shaped concrete ceiling that acts rather like a storage heater in reverse.

At the start of the day the ceiling is cool. It takes energy from the room by absorbing radiation from the floor and by being heated by hot air rising to it. During the night, this stored energy is removed by cooling the ceiling with colder night air drawn across both sides of the ceiling by low-power fans and vented from the building. This is how the ceiling starts the day cool, allowing a net thermal flow of energy into it from the room as the day progresses. For this cooling method to be effective, the ceiling must have a large thermal capacity ($\mathrm{J\,K^{-1}}$), a large surface area (the wavy shape increases the area by 25%) and must be a good absorber of infrared radiation. So the ceiling is actually a "storage cooler".

Additional cooling is provided by water flowing

These tilting blinds deflect direct sunlight and, being translucent, they diffuse light into the interior. Their angle is controlled automatically to track the angle of the Sun.

These tall cylindrical ventilation stacks carry hot air up and out of the building, bringing in cooler air at the bottom. They are designed to achieve one complete air change per hour. Conventionally, large buildings are ventilated by electric fans, which draw stale air out of the building, replacing it with fresh air. But here the ventilation system is solar powered. On a sunny day the air inside the stacks is heated by the sunlight and rises up the stack, driven by the difference in density between warm and cool air. The rising warm air carries stale air up the stacks and out of the building, creating a current drawing cool fresh air into the building. Electric fans are still needed when the solar power is not sufficient, but the system makes a considerably reduced overall power demand.

through pipes under the floor. The water is pumped out of the ground at 10 °C, from a depth of 70 m under the car park. After flowing through the building it is returned to the ground through a 20 m borehole, into porous chalk so that it can flow away freely. During the winter, the same pipes under the floor can be used to heat the building.

Natural ventilation

The EOF uses solar power to ventilate the building. Parts of the south-facing wall are glazed, creating pockets of hot air in ventilation stacks running up the building. Being less dense than cooler air, the hot air rises up the ventilation stacks and is replaced by cool air drawn through the building via automatically controlled windows that open directly into the stacks. This is particularly effective when it is hot but there is no wind. Fans in the stacks can assist ventilation during the winter when there is less solar energy to drive the system.

Lighting

Half of the wall space of the EOF is glazed, with high-level windows. This lets lots of light into the building, reducing the need for electric lighting. The open-plan design of the office space helps by allowing each part of it to get light from the windows. However, direct sunlight is uncomfortable so special louvred blinds are placed in front of the windows on the south side.

The angle of the blinds changes automatically during the day, tracking the motion of the Sun. Not only do the blinds reflect some of the light onto the office ceilings but they are also translucent, letting diffuse light through.

The electric lighting automatically compensates for daylight level and occupancy. Each lamp is

controlled separately by the building-management system operated by a computer. Sensors for daylight, infrared and motion nearby have been built into each lamp so that it can be turned up or down to keep a constant light level, or turned off when there is nobody around. Of course, each lamp uses energy saving light bulbs.

Personal controls

People are comfortable with a range of temperatures and lighting conditions. Automatic systems, which sense and control these conditions, produce energy efficiencies by operating only to preset maximum levels and only when people are present. Less well known is that people feel happier over a wider range of heating and lighting conditions when they can control it themselves. So the EOF allows its occupants to override the automatic settings and alter local lighting levels and temperature as they wish.

Experience suggests that, although most people are in favour of energy saving measures, they sacrifice these principles if the result interferes with their own comfort. The EOF demonstrates that appropriate use of technology can result in a low-energy environment without significant sacrifice. We will all probably have to make do with less energy in the future, but that doesn't inevitably mean a lower quality of life. This is what the EOF achieves. It cuts energy use and carbon emission by a factor of about four, using mainly simple and cheap technologies. And it respects the feelings of its users.

Questions

Conduction losses
Useful data: the EOF is approximately 12 m high, 60 m long and 11 m deep.

1. If a rate of thermal flow of energy of 10 kW (100 people at 100 W each) maintains a temperature difference of 3°C between inside and outside, estimate the thermal resistance of the building.

2. The thermal conductivity of glass is $1.1 \, W \, m^{-1} K^{-1}$. Double glazing has two 3 mm panes of glass that trap a 10 mm layer of air (thermal conductivity $0.024 \, W \, m^{-1} K^{-1}$). Calculate the thermal resistance of a $1 \, m^2$ pane of double-glazing. Calculate its thermal conductance, the reciprocal of thermal resistance.

Photovoltaics
Useful data: the $47 \, m^2$ of photovoltaic cells on the outside wall of the EOF produces 20 kW h a day when the daily solar energy input is $5 \, kW \, h \, m^{-2}$.

1. Show that the cells have an efficiency of about 10%.

2. The fans used to draw night air through the ceilings have a power of 80 W. If the photovoltaic cells charge up batteries during the day with an efficiency of 70%, how many fans can run off the batteries all night?

3. Sunlight has an average wavelength of about 550 nm. Photons of sunlight absorbed by a photovoltaic cell deliver their energy to electrons. Calculate that energy in electron volts (eV).

4. The typical voltage drop across a single photovoltaic cell is 0.5 V. How much energy (in eV) do electrons gain from the cell? From this determine the maximum theoretical efficiency of the cell.

Water cooling
Useful data: the temperature of the groundwater rises by 5 °C as it is pumped though the building, extracting energy at a rate of 35 kW.

1. The specific thermal capacity and density of water are $4.2 \times 10^3 \, J \, kg^{-1} K^{-1}$ and $1 \times 10^3 \, kg \, m^{-3}$. Calculate the volume flow rate (in $m^3 \, s^{-1}$) of the cooling water.

2. Suggest why the water is extracted at a depth of 70 m but returned to a depth of only 20 m.

Light sensors
Useful data: a light-dependent resistor (LDR) is a semiconductor for which resistance drops rapidly with increasing illumination. An LDR and a fixed resistor can be used to create a potential difference that increases from 0 V to 12 V as the light intensity increases.

1. Draw a circuit diagram for the system.

2. Explain how the circuit operates.

Go to the *Advancing Physics A2* students' CD-ROM for a larger set of questions.

Case study: dark matter

In the 1960s, astronomers were arguing bitterly about red shifts and the distance scale of the universe. Vera Rubin was fed up. "I wanted a problem that nobody would bother me about," she said. Little did she know that the "safe" problem she chose would lead to the radical suggestion that the universe, like an iceberg, has 90% of its mass hidden from view.

A safe problem?

Rubin and her colleague Kent Ford had built one of the most sensitive spectrometers in existence, to measure galactic red shifts. So the "safe" problem they chose was to measure how the Doppler shift of light from nearby galaxies varied at different places across a galaxy. They hoped to see evidence of the galaxies rotating, because stars on one side would be coming towards us while stars on the other side would be going away. It hadn't been done before because previous spectrometers needed the light from the whole galaxy to get a result, thus averaging out internal movements. Rubin and Ford's spectrometer was so sensitive that it could record the spectrum from a galaxy bit by bit. They set about taking spectra across the width of the nearest large galaxy, M31 in Andromeda.

Where is the mass?

Knowing the radius and speed of a star's orbit round a galaxy provides information about the mass of that galaxy. You can find out not only what the mass is but also how it is distributed in the galaxy. For speed v and radius r the acceleration towards the centre $\frac{v^2}{r}$ measures the gravitational field $\frac{GM}{r^2}$. Thus the mass M is just

$$M = \frac{v^2 r}{G}$$

and the quantity $v^2 r$ keeps track of the mass inside radius r.

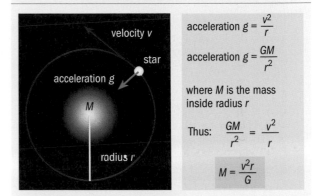

velocity v

star

acceleration g

M

radius r

acceleration $g = \frac{v^2}{r}$

acceleration $g = \frac{GM}{r^2}$

where M is the mass inside radius r

Thus: $\frac{GM}{r^2} = \frac{v^2}{r}$

$M = \frac{v^2 r}{G}$

The mass M inside radius r is tracked by the quantity $v^2 r$, for varying radius r

This is Vera Rubin who, with Kent Ford, made the first studies of Doppler shifts from different parts of distant galaxies, using a very sensitive spectrometer they had built. They found evidence of large amounts of invisible matter in the universe.

The argument depends on the fact that the gravitational pull anywhere outside a sphere of matter is the same as if the mass were concentrated at the centre, as long as the distribution is spherically symmetric. Also, inside such a spherical shell of matter there is no resultant gravitational pull from the shell. (Very similar results apply to distributions of electric charge, because the arguments depend only on the inverse square law for the field.)

Dark matter appears

Most of the visible matter in a galaxy is clustered at its centre. This is where much of the light comes from. So Rubin and Ford expected a slowing-down at large distances from the galactic centre (as

Key summary: models of distribution of matter in galaxies

Mass concentrated at centre

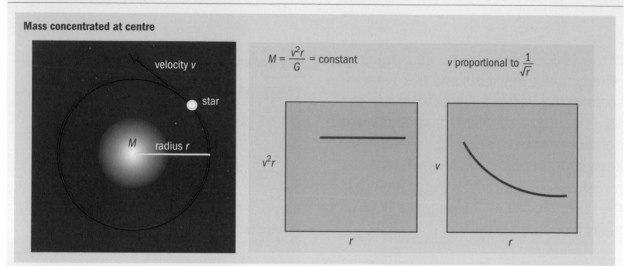

$M = \dfrac{v^2 r}{G} = \text{constant}$

v proportional to $\dfrac{1}{\sqrt{r}}$

Mass spread with uniform density ρ

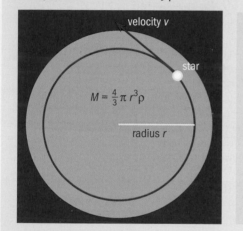

$M = \dfrac{v^2 r}{G} = \dfrac{4}{3}\pi r^3 \rho$

mass proportional to r^3

v proportional to r

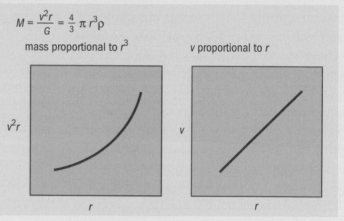

Gas model: density varies as $1/r^2$

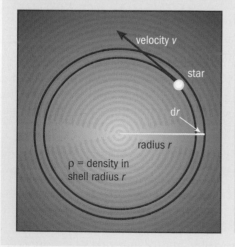

mass of shell = $4\pi r^2 \rho \, \mathrm{d}r$

If ρ varies as $1/r^2$, then mass of each shell is same. Thus mass inside radius r increases uniformly with distance

v constant with r

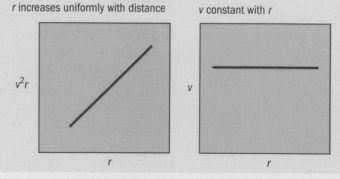

Measurements of the velocity of stars at different distances from the centre of spiral galaxies suggest the existence of a halo of dark matter surrounding each galaxy

Tracing how mass in a galaxy is spread out

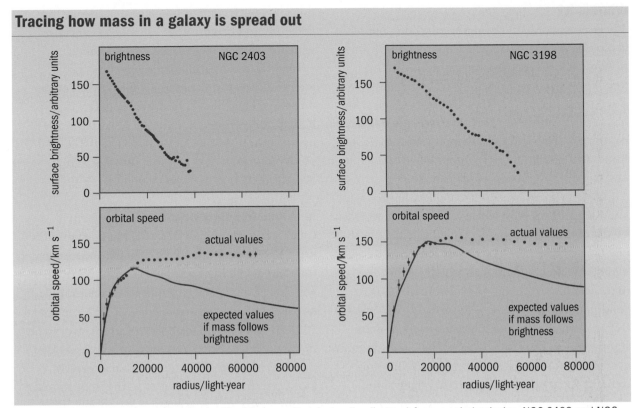

Graphs of brightness against radius (top) and orbital speed against radius (bottom) for two spiral galaxies, NGC 2403 and NGC 3198. The brightness drops steadily with radius, indicating that luminous matter is concentrated towards the centre. The orbital speed increases with radius at first, but then remains almost constant – even out to where there is no luminous matter.

also happens in the solar system). Their astonishing result was that the speed of rotation did no such thing – it stayed more or less constant. This meant that the mass $M = \frac{v^2 r}{G}$ inside radius r must be increasing in proportion to r. Although visible stars thinned out away from the centre of a galaxy, the invisible mass went on increasing. Dark matter had been discovered, and in large quantities.

Rubin and Ford's rotation curves, their plots of speed v against radius r, have become a standard tool for tracing how the mass in a galaxy is spread out. Again and again, the flat curves at large radii point to a halo of unseen matter filling the outer reaches of galaxies. Radio astronomers soon used Doppler shifts from 21 cm hydrogen emission to show that our own galaxy has such a halo too.

How much?

The amount of dark matter in galactic haloes is staggering. Consistently it comes out at at least 10 times the mass associated with all the visible stars in a galaxy. Not just some, but most of a galaxy seems to be invisible dark matter in some

form. A similar idea had been proposed 30 years before Rubin and Ford, by the Swiss-American astronomer Fritz Zwicky. He had argued that galaxies in clusters were moving too fast to be held together by the observable mass in their cluster (their kinetic energy was much more than half the estimated gravitational potential energy). The clusters should have flown apart long ago. But Zwicky was notorious for his wild ideas and nobody had taken much notice.

The density variation as $\frac{1}{r^2}$, indicated by the flat rotation curve, is what you expect to find if the halo of dark matter behaves like a gas of randomly moving particles. The outward diffusing effect of their random motions is restricted by the inward pulling effect of their own gravity.

Yet another reason for believing in dark matter comes from computer simulations of stars moving in a spiral galaxy. Without a massive halo around them, the model galaxies are unstable, collapsing to a rotating bar. With a halo, the spiral form persists. As Rubin said: "Nobody ever told us all matter radiated. We just assumed it did."

What is it?

There is certainly dark matter out there, but what is it made of? Cambridge astronomer Malcolm Longair offers the following list of possibilities: interstellar planets, brown dwarfs, very-low-mass stars, isolated neutron stars, little black holes, big black holes, supermassive black holes, massive neutrinos, unknown weakly interacting particles, standard household bricks, abandoned spaceships and copies of the *Astrophysical Journal*.

The jokey examples underline how hard it is to know what things are when you detect them only by the effect of their mass. Longair calculates that one house brick per cube of side length 500 million kilometres is enough to provide all the needed mass and that, spread uniformly, these bricks wouldn't obscure the view. It probably isn't mostly stellar remnants (brown dwarfs, very-low-mass stars, neutron stars, black holes), because too many generations of stars would have had to live and die to make them all, and there hasn't been time.

It isn't protons and neutrons

There's a reason to think that most of the dark matter must be in some peculiar form. The reason is that the observed proportion of hydrogen to helium in the universe is just what would be expected to be "cooked up" in the Big Bang. There is about 75% hydrogen to 24% helium by mass, so neutrons make up about 12% of all nucleons.

The neutrons are scarcer because they are more massive than protons, by a mass equivalent to rest energy $E = mc^2 = 1.3$ MeV. In the early stages of the Big Bang, while the temperature is high enough, protons and neutrons convert freely into one another, emitting or absorbing electrons, positrons and neutrinos. The electrons and positrons are continually created in pairs from photons with an energy of at least 1 MeV, corresponding to the rest energy $E = mc^2$ required to make a positron and an electron.

This energy corresponds to an average energy per particle kT at a temperature of about 1×10^{10} K. As the universe cools past this temperature, the electrons and positrons annihilate each other, freezing the ratio of neutrons to protons. The Boltzmann factor at this freezing point fixes the ratio. Finally, about 100 s into the Big Bang, before

the neutrons have had much time to decay, the neutrons and protons fuse to make deuterium, which is used immediately to make helium.

This process, mainly determined by the temperature, also depends on the density of the universe. The result is that the observed ratio of helium to hydrogen is compatible with a mass of neutrons and protons not much larger than the observed visible matter in the universe. An amount 10 times larger or more is ruled out.

Is it massive neutrinos?

Enter a Russian pursued by two Hungarians. The great Russian astrophysicist Yakov Zel'dovich noted that, just as the Big Bang left the universe flooded with the photons of the cosmic ray background, it also left it flooded with neutrinos. There are about 100 000 neutrinos in each litre of space, everywhere in the universe. Neutrinos interact so weakly that this level is completely undetectable at present. However, they each only need a tiny mass to make an observable difference to the expansion of the universe. The Hungarian physicists George Marx and his student Sandor Szalay calculated that a rest energy (mass) of between 25 and 100 eV per neutrino would be enough.

The story has a further twist. If neutrinos have mass, they can turn into other kinds of neutrino. This might have explained why experiments were picking up too few neutrinos from the Sun. But the Hungarian physicists were refused permission to check their neutrino predictions on US models of the Sun. It was during the Cold War and the models were also used in designing hydrogen bombs.

Marx and Szalay's idea stimulated people to make better measurements of the maximum neutrino mass. Unfortunately, the experiments currently put the largest possible neutrino mass too low to do the job. Also, fast-moving neutrinos would clump matter into galaxies on too large a scale. So the idea that dark matter is made from neutrinos has dropped out of favour. But particle physicists are inventive characters and have suggested all sorts of exotic possibilities.

It still isn't understood

As we write these words in the year 2007, the question of the nature of 90% of the universe is far

Cooking helium in the Big Bang universe

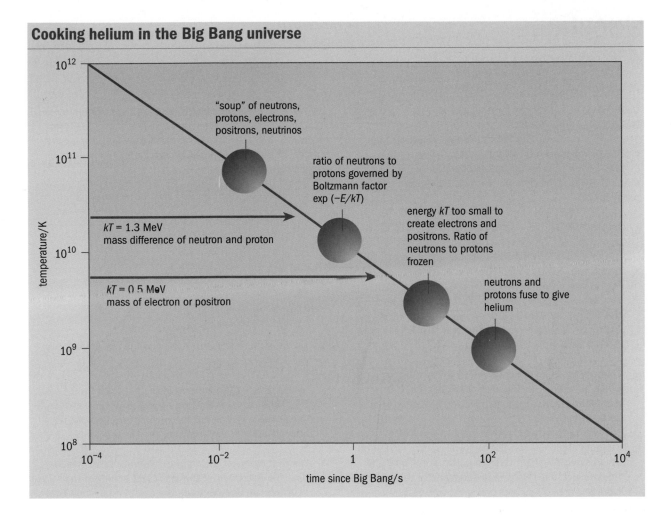

"soup" of neutrons, protons, electrons, positrons, neutrinos

ratio of neutrons to protons governed by Boltzmann factor exp ($-E/kT$)

energy kT too small to create electrons and positrons. Ratio of neutrons to protons frozen

neutrons and protons fuse to give helium

kT = 1.3 MeV mass difference of neutron and proton

kT = 0.5 MeV mass of electron or positron

temperature/K

time since Big Bang/s

from settled. Candidates that might be responsible for some of the missing mass are whimsically described as MACHOs (massive astrophysical compact halo objects), alias lumps of ordinary matter or black holes, and WIMPs (weakly interacting massive particles), alias the neutrino idea in other forms. Those who fancy MACHOs are looking to see if their gravitational fields might act like a lens on light going past them. (You can see the effect of a lens as it passes between you and a lamp as a fluctuation in brightness.)

Those who fancy WIMPs are burying large detectors underground to try to detect them hitting a nucleus head on and making it recoil. In both cases, significant events are expected to be very rare and easily masked by others. Results to date are inconclusive. By the time you read this there may be more substantial evidence to hand. Or there may not.

Of course it is just conceivable that part of the problem lies with the Newtonian physics used

to show that the dark matter must be there. As the astrophysicist Jim Peebles wrote: "Since the subject is still being explored, it is as well to bear in mind the alternative that we are not using the right physics." But he wasn't betting on it: you only change the foundations if you absolutely have to.

However, there are those who think that the foundations will have to be changed in some way. Recent evidence strongly suggests that the expansion of the universe is increasing. If so, something unlike gravity that acts as a repulsive force is needed. Since energy, including rest energy (mass), exerts a gravitational attraction, a way to fill the universe with "dark energy" is required. Other ideas involve changing the gravitational inverse square law at small distances. Nobody knows at present which of these ideas will win out. It is just as likely that the answer, if found, will be none of them. Whatever happens, it seems a good bet that an understanding of dark matter and dark energy will change ideas about the universe.

Questions

Orbits in galaxies

Useful data: a galaxy can be modelled as a disc of stars in circular orbit about a common centre, with gravity providing the centripetal force required for the circular motion of the stars. The orbital speed v of a star at a distance r from the galaxy centre depends only on the mass M of the stars that are less than r from the centre of the galaxy.

1. By considering the centripetal acceleration of the star, show that $M = \dfrac{v^2 r}{G}$. Suggest why the mass of the rest of the galaxy (more than r from the centre) does not affect the acceleration of the star.

2. The Sun is 33 000 light-years from the centre of the Milky Way, with an orbital speed of $260 \, \text{km s}^{-1}$. Calculate the mass of the Milky Way that is less than 33 000 light-years from its centre.

3. The average thickness of the Milky Way's disc is 100 light-years. Use the answer to question 2 to calculate the average density of the galaxy up to 33 000 light-years.

4. Suppose the mass in a galaxy were evenly distributed at constant density ρ through a disc of radius r and thickness d. Use this idea to show that in this case the equation $v^2 = \dfrac{GM}{r}$ means that $v = \sqrt{\pi r \rho d G}$.

Photon conversion

Useful data: electrons and positrons have the same mass of $9.11 \times 10^{-31} \, \text{kg}$ but equal and opposite electric charges.

1. Show that the mass of an electron is equivalent to a rest energy of 0.51 MeV.

2. A photon can decay into a positron and an electron. Why can it not decay into a single electron? Calculate the kinetic energy of each particle if the photon has an energy of 5.0 MeV. How can you be sure that their speed is close to the speed of light?

3. The case study suggests that photons can only decay into particles of total rest energy mc^2 if the energy kT is large enough. Explain this suggestion.

4. Estimate the temperature of the universe at which photons can no longer create positron–electron pairs.

Protons, neutrons and neutrinos

Useful data: neutrons and protons are converted into one another in three reactions involving electrons, positrons, neutrinos and antineutrinos. The table gives data for the masses of the various particles involved. (The neutrinos can be considered as massless.)

Particle	Mass/kg
electron	9.11×10^{-31}
neutron	1.6748×10^{-27}
positron	9.11×10^{-31}
protron	1.6725×10^{-27}

$$n + \nu_e \leftrightarrow p + e^-$$
$$n + e^+ \leftrightarrow p + \overline{\nu}_e$$
$$n \leftrightarrow p + e^- + \overline{\nu}_e$$

1. Taking each reaction in turn, state in words the process it represents.

2. By considering the masses of the particles involved in the reaction, explain why a neutron will spontaneously decay into a proton but the reverse reaction does not happen at room temperature.

3. The first reaction shown above forms a basis for neutrino detection. If the mass and energy of the neutrino can be ignored in the calculation, show that the electron emerges with an energy of 780 keV.

Go to the *Advancing Physics* A2 students' CD-ROM for a larger set of questions.

Case study: profiles from space

Plenty of systems find the distance between two points by measuring the time it takes for waves to travel between them. For example, bats and porpoises use the time delay between emitting a pulse of ultrasound and receiving its echo to work out how far away their lunch is. Similarly, radar systems track aircraft and ships by timing microwaves reflected from them. More recently, laser light has been used to map the mountains of Mars, and the same technique is now being used to study the structure of forest canopies on Earth.

Distance in light-nanoseconds

A light-year is a useful measure of distance for astronomy. This is because distances measured in light-years are manageable numbers greater than 1. For example, the nearest star is 4.2 light-years away, the galaxy is 100 000 light-years across and the distance scale of the visible universe is (only) 15 000 000 000 light-years. A light-second is 300 000 km, more than halfway to the Moon, so is a useful measure of distance within the solar system. The Sun is 500 light-seconds away from Earth.

But what about measuring more familiar distances, such as the distance across a room? The time taken for a pulse of light to travel across a typical laboratory and back is about 100 ns. This time interval can be measured with reasonable precision using a fast oscilloscope, with an uncertainty of 1 ns translating into a distance of only 300 mm. The light-nanosecond is 300 mm, ideally suited to measurement on a human scale. It provides the basis for a quick but fairly rough surveying method – yet it would not be difficult to improve on this precision using a humble tape measure. However, if the distance to be measured is to an inaccessible place, like the vault of a cathedral or the surface of Mars, then 300 mm precision using a simple spot of light might seem a very attractive proposition. This is the basis of laser ranging.

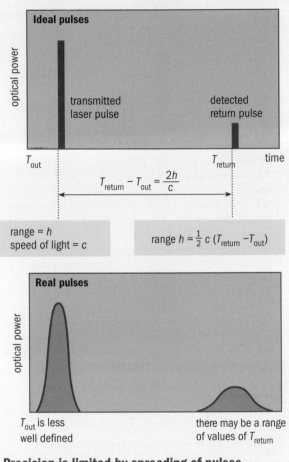

Key summary: laser ranging

Ideal pulses

optical power

transmitted laser pulse   detected return pulse

T_{out} T_{return} time

$$T_{return} - T_{out} = \frac{2h}{c}$$

range = h
speed of light = c

range $h = \frac{1}{2} c \, (T_{return} - T_{out})$

Real pulses

optical power

T_{out} is less well defined   there may be a range of values of T_{return}

Precision is limited by spreading of pulses

Lidar

Lidar (**li**ght **d**etection **a**nd **r**anging) is the optical equivalent of the microwave pulse–echo technique known as radar (**ra**dio **d**etection **a**nd **r**anging). It has a become a practical proposition since the development of two pieces of technology:

- lasers able to generate short powerful pulses of light;
- photon detectors with fast enough response times.

The distance h to be measured is found from the difference between the time t_0 when the pulse was sent and the time t_r when it returns:

$$h = \frac{c}{2}(t_r - t_0)$$

Key summary: optics of laser ranging

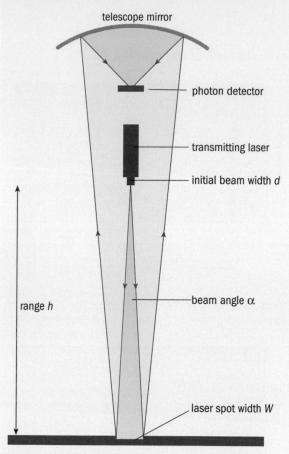

telescope mirror

photon detector

transmitting laser

initial beam width d

range h

beam angle α

laser spot width W

The laser spot width is limited by diffraction

compared with the lateral scale of the surface roughness. However, it may very well not be. The angular width α of the laser beam is inversely proportional to its initial width d:

$$\alpha = \frac{\lambda}{d}$$

where λ is the wavelength of the light. The spot diameter W will therefore approximate to $W = \alpha h$, where h is the distance of the laser above the surface being studied. Spot diameters of less than 100 m can be achieved at distances of 100 km.

Profiles from orbit

The first application of lidar to the detailed mapping of the shape of a whole planet took place during 1998 and 1999. The Mars orbital laser altimeter (MOLA) aboard the spacecraft *Mars Global Surveyor* carried out a topographical survey of Mars from an orbital height of about 400 km.

The orbital path of MOLA is almost circular, and can be determined to an uncertainty of a few metres. So a succession of range measurements from spacecraft to planetary surface, obtained from a sequence of laser pulses, builds up a profile of the local deviation of the surface from the mean planetary radius. Features like mountains and valleys can be mapped. The width of the laser beam at the surface is quite small (about 100 m), but a dense mesh of sampling points can be built up over a large number of successive orbits.

The number N of samples per kilometre on the planetary surface is determined by the rate n at which the laser produces pulses and by the orbital velocity v:

$$N = \frac{nR}{vR_0}$$

where R is the orbital radius of the spacecraft and R_0 is the mean planetary radius. In practice the pulse rate is $10\,\mathrm{s}^{-1}$ and the orbital velocity is $3\,\mathrm{km\,s}^{-1}$, giving about three samples per kilometre. Since at least two samples are required per cycle, this means that MOLA cannot detect surface variations with a spacing of less than $2/3$ km. Sampling of variations in surface height is the same problem as the sampling of a waveform in time (chapter 3, *Advancing Physics AS*); sampling frequency must be at

The problem is measuring the times t_r and t_0 accurately. Ideally the transmitted and returned pulses would have sharply defined edges, but in practice they do not. The transmitted laser pulse is usually bell-shaped and the return pulse from a flat surface would normally be similar but less intense (most surfaces are far from perfectly reflecting). However, if the surface is not flat within the illuminated area, the return pulse will be wider than the transmitted pulse. Photons reflected from the peaks of the surface will arrive earlier than those reflected from the valleys.

Measuring the profile of a surface with lidar is not just a question of how accurately the round trip can be measured. The effect of the roughness of the surface on the shape of the return pulse, and the effect of diffraction in determining the size of the laser spot on the surface, must both be considered. Ideally, the spot should be small

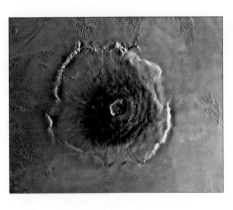

A *Viking* orbiter image of Mount Olympus on Mars.

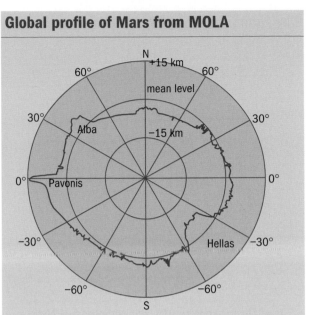

An illustration of the global profile of Mars from MOLA, crossing longitudes 52°E and 247°E. The north and south polar caps are at the top and bottom. The Pavonis Mons and Alba Patera volcanic shields, both within the Tharsis province, appear top left, and the Hellas impact basin is situated bottom right.

least twice the highest frequency. But now these are frequencies in space, not in time.

MOLA's orbit passes over the planet's north and south polar ice caps and is completed in just under two hours. Mars revolves on its axis underneath the orbiting spacecraft so that successive orbits sample different parts of the planet's surface. In a mapping mission lasting one Martian year (686 Earth days), MOLA accumulated profile data over nearly 10 000 orbits.

Getting there

Over the past half-century there have been more than 30 attempts to get a spacecraft close to Mars, and the large majority of these have failed. Many of these failures were during the period of the space race between the USA and the former Soviet Union, resulting in the waste of a huge amount of money. The current cost of the failure of a Mars mission would be about a billion dollars.

To get to Mars a spacecraft must escape from Earth's gravity and become a satellite of the Sun on a carefully chosen elliptical orbit that intercepts the orbit of Mars. Then the spacecraft must be captured by Martian gravity. For this to happen, the craft is slowed down by firing its main rocket engine. (The 1999 *Mars Climate Orbiter* was lost because of a faulty command at this point.) This leaves it in a highly elliptical orbit whose low point enters the upper atmosphere of Mars. Each pass through this low point takes energy from the orbiting craft and, after four months of such "aero-braking", the orbit is nearly circular. A final rocket thrust then establishes the very nearly circular orbit needed for the mapping operation. You can see that there are plenty of chances for things to go wrong, as indeed they have.

Mountains on Mars

The MOLA profiles are a spectacular improvement on our knowledge of the topography of Mars. Previously this had been deduced from Earth-based radar and the interpretation of images from the *Mariner* and *Viking* missions. The uncertainty in the surface height was typically 2 km, compared with about 10 m for MOLA. We now know the profile of some parts of Earth less accurately than we know that of the whole of Mars.

Results confirm that the mountains of Mars are far higher than those on Earth. For example, the Martian Mount Olympus rises 20 km above the mean surface level compared with 9 km for Mount Everest on Earth. An enormous impact crater called Hellas is about 9 km deep and is surrounded by mountainous ridges of ejected material.

Early astronomers thought they could see canals on Mars and imagined a civilization existing there. The differences between the northern and southern hemispheres do suggest the presence of a huge ocean at some point in the past. The northern hemisphere is mainly a vast smooth-surfaced depression, possibly the site of an ancient northern ocean, whereas the southern hemisphere largely consists of heavily cratered uplands with

Key summary: scanning lidar imager of canopies by echo recovery (SLICER)

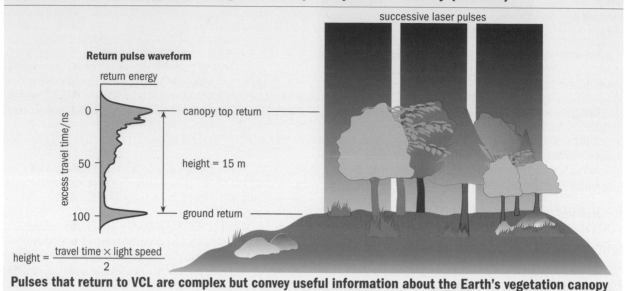

successive laser pulses

Return pulse waveform

return energy

excess travel time/ns

0 — canopy top return

50 — height = 15 m

100 — ground return

$$height = \frac{travel\ time \times light\ speed}{2}$$

Pulses that return to VCL are complex but convey useful information about the Earth's vegetation canopy

a rough surface. Many valleys and channels also suggest that water once flowed on the surface.

Back to Earth: looking into the forest canopy

A more elaborate version of the MOLA instrument has now been developed to map the surface topography of Earth. Known as the vegetation canopy lidar (VCL), it was launched in February 2000. Whereas Mars is dry with a well defined, hard surface, Earth is a soft, wet planet, the surface of which is mostly watery, vegetated or covered with clouds. The lidar pulses that return to VCL are therefore complex but convey much useful information about the soft outer edge of our planet.

The idea of mapping the Earth from satellites is not new. Since 1978, radar altimeters have been sent aloft. Microwave wavelengths are typically 10 mm, many orders of magnitude larger than optical wavelengths. Like laser light, a radar beam spreads out through diffraction. Since the largest radar aerial that can be carried into orbit is only a few metres across, the width of the radar spot on the ground is quite large (about 1 km). Lidar spot sizes are much smaller (about 25 m), allowing a much higher resolution of surface variation. On the other hand, radar penetrates clouds and light does not.

VCL is a more powerful instrument than MOLA. It has five lasers that produce 25 m spots from an orbital height of 400 km. As implied by

its name, VCL is chiefly intended for the study of forested regions. It can solve two problems:

- accurately locating ground height under forest canopies (not possible using conventional stereophotographic techniques);
- generating data about the three-dimensional distribution of the forest canopy (differently shaped trees produce different return pulses).

The return lidar pulse from a forest canopy consists of two main sections. The shortest round trip is to the canopy top, so this generates the leading edge of the return pulse. The shape and length of this section will depend on the size and type of tree. The vertical interval where trunks are bare will produce no return. Finally, there will be a relatively short return pulse from ground level, although its size will depend on the luxuriance of the tree canopy. The time interval between the canopy-top pulse and the ground-return pulse gives a means of estimating the tree height. The density of the trees and how much foliage they carry can also be estimated in this way.

Of course, specifying tree heights in light-nanoseconds instead of metres is more popular with physicists than with foresters. However, the latter will appreciate the copious amounts of information that VCL is giving them about the make-up and maturity of their forests without the need to scramble through them with a tape measure.

Questions

Laser ranging

Useful data: the average height of the circular orbit of a laser altimeter above a planetary surface is 400 km.

1. Estimate a typical round-trip time for a light pulse.

2. If the radius of the planet varies by up to 10 km either side of the average value, estimate the range of round-trip times for pulses reflected from the planet's surface.

3. How high is a typical tree in light-nanoseconds?

Laser beams

Useful data: the laser on the VCL satellite produces infrared light of wavelength 1064 nm. The satellite is, on average, 400 km above the surface of Earth. The beam width is 25 m when it reaches the surface of Earth.

1. Estimate the angular width of the laser beam due to diffraction. Hence estimate the width of the beam as it emerges from the laser.

2. Suppose that the energy of each laser pulse is 10 mJ. Neglecting any absorption by the atmosphere, calculate the number of photons that hit Earth in each pulse.

3. If half of the photons are absorbed by the surface and the others are reflected in all directions, estimate how many are collected by the satellite mirror (diameter 0.90 m).

Mars orbits

Useful data: here are some data for Mars.

mass /kg	mean radius/m	Martian day/s	Martian year/s
6.45×10^{23}	3.38×10^{6}	8.86×10^{4}	5.95×10^{7}

1. Write down an expression for the centripetal force required to keep a satellite of mass m in a circular orbit of radius r about Mars. By equating this to the gravitational force on the satellite, show that the orbital time T of the satellite is given by $T^2 = \dfrac{4\pi^2 r^3}{GM}$, where M is the mass of Mars.

2. Use the data in the table to show that the orbital time of the MOLA satellite is just under two hours. The average height of the satellite above the surface is 400 km.

3. Each successive orbit of the satellite maps a different part of the surface. Explain why.

Mountain height

Useful data: the height of the highest mountain on a planet can be used to estimate its gravitational field strength. The model assumes that the maximum pressure at the base of a mountain is the same on all planets. This is because the rock under the mountain will flow away to the sides if the pressure is great enough.

1. Model Mount Everest as a cylinder of height 8.8 km with a density of 2.6×10^3 kg m^{-3}. Calculate the pressure it exerts on the rocks under it.

2. Mount Olympus on Mars has a height of 20 km. Use this to estimate the gravitational field strength on Mars, assuming that the crust of both planets is made of material with similar yield behaviour under stress.

3. Use the mass and radius of Mars (given in the table on this page) to calculate the value of g on its surface. Compare its value with your answer to question 2. Suggest reasons why they are different.

Go to the *Advancing Physics* A2 students' CD-ROM for a larger set of questions.

Case study: hybrid vehicles

Pollution from vehicles has a major impact on health. City centres choke with the pollution that emerges from vehicle exhausts. Cars, vans, lorries and buses all remorselessly push a range of pollutants into the atmosphere, including:

- carbon dioxide, which contributes to global warming
- carbon monoxide, which poisons people
- oxides of nitrogen, which lower the pH of rain, making it acid
- specks of carbon compounds, which impair the function of our lungs
- volatile organic compounds (mostly hydrocarbons), which are toxic to humans

Pollutants are pushed into the atmosphere during a traffic jam. A bicycle can be a quicker way through the jam and doesn't contribute to pollution, but the cyclist still breathes it in.

It can't go on. Despite the major reductions achieved with the latest catalyst technologies, there seem to be only two ways forward for cities of the future: either ban personal vehicles completely and force everyone to use public transport, or replace the internal combustion engine with something else. Neither of these is likely to happen soon, for a variety of political and practical reasons. However, the hybrid vehicle offers a partial solution to the problem, which is sensible in both economic and technological terms.

Public transport

As you can see from the table opposite, the most energy-efficient way of getting from one place to another is by bicycle.

The motor car is inefficient, mainly because most journeys have only the driver in the vehicle. Typically 1 tonne of hardware is being used to move 75 kg of person. The efficiency rises as the number of occupants increases. A full car becomes as efficient as a train. However, energy efficiency is only one measure of a car's performance – exhaust emissions have to be taken into account, too.

Energy efficiency of modes of transport

Mode of transport with typical loading	Approximate energy per passenger kilometre/kJ
bicycle	100
walking	200
long distance coach	400
train	600
bus	700
motor car	1600
jet airplane	2500

The energy needed to run an electric train is obtained by burning a fuel in a power station. The combustion of the fuel is constantly monitored, reducing the combustion products other than carbon dioxide and water, which emerge from the chimney. In contrast, the burning of fuel in an internal combustion engine is usually far from optimal. Careful design and catalytic converters in the exhaust pipe reduce emissions of toxic by-products but do not eliminate them.

Electric cars

The state of California has had enough. By 2004 it required that one in 10 of the vehicles sold could release no polluting gases into the atmosphere. The initial response to this zero-emissions policy from car manufacturers was a heavy investment in research programmes to develop electric cars.

Electric traction has several advantages over internal combustion for a car. To start with, you can dispense with a lot of gearing and shafts simply by connecting a motor directly to each front wheel so there is an immediate saving in weight and space. Electric motors can produce a large torque at low speeds, giving the car a big acceleration without the need for a gearbox. Then there is the possibility of regenerative braking, using the motors as generators to take energy out of the motion of the vehicle and put it back into the battery as the car slows down. At least 50% of the energy can be recycled with regenerative braking, compared with 0% for normal friction braking systems, which simply use the energy taken from the motion to make the brakes hot. Transferring energy to and from each wheel by electricity makes computer control relatively straightforward, so electric vehicles can easily be provided with anti-skid braking.

Energy density

So why aren't electric cars everywhere? The problem is storing the energy. Despite massive research effort into developing new battery technology, the lead-acid battery still reigns supreme for storing energy. It can hold 20 times more energy per kilogram ($kWh\,kg^{-1}$) than its nearest popular competitor, the nickel-cadmium rechargeable cell. A typical lead-acid car battery has the following properties:

- storage capacity of $750\,Wh$ (or $2.7\,MJ$) of energy;
- useful lifetime of several years if not discharged completely;
- mass of about $16\,kg$;
- volume of about 7 litres;

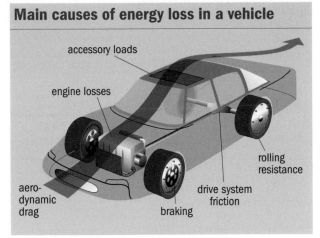

Main causes of energy loss in a vehicle

accessory loads

engine losses

rolling resistance

aero-dynamic drag

drive system friction

braking

- operates at everyday temperatures.

But this compares very unfavourably with a liquid fuel, such as petrol. The same amount of energy requires only $60\,g$ of petrol, taking up $80\,ml$ of space. Petrol has an energy density of $5 \times 10^7\,J\,kg^{-1}$; the same figure for lead-acid batteries is a miserable $2 \times 10^5\,J\,kg^{-1}$. Of course, the figures for petrol are improved by not also having to cart around the oxygen with which to burn it.

Range and speed

Unless there is a startling improvement in the energy-mass ratio of batteries, electric vehicles are not going to happen without a lot of protest from the general public. The following calculation should make this clear.

A top speed v of 50 mph is about $25\,m\,s^{-1}$, more than adequate in an urban environment. At that speed the traction force F from the wheels has to balance the counter force of wind resistance. For a family car with a drag coefficient of 0.25, this could easily be $300\,N$. So the maximum traction power Fv of an electric car will need to be about $300\,N \times 25\,m\,s^{-1} = 7.5\,kW$. A single car battery could only manage this for about five minutes.

Six hours at top speed would require enough batteries to add about 1 tonne to the mass of the vehicle. Add the driver, passengers and the rest of the structure and the total mass of the car could easily be 1500 kg. This gives the vehicle a maximum speed of less than $5\,m\,s^{-1}$ (about 10 mph) when moving up a gradient of only 10%.

You will be familiar with massive lorries crawling up the slow lane of a long hill. The need

Key summary: drag force on a van

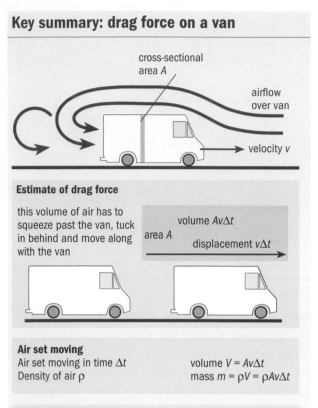

Estimate of drag force

this volume of air has to squeeze past the van, tuck in behind and move along with the van

Air set moving
Air set moving in time Δt

Density of air ρ

volume $V = Av\Delta t$

mass $m = \rho V = \rho Av\Delta t$

Increase of momentum
In time t mass m of air increases in momentum by mv

momentum increase $\Delta p = mv = \rho Av\Delta t \times v = \rho Av^2\Delta t$

Force on van
Force F is rate of change of momentum $\dfrac{\Delta p}{\Delta t}$

$$F = \frac{\Delta p}{\Delta t} = \frac{\rho Av^2\Delta t}{\Delta t} = \rho Av^2$$

$$F = \rho Av^2$$

Drag coefficient C_D
Streamlining reduces the mass of air dragged along by the van. Drag coefficient C_D is less than 1

The drag force on a vehicle is proportional to the square of the vehicle speed.

$$F = C_D\rho Av^2$$

Drag force is proportional to the speed squared

Power required for a typical car

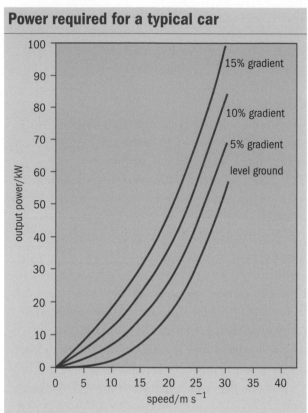

This graph shows how the power required for a typical car varies with the gradient of the road and the speed of the car.

Hybrid vehicles

The power system of a hybrid vehicle has three distinct parts:

- a liquid/gas fuel engine that turns the shaft of a generator;
- a battery that stores the electrical energy from the generator;
- an electric traction system that drives the vehicle forwards, using energy stored in the battery.

The engine is run at a speed that maximises its fuel efficiency. Since most of the noxious emissions from internal combustion engines occur when they are running slowly, this also drastically reduces the amount of pollution present in the exhaust gases. The engine only comes on when the energy stored in the battery (as indicated by the battery voltage) goes below a certain level, and goes off as soon as the battery is fully charged.

A hybrid vehicle has all of the advantages of electric traction. Its technology is proven – it provides regenerative braking, there are no gear trains and you get high acceleration and computer control – together with all of the advantages of

to lift a massive vehicle uphill severely limits the maximum speed for a given power.

The final blow for the future of electric vehicles is delivered by the recharging arrangements. Whereas you can refill a petrol tank in a matter of minutes, recharging lead-acid batteries takes hours and your car stays idle, plugged into the supply. You could, of course, exchange dead batteries for fully charged ones, but imagine the problems of replacing a 1 tonne battery pack quickly.

How a hybrid car works

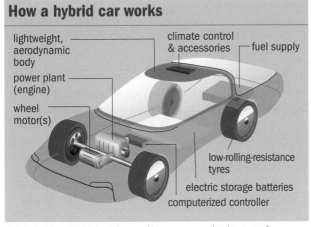

- lightweight, aerodynamic body
- climate control & accessories
- fuel supply
- power plant (engine)
- wheel motor(s)
- low-rolling-resistance tyres
- electric storage batteries
- computerized controller

The hybrid petrol/electric car. You can see the layout of component systems in the vehicle.

The Toyota Prius was an early commercial example of the hybrid car concept. Other makers have since followed suit.

liquid fuels (high energy–mass ratio, low cost, wide availability). The battery can be relatively small, since it only has to provide large amounts of energy in bursts for acceleration or climbing hills.

Consider a hybrid car that needs a traction power of 7.5 kW to cruise at 50 mph. Generators and electric motors have typical efficiencies of 90%, so the combustion engine has to be able to deliver a continuous power of, say, 10 kW. Assuming a loaded mass of 500 kg, the vehicle will need a traction power of 20 kW to climb a 10% gradient without slowing down. Taking account of the motor efficiency and the power from the generator, this means that the battery has to supply energy at a rate of 14 kW to the electric motors to keep the car going. Four standard lead-acid car batteries could manage this for at least 10 minutes, contributing a mass of only 64 kg to the vehicle, which is equivalent to the mass of one passenger.

Evolution, not revolution

For many years car manufacturers have tried to develop satisfactory, purely electric cars. Prototypes of all kinds have been built and tried out. None has made it to full-scale mass-production and electric cars have not won widespread consumer acceptance. When asked, the public is in favour of cleaner, quieter cars, though not so much in favour of slower ones. The speed and range of electric cars, limited as they are for the reasons explored above, have always counted against them. But new anti-pollution laws provide a new incentive, and manufacturers have now turned to the seemingly

more complicated solution of the hybrid vehicle as a more practical, realistic and acceptable proposition.

All of the major car manufacturers are taking the hybrid car very seriously. Billions of dollars have already been invested in research and production, and hybrid cars are already on the streets in respectable numbers. Since it only uses proven technology (combustion engines, generators, batteries and electric motors), the hybrid car is hardly more expensive than other cars. Its widespread adoption offers the prospect of reduced emissions and improved fuel economy (up to 70 miles per gallon) without any significant loss of performance. Furthermore, it is amenable to the introduction of revolutionary new systems that promise to improve fuel efficiency and reduce emissions even further. As improved technologies become available for each of its component parts, they can easily be slotted into a hybrid car. These include the following possible innovations:

- flywheels to store energy instead of batteries;
- gas turbines to replace internal combustion engines, saving weight and increasing fuel efficiency;
- engines that use non-fossil fuels, such as alcohol;
- fuel cells that combine hydrogen with oxygen directly to make electricity, resulting in no emissions at all apart from harmless H_2O.

This last item is the Holy Grail of green transport, particularly if the hydrogen can be generated directly from sunlight, perhaps using photovoltaic cells to electrolyse water. The zero-emission hybrid car may be a real possibility.

Questions

Regenerative braking

Useful data: the case study claims that 50% of the kinetic energy of a vehicle can be stored in the battery of an electric vehicle each time that it is stopped.

1. Use the idea of electromagnetic induction to explain how an electric motor is able to charge the battery.

2. The mass of a typical car is 1000 kg. The maximum legal speed of a car in a town is 50 km h^{-1}. If the car has to come to a stop from this speed for a short time once every two minutes, using friction brakes, estimate how much energy is wasted in heating up the brakes and the air around them every half an hour.

3. The traction force required to keep a typical car moving at a steady speed of 50 km h^{-1} is 100 N. Calculate how much work is done by the car engine to keep it going at 50 km h^{-1} for half an hour.

4. Use your answers to questions 2 and 3 to estimate the percentage fuel saving of a car driving through a town when it uses regenerative braking.

Electric motors

Useful data: the construction of a DC electric motor includes several features that maximise its performance. The rotor spins in the magnetic field produced by the stator, with a very small air gap between the two.

1. The rotor is made from thin sheets (laminations) of iron. Explain why this improves the performance of the motor.

2. Explain why reducing the gap between the rotor and the stator improves the performance of the motor.

3. Some electrical energy is wasted in heating the coils of the stator, which generate the magnetic field around the rotor. Explain why increasing the thickness of wire in the coils improves the performance of the motor.

Traction power

Useful data: the drag force F on a vehicle of cross-sectional area A moving at a steady speed v on level ground through air of density ρ is given by $F = C_D \rho A v^2$. The quantity C_D is the drag coefficient of the vehicle.

1. The case study suggests that a car that has a drag coefficient of 0.25 has a drag force of 300 N at a speed of 25 m s^{-1}. Calculate the cross-sectional area of such a car. The density of air is 1.2 kg m^{-3}.

2. Show that the power P required to keep a car moving at a steady speed v is proportional to v^3.

3. Estimate the maximum power of a car that has a top speed of 40 m s^{-1} (about 85 mph).

Battery power

Useful data: a typical lead-acid battery has the following properties: a storage capacity of 750 W h, a volume of 7×10^{-3} m^3, a mass of 16 kg and a terminal voltage of 12 V.

1. The capacity of a battery is often quoted in ampere-hours. This is the number of hours for which a fully charged battery can supply a current of 1 ampere. Use the data to estimate the capacity in ampere-hours of a typical lead-acid battery.

2. If the drag force on a car at 25 m s^{-1} is 300 N, estimate how far it could travel on a level road using the energy stored in a bank of lead-acid batteries that has a mass of 1 tonne.

3. Petrol provides energy by burning with air. If you only have to carry the petrol, you could think of the mass of the petrol as providing energy of 50 MJ kg^{-1}. The density of the petrol is 700 kg m^{-3}. Calculate the mass and volume of petrol that provides the same energy as the lead-acid batteries of question 2. Comment on the significance of your answers for the future of electric vehicles.

Go to the *Advancing Physics A2* students' CD-ROM for a larger set of questions.

Acknowledgments

The publishers are grateful to the following for permission to reproduce images and poetry in this book.

1   Scott Camazine/K Visscher/Science Photo Library
5   Courtesy of Reed College
9   (two images) Science Museum/Science & Society Picture Library
13 (right) Courtesy of MetroCentre, Gateshead
15 (top) Courtesy of Cambridge University Press
15 (bottom) By permission of the Syndics of Cambridge University Library and the Particle Physics & Astronomy Research Council
17 (left) Science Museum/Science & Society Picture Library
17 (right) National Maritime Museum, Greenwich, London
23 ©The Board of Trustees of the Armouries, XI.10, 23
27 (top left) ©The British Museum
27 (bottom right) Marconi plc
31 (top) Anglo-Australian Observatory/David Malin Images
31 (bottom right) David Parker/Science Photo Library
33 R Sheridan ©Ancient Art & Architecture Collection
35 (five images) Courtesy of Lowell Observatory
45 (four images) NASA
51 (top) Courtesy of Spalding Sports Worldwide Inc
51 (bottom) ©Raj Kamal/www.osf.uk.com
52 Courtesy of Toyota Europe
56 (two images) NASA
58 Courtesy of Kitsou Dubois, Ki Productions and the Arts Catalyst
65 (top) Philip Perkins, www.astrocruise.com/m31.htm
63 (middle and bottom) Anglo-Australian Observatory/David Malin Images
66 RGO/IAC/INT/David Malin Images
67 (bottom) JPL/NASA
68 AIP Emilio Segrè Visual Archives
69 National Maritime Museum, Greenwich, London
70 Courtesy of the American Association of the Variable Star Observers (AAVSO)
72 Gary Bower, Richard Green (NOAO), the STIS Instrument Definition Team and NASA
73 (left) Dave Finley, NRAO/AUI/NSF
84 The Observatories of the Carnegie Institution of Washington
86 (top) R Williams and the HDF Team (STScI) and NASA
86 (bottom) Jeff Hester & Paul Scowen (Arizona State University) and NASA
88 Lucent Technologies
89 NASA
93 (three images) © Breitling
95 By permission of The British Library, Section 8 1601/576 or 536.i.14
101 (left) Courtesy of the Archives, California Institute of Technology
101 (right) AIP Emilio Segrè Visual Archives, Segrè Collection
109 (top) Science Museum/Science & Society Picture Library
109 (bottom, two images) Courtesy of Ray Knowles, Crofton Pumping Station
110 Photograph by Lady Roscoe, courtesy of Caltech Archives
117 (top) François Gohier/Science Photo Library
117 (middle) Gianni Tortoli/Science Photo Library
117 (bottom left) Matt Meadows/Science Photo Library
117 (bottom right) James Holmes/Science Photo Library
117 The poem "Fire and ice" from *The Poetry of Robert Frost*, edited by Edward Connery Latham and published by Jonathon Cape, is reprinted with permission of the Random House Group Ltd
120 Alfred Pasieka/Science Photo Library
121 (left) Dr Jeremy Burgess/Science Photo Library
121 (right) By courtesy of the Percival David Foundation of Chinese Art
123 Courtesy of Dr Peter Ford, formerly of the Department of Physics, University of Bath
130 Richard Boohan & Isabel Martins *School Science Review* **77** 281
131 Courtesy of Frantisek Zboray
135 WT Sullivan III/Science Photo Library
136 (left) Courtesy of Brush Electrical Machines Ltd, Loughborough
136 (right) Electrical Machines and Drives Group, University of Sheffield
137 (four images) Used with permission of the Physical Science Study Committee (PSSC)
140 (top right) National Trust Regional Libraries and Archives
140 (bottom, two images) The Royal Institution, London, UK/Bridgeman Art Library
144 Courtesy of Ian Lawrence
145 Courtesy of Richard Wilson (Dencol) Ltd
147 Courtesy of Edison Mission Energy
149 Maximilian Stock Ltd/Science Photo Library
152 (inset) Courtesy of Iowa State University
158 (six images) Electrical Machines and Drives Group, University of Sheffield
163 (top) Samuel Ashford/Science Picture Library
163 (bottom) Mauro Fermariello/Science Picture Library
164 Dominik Friedrichs, www.forlix.org
169 Courtesy of SLAC
171 (inset) Reproduced by permission of Ordnance Survey on behalf of HMSO. ©2008. Ordnance Survey Licence number 100048011
175 (two images) Courtesy of Jon Foyle, King Edward VI School, Southampton
176 Tina Carvalho/Oxford Scientific Films

177 (left) CERN, P Loiez/Science Photo Library
177 (right) CERN/Science Photo Library
178 (left) CERN
178 (right) David Parker/Science Photo Library
180 Stanford Linear Accelerator Centre/Science Photo Library
184 Courtesy of Ian Lawrence
189 (bottom) Dr John Mazziotta et al/Science Photo Library
190 Wellcome Dept of Cognitive Neurology/Science Photo Library
191 Lawrence Berkeley Laboratory/Science Photo Library
196 Science Photo Library
206 Courtesy of Brookhaven National Laboratory
207 CERN
209 (four images) IBM, Crommie, Lutz and Eigler
211 Department of Physics, Imperial College/Science Photo Library
218 (left) ©R Williams (STScI), the HDF-South Team, and NASA
218 (right) CERN
221 (top left) Courtesy of Martin Lewis
221 (top right) Courtesy of NASA/JPL/Caltech
221 (bottom left) Courtesy of Konrad Spindler
221 (bottom right) Hank Morgan/Science Photo Library
222 (top) US Department of Energy, Oak Ridge National Laboratory
222 (bottom) Custom Medical Stock/Science Photo Library
223 (right) W Gentner, H Maier-Liebnitz and W Bothe *An Atlas of Typical Cloud Chamber Photographs* (Pergamon Press, 1954) Fig 10, p13
225 (left) Niewenglowski G N 1896 *La photographie de l'invisible au moyen des rayons x, ultra violets, de la phosphorescence et de l'effluve électrique: Historique thérie pratique* (pamphlet) (Paris: Société d'Editions Scientifiques)
225 (middle) Hempelmann L H 1949 Potential dangers in the uncontrolled use of shoe-fitting fluoroscopes *New England J. Medicine* **241** 335–7
225 (right) Scott Camazine/Science Photo Library
226 Courtesy of the Gray Laboratory Cancer Research Trust
236 Courtesy of the Archives, California Institute of Technology
241 EFDA-JET
242 (top left) Bernard Edmaier/Science Photo Library
242 (top right) John Mead/Science Photo Library
242 (middle left) ©Charles E Rotkin/CORBIS
242 (middle right) ©Charles E Rotkin/CORBIS
242 (bottom left) Ria Novosti/Science Photo Library
242 (bottom right) *Encyclopedia of the Atmospheric Environment* aric circa 2000: Acid rain, Trees: www.doc.mmu.ac.uk/aric/eae/Figures/trees.html
248 Courtesy of Building Research Establishment
251 (two images) Courtesy of Building Research Establishment
253 AIP Emilio Segrè Visual Archives, John Irwin Slide Collection
261 NASA/NSSDC
264 (main) ©Tony Stone Images; Martine Mouchy
264 (inset) Adrienne Hart/Davis/Science Photo Library

Every effort has been made to trace the holders of copyright. If any have been overlooked, the publishers will be pleased to make the necessary amendments at the earliest opportunity.

Answers

Chapter 10

1 a approximately 500 000, b plot log count rate against time, c 2.28 hours, d $8.4\times10^{-5}\,\text{s}^{-1}$, e 84, f about 1.6 million

2 a $3.2\times10^{-5}\,\text{s}^{-1}$, b 3.1×10^{7}, c $7\times10^{-7}\,\text{J}$, d $5\times10^{-9}\,\mu\text{g}$, e about $4\,\text{s}^{-1}$

3 a $0.06\,\text{C}$, b $1\,\text{mA}$, c about $\frac{1}{6}$, d $60\,\text{s}$, the time constant, e $0.18\,\text{J}$, f $36\,\Omega$, g time constant $=0.36\,\text{s}$

4 a $8\times10^{4}\,\text{N}\,\text{m}^{-1}$, b $63\,\text{mm}$, c about $400\,\text{J}$, d about $1.3\,\text{m}\,\text{s}^{-1}$

5 a $5\times10^{12}\,\text{Hz}$, b $5\times10^{12}\,\text{Hz}$, c $49\,\text{N}\,\text{m}^{-1}$

Chapter 11

1 a $8.0\,\text{m}\,\text{s}^{-1}$, b $10/\pi\,\text{s}^{-1}$

2 a $8.1\times10^{-8}\,\text{N}$, b no numerical answer

3 1:19 000

4 Mars $3.5\times10^{-9}\,\text{N}$, midwife $8.5\times10^{-8}\,\text{N}$, ratio 0.04

5 $1.9\times10^{27}\,\text{kg}$

6 a $26\,600\,\text{km}$, b $3.9\,\text{km}\,\text{s}^{-1}$, c $20\,200\,\text{km}$, d $67\,\text{ms}$

7 a $3.8\,\text{N}\,\text{kg}^{-1}$, b no numerical answer

8 a $1.5\times10^{4}\,\text{kg}\,\text{m}\,\text{s}^{-1}$, b $75\,\text{kN}$

9 a $50\,\text{kJ}$, $200\,\text{kJ}$, $-50\,\text{kJ}$, b $50\,\text{J}\,\text{kg}^{-1}$, $200\,\text{J}\,\text{kg}^{-1}$, $-50\,\text{J}\,\text{kg}^{-1}$, c $50\,\text{J}\,\text{kg}^{-1}$, $200\,\text{J}\,\text{kg}^{-1}$, $-50\,\text{J}\,\text{kg}^{-1}$, d $150\,\text{J}\,\text{kg}^{-1}$, $100\,\text{J}\,\text{kg}^{-1}$, e $120\,\text{kJ}$

Chapter 12

1 a $438\,\text{nm}$, b $316\,\text{nm}$, c 2.6, d 3.6, e $1.6\times10^{-18}\,\text{J}$, $0.45\times10^{-18}\,\text{J}$, f 3.6, same as d

2 a $30\,\text{km}\,\text{s}^{-1}$, b $5.9\times10^{-3}\,\text{m}\,\text{s}^{-2}$, c $5.9\times10^{-3}\,\text{N}\,\text{kg}^{-2}$, d $g=GM/R^{2}$, e $2.0\times10^{30}\,\text{kg}$

3 a $v=210\times100\,\text{km}\,\text{s}^{-1}$, of the order $1/10\,c$, b if
$$v/c=\gamma=\frac{1}{\sqrt{1-v^{2}/c^{2}}}=1.005 \text{ approximately}$$

or an error of 0.5% in the uncorrected equation

4 a $1/30$ radian, b $1.9°$, d $2.6\,\text{s}$, e perhaps one wavelength, $0.3\,\text{m}$

5 a no numerical answer, b about $10\,\text{pm}$ $(10^{-11}\,\text{m})$, c about 10^{5}, d $0.12\,\text{MeV}$

Chapter 13

1 no numerical answer

2 a $500\,\text{kg}$, b $1.0\,\text{kg}\,\text{m}^{-3}$, c $(293\,\text{K}\times1.2\,\text{kg}\,\text{m}^{-3})/1.0\,\text{kg}\,\text{m}^{-3}=352\,\text{K}=79\,°\text{C}$

3 a 5.7, b no numerical answer

4 a $63\,\text{kJ}$, b $52\,\text{W}$

5 a 2.7×10^{25} molecules, b $3.7\times10^{-26}\,\text{m}^{3}$ per molecule, c $9.1\times10^{-4}\,\text{m}^{3}$ (about 0.9 litre)

6 a $0.25\,\text{kg}$, b and h see table below, c, d, e no numerical answers, f $490\,\text{m}\,\text{s}^{-1}$, g no numerical answer

Volume /m³	Density at 27 °C /kg m⁻³	Density at 127 °C /kg m⁻³
0.40	0.62	0.46
0.34	0.73	0.55
0.29	0.86	0.64
0.25	0.99	0.74
0.20	1.24	0.93
0.18	1.38	1.03
0.17	1.46	1.09

Chapter 14

1 a $10^{9}\,\text{K}$, b $0.1\,\text{K}$, c $5\times10^{4}\,\text{K}$, d $6\times10^{9}\,\text{K}$

2 a $1.4\times10^{-8}\,\text{J}$, b $\text{KE}=2\times10^{-8}\,\text{J}$, c $1.5\times10^{-25}\,\text{kg}=91$ proton masses

3 a $3.3\times10^{-20}\,\text{J}$, b $kT=4.1\times10^{-21}\,\text{J}$, $\varepsilon/kT=8.0$, c $2400\,\text{K}$

4 no numerical answer

5 a $20\,\text{km}$, b $14\,\text{km}$, c $290\,\text{km}$, d $120\,\text{km}$

6 a, b, c no numerical answer, d $3.4\,\text{eV}$

7 a, b no numerical answer, c $4.6\,kT$

Chapter 15

1 no numerical answer

2 a, b, c, d no numerical answers, e $3\,\text{V}$, f 70, g, h no numerical answer

3 no numerical answers

4 a about $0.5\,\text{Wb}$, b about $1/4$ turn, $1/200\,\text{s}$, c $100\,\text{Wb}\,\text{s}^{-1}$, d 100

Chapter 16

1 a $1\,\text{keV}$, b $0.51\,\text{MeV}$, KE is 500 times smaller than rest energy, c no numerical answer, d $1.9\times10^{7}\,\text{m}\,\text{s}^{-1}$, e $4.4\times10^{5}\,\text{m}\,\text{s}^{-1}$

2 no numerical answer

3 no numerical answer

4 a field $=5.7\times10^{11}\,\text{V}\,\text{m}^{-1}$, force $=9.1\times10^{-8}\,\text{N}$, potential $=29\,\text{V}$, b field $=1.4\times10^{19}\,\text{V}\,\text{m}^{-1}$, force $=2.3\,\text{N}$, potential $=140\,\text{kV}$, c field $=2.5\times10^{7}\,\text{V}\,\text{m}^{-1}$, force $=4.0\times10^{-12}\,\text{N}$, potential $=5\,\text{kV}$

5 no numerical answer

6 no numerical answer

7 a, b no numerical answers, c $3.3\,\text{m}$

Chapter 17

1 estimates – no answers given

2 no numerical answer

3 a $3.9\times10^{-11}\,\text{m}$, b $6.3\times10^{-15}\,\text{m}$, c $1.2\times10^{-16}\,\text{m}$

4 a $206/88=2.34$, b $82/38=2.16$, c 1.0 approximately, d $124/50=2.48$, e $(206/88)^{1/3}=1.33$

5 no numerical answers

6 no numerical answers

7 no numerical answers

8 a $1.02\,\text{MeV}$, b $212\,\text{MeV}$, c $1.88\,\text{GeV}$

9 a $4\,\text{nm}$, $2\,\text{nm}$, b $4.5\times10^{-20}\,\text{J}$, c $6.8\times10^{13}\,\text{Hz}$, d $4.4\times10^{-6}\,\text{m}$

Chapter 18

1 no numerical answer

2 a, b no numerical answers, c, d see table below

Nucleus	Energy difference /MeV	Energy difference per nucleon /MeV
$^2_1$H	−2.23	−1.1
$^4_2$He	−28.4	−7.1
$^3_1$H	−8.5	−2.8
$^6_3$Li	−32.1	−5.3

3 no numerical answers

4 no numerical answers

5 no numerical answers

6 a 2.3×10^{-14} J = 0.14 MeV, b 296 MeV

Case studies: advances in physics

Energy efficient offices

Conduction losses

1 3.0×10^{-4} K W^{-1}

2 0.42 K W^{-1} per square metre panel, 2.4 W K^{-1} per square metre panel

Photovoltaics

1 no numerical answer

2 14 fans

3 2.26 eV

4 22%

Water cooling

1 1.67×10^{-3} m^3 s^{-1}

2 no numerical answer

Light sensors

1 no numerical answer

2 no numerical answer

Dark Matter

Orbits in galaxies

1 no numerical answer

2 3.2×10^{41} kg

3 1.1×10^{-18} kg m^{-3}

4 no numerical answer

Photon conversion

1 no numerical answer

2 4.0×10^{-13} J

3 no numerical answer

4 1.2×10^{10} K

Protons, neutrons and neutrinos

1 no numerical answer

2 no numerical answer

3 no numerical answer

Profiles from space

Laser ranging

1 2.67 ms

2 2.60–2.73 ms

3 50 light-ns

Laser beams

1 6.25×10^{-5} radian, 17 mm

2 5.35×10^{16} photons per pulse

3 1.7×10^{4} photons per pulse

Mars orbits

1 no numerical answer

2 no numerical answer

3 no numerical answer

Mountain height

1 2.24×10^{8} Pa

2 4.3 N kg^{-1}

3 3.8 N kg^{-1}

Hybrid vehicles

Regenerative braking

1 no numerical answer

2 1.5×10^{6} J

3 2.5×10^{6} J

4 37%

Electric motors

1 no numerical answer

2 no numerical answer

3 no numerical answer

Traction power

1 1.6 m^2

2 no numerical answer

3 31 kW

Battery power

1 63 ampere-hours

2 570 km

3 3.4 kg, 4.9×10^{-3} m^3

Index

Page numbers in *italics* refer to entries mainly or wholly, in photographs, diagrams or summaries